Fundamentals of Electric Machines

Fundamentals of Electric Machines

A Primer with MATLAB®

Warsame Hassan Ali
Samir Ibrahim Abood
Matthew N. O. Sadiku

CRC Press is an imprint of the
Taylor & Francis Group, an **informa** business

CRC Press
Taylor & Francis Group
6000 Broken Sound Parkway NW, Suite 300
Boca Raton, FL 33487-2742

First issued in paperback 2021

ISBN 13: 978-0-367-25098-0 (hbk)
ISBN 13: 978-1-03-224286-6 (pbk)

Library of Congress Cataloging-in-Publication Data

Names: Ali, Warsame Hassan, author. | Sadiku, Matthew N. O., author. | Abood, Samir, author.
Title: Fundamentals of electric machines : a primer with MATLAB / Warsame Hassan Ali, Matthew N. O. Sadiku and Samir Abood.
Description: Boca Raton : Taylor & Francis, a CRC title, part of the Taylor & Francis imprint, a member of the Taylor & Francis Group, the academic division of T&F Informa, plc, 2019.
Identifiers: LCCN 2019007098 | ISBN 9780367250980 (hardback : acid-free paper) | ISBN 9780367250980 (ebook)
Subjects: LCSH: Electric machinery. | Electric machines. | MATLAB.
Classification: LCC TK2000 .A566 2019 | DDC 621.31/042--dc23
LC record available at https://lccn.loc.gov/2019007098

Visit the Taylor & Francis Web site at
http://www.taylorandfrancis.com

and the CRC Press Web site at
http://www.crcpress.com

Publisher's Note
The publisher has gone to great lengths to ensure the quality of this reprint but points out that some imperfections in the original copies may

My parents, Fadma Ibrahim Omer and Hassan Ali Hussein (Shigshigow), for their unconditional love and support, to my beloved wife, Engineer Shukri Mahdi Ali, and my children, Mohamed, Faduma, Dahaba, Hassan, and Khalid, for their patience, and to all my caring siblings.

Warsame Hassan Ali

My great parents, who never stop giving of themselves in countless ways, and my beloved brothers and sisters.

My dearest wife, who offered me unconditional love with the light of hope and support.

My beloved kids: Daniah, and Mustafa, whom I can't force myself to stop loving.

To all my family, the symbol of love and giving,

Samir Ibrahim Abood

My parents, Ayisat and Solomon Sadiku, and my wife, Kikelomo.

Matthew N. O. Sadiku

Contents

Preface

An electric machine is a device that converts mechanical energy into electrical energy or vice versa. It can take the form of an electric generator, electric motor, or transformer. Electric generators produce virtually all electric power we use all over the world.

Electric machine blends the three major areas of electrical engineering: power, control, and power electronics. This book presents the relation of power quantities for the machine as the current, voltage power flow, power losses, and efficiency. The control condition presents the methods of speed control and electrical drive. Power electronics is important to machine control and drive.

The purpose of this book is to provide a good understanding of the machine behavior and its drive. The book begins with the study of salient features of electrical DC and AC machines. Then it presents their applications in the different types of configurations in lucid detail. This book is intended for college students, both in community colleges and universities.

This book organized into 11 chapters. With a short review of the basic concept of magnetism in Chapter 1, it starts with a discussion of magnetism and electricity, history of magnetism, types of magnets, and magnetic materials. It also discusses the lines of magnetic forces and force generated in the field. Chapter 2 presents magnetic circuits and quantities. The description of electrical circuit elements, its connection in series and parallel, and its phasor diagram are discussed in Chapter 3 under the title of Alternating Current Power.

Chapter 4 deals with an electric transformer, installation of the transformer, core shape, principle of operation, and the operation of the transformer under no load and full load conditions. Chapter 5 deals with the transformer design techniques. It includes the conventional design of core and shell type for single- and three-phase transformers. It illustrates all the calculation that is required to design a transformer. The description details of all types of DC machine and related voltage, current, power, and efficiency are covered in Chapter 6. All types of AC machines are also covered in Chapter 7. The chapter also elaborates the related voltage, current, power, and efficiency.

In Chapter 8, the principles of conversion from AC to DC involving single-phase as well as three-phase are presented. DC choppers and the study of several applications of power electronics are also mentioned. Chapter 9 discusses electric drives in general and concept of DC drive in particular. Chapter 10 describes the basic principles of speed control techniques employed in three-phase induction motors using power electronics converters.

Chapter 11 introduces some machines that have special applications. The examples explained in this chapter include stepper motors, brushless DC motor, switched reluctance motor, servomotors, synchro motors, and resolvers.

Several problems are provided at the end of each chapter. The answers to odd-number problems appear in Appendix D.

It is not necessary that the reader has previous knowledge of MATLAB®. The material of this text can be learned without MATLAB. However, the authors highly recommend that the reader studies this material in conjunction with the MATLAB Student Version. Appendix C of this text provides a practical introduction to MATLAB.

MATLAB® and Simulink® are registered trademarks of The MathWorks, Inc. For product information, please contact:

The MathWorks, Inc.
3 Apple Hill Drive
Natick, MA 01760-2098 USA
Tel: 508 647 7000
Fax: 508-647-7001
E-mail: info@mathworks.com
Web: www.mathworks.com

Acknowledgments

We are indebted to Dr. John Fuller and Dr. Penrose Cofie who reviewed the manuscript. We are very grateful to Dr. Kelvin K. Kirby, the interim head of the Department of Electrical and Computer Engineering for his support. Dr. Pamela Obiomon, the Dean of the Roy G. Perry – College of Engineering, for providing a sound academic environment at Prairie View A&M University, We are very thankful to Dr. Ruth Simmons, the President of Prairie View A&M University, who is consistently encouraging faculties to write books. Special thanks to our graduate students, Nafisa Islam, Faduma Sheikh Yusuf, Yogita Akhare, Kenechukwu Victor Akoh, Chandan Reddy Chittimalle, and, Akeem Green, for going over the manuscript and pointing out some errors.

Finally, we express our profound gratitude to our wives and children, without whose cooperation this project would have been difficult if not impossible. We appreciate feedback from students, professors, and other users of this book. We can be reached at whali@pvamu.edu, sabood@student.pvamu.edu, and sadiku@ieee.org.

Authors

Warsame Hassan Ali received his BSc from King Saud University Electrical Engineering Department, Riyadh, Saudi Arabia, and his MS from Prairie View A & M University, Prairie View, Texas. He received his PhD in Electrical Engineering from the University of Houston, Houston, Texas. Dr. Ali was promoted to associate professor and tenured in 2010 and professor in 2017. Dr. Ali joined NASA, Glenn Research Center, in the summer of 2005, and Texas Instruments (TI) in 2006 as a faculty fellow.

Dr. Ali has given several invited talks and is also the author of more than 100 research articles in major scientific journals and conference. Dr. Ali has received several major National Science Foundation (NSF), Naval Sea Systems Command (NAVSEA), Air Force Research Laboratory (AFRL) and Department of Energy (DOE) awards. At present, he is teaching undergraduate and graduate courses in the Electrical and Computer Engineering Department at Prairie View A & M University. His main research interests are the application of digital PID Controllers, digital methods to electrical measurements, and mixed-signal testing techniques, power systems, High Voltage Direct Current (HVDC) power transmission, sustainable power and energy systems, power electronics and motor drives, electric and hybrid vehicles, and control system.

Samir Ibrahim Abood received his BSc and MSc from the University of Technology, Baghdad, Iraq in 1996 and 2001, respectively. From 1997 to 2001, he worked as an engineer at the same university. From 2001 to 2003, he was an assistant professor at the University of Baghdad and at AL-Nahrain University, and from 2003 to 2016, Mr. Abood was an assistant professor at Middle Technical University/Baghdad – Iraq. Presently, he is doing his PhD in an electrical power system in the Electrical and Computer Engineering Department at Prairie View A & M University. He is the author of 25 papers and 3 books. His main research interests are in the area of sustainable power and energy system, microgrid, power electronics and motor drives, and control system.

Matthew N. O. Sadiku received his BSc in 1978 from Ahmadu Bello University, Zaria, Nigeria, and his MSc and PhD from Tennessee Technological University. He was an assistant professor at Florida Atlantic University, where he did graduate work in computer science. From 1988 to 2000, he was at Temple University, Philadelphia, Pennsylvania, where he became a full professor. From 2000 to 2002, he worked with Lucent/Avaya, Holmdel, New Jersey, as a system engineer and with Boeing Satellite Systems as a senior scientist. At present, he is a professor at Prairie View A & M University.

He is the author of over 500 professional papers and over 80 books, including *Elements of Electromagnetics* (Oxford University Press, 7th ed., 2018), *Fundamentals of Electric Circuits* (McGraw-Hill, 6th ed., 2017, coauthored with C. Alexander), *Numerical Techniques in Electromagnetics with MATLAB®* (CRC Press, 3rd ed., 2009), and *Metropolitan Area Networks* (CRC Press, 1995). Some of his books have been translated into French, Korean, Chinese (and Chinese long form in Taiwan), Italian, Portuguese, and Spanish. He was the recipient of the McGraw-Hill/Jacob Millman Award in 2000 for an outstanding contribution in the field of electrical engineering. He was also the recipient of the Regents Professor award for 2012–2013 given by the Texas A&M University System.

His current research interests are in the area of computational electromagnetics and computer communication network. He is a registered professional engineer and a fellow of the Institute of Electrical and Electronics Engineers (IEEE) "for contributions to computational electromagnetics and engineering education." He was the IEEE Region 2 Student Activities Committee chairman. He was an associate editor for IEEE Transactions on Education. He is also a member of the Association for Computing Machinery (ACM).

1

Basic Concepts of Magnetism

I am a slow walker, but I never walk backwards.

Abraham Lincoln

Magnetism is a force generated in the matter by the motion of electrons within its atoms. Magnetism and electricity represent different aspects of the force of electromagnetism, which is one part of nature's fundamental magnetic force. The region in space that is penetrated by the imaginary lines of magnetic force describes a magnetic field. The strength of the magnetic field is determined by the number of lines of force per unit area of space. Magnetic fields are created on a large scale either by the passage of an electric current through magnetic metals or by magnetized materials called *magnets*. The elemental metals—iron, cobalt, nickel, and their solid solutions or alloys with related metallic elements—are typical materials that respond strongly to magnetic fields. Unlike the all-pervasive fundamental force field of gravity, the magnetic force field within a magnetized body, such as a bar magnet, is polarized—that is, the field is strongest and of opposite signs at the two poles of the magnet.

1.1 History of Magnetism

The history of magnetism was dated to earlier than 600 B.C., but it is only in the twentieth century that scientists have begun to understand it and develop technologies based on this understanding. Magnetism was most probably first observed in a form of the mineral magnetite called lodestone, which consists of an iron oxide—a chemical compound of iron and oxygen. The ancient Greeks were the first known to have used this mineral, which they called a magnet because of its ability to attract other pieces of the same material and iron.

The British physicist William Gilbert (1600 B.C.) explained that the earth itself is a giant magnet with magnetic poles that are somewhat distracted from its geographical poles. The German scientist Gauss then studied the nature of earth's magnetism, followed by the French scientist Koldem (1821 A.C.) known that the magnet is a ferrous material only.

Quantitative studies of magnetic phenomena initiated in the eighteenth century by Frenchman Charles Coulomb (1736–1806), who established the inverse square law of force, which states that the attractive force between two magnetized objects is directly proportional to the product of their individual fields and inversely proportional to the square of the distance between them. Danish physicist Hans Christian Oersted (1777–1851) first suggested a link between electricity and magnetism. Experiments involving the effects of magnetic and electric fields on one another were then conducted by Frenchman Andre Marie Ampere (1775–1836) and Englishman Michael Faraday (1791–1869), but it was the Scotsman, James Clerk Maxwell (1831–1879), who provided the theoretical foundation to the physics of electromagnetism in the nineteenth century by showing that electricity and magnetism represent different aspects of the same fundamental force field. Then, in the late 1960s, American Steven Weinberg (1933–) and Pakistani Abdus Salam (1926–), performed yet another act of theoretical synthesis of the fundamental forces by showing that electromagnetism is one part of the electroweak force.

The modern understanding of magnetic phenomena in condensed matter originates from the work of two Frenchmen: Pierre Curie (1859–1906), the husband and scientific collaborator of Madame Marie Curie (1867–1934) and Pierre Weiss (1865–1940). Curie examined the effect of temperature on magnetic materials and observed that magnetism disappeared suddenly above a certain critical temperature in materials like iron. Weiss proposed a theory of magnetism based on an internal molecular field proportional to the average magnetization that spontaneously aligns the electronic micro magnets in the magnetic matter. The present-day understanding of magnetism based on the theory of the motion and interactions of electrons in atoms (called quantum electrodynamics) stems from the work and theoretical models of two Germans, Ernest Ising (1900–) and Werner Heisenberg (1901–1976). Werner Heisenberg was also one of the founding fathers of modern quantum mechanics.

1.2 The Cause of Magnetism

The prime reason of magnetism is the movement of electrons in a particular object. The movement of electrons in specific orbits around the nucleus is as same as the motion of the planets around the earth and the rotation of the earth around its axis. In this case, there are two movements: orbital and rotational. Due to these movements, a magnetic moment is produced on each electron that behaves like an intricate magnet. In a particular matter, the numbers of all electrons of an atom are equal. In each orbit, there are many pairs of electrons that are usually remaining as many pairs in each orbit and each electron is rotating in two opposite direction. As a result, the total torque that was created by electrons in each orbit is zero.

The metallic bond is usually seen in pure metal and some metalloids. Most of the metal is symbolized by high electrical and thermal conductivity as well as by malleability, materials contain transition atoms. On the other hand, there are some metal that contains an odd number of an electron at their outer shell such as—iron, cobalt, and nickel. These metals have their own specific value for the sum of the torque. Some of which contain an odd number of electrons such as iron, cobalt, and nickel. The sums of torque in these orbits contain a specific value. The positively charged particles of the magnetic materials gain the magnetic effect generated by the electrons which are revolving around them. Positively charged ions or group of atoms (about 10^5 atoms) that are directed toward the same direction, is called a domain. Each field has its magnetic field with its northern and southern poles. Non-magnetized field of iron pieces is oriented in random, non-uniform directions. When placing the iron piece within a relatively weak external magnetic field, some iron fields will align with some and with the outer magnetic field. As the external magnetic field increases, lined and outward-facing fields will increase remarkably. As the strength of the outer field increases, we retain a situation in what all the fields of the iron segment are heading toward this area. Consequently, the object is perceived as being saturated magnetically, whereas all the fields have been lined up in the desired direction. In either case, lifting the effect of the magnetic field outside from the iron piece, the fields will return to the previous random states.

1.3 Types of Magnets

Magnets are divided into two types according to the composition method, namely, natural and artificial magnets. There are confined to the natural magnets in those present in the form of natural rocks such as the ancient Greeks and China, and the characteristics of these magnets to the rocks produced by nature and no human being. Natural rock or magnets are currently present in the United States of America, Sweden, and Norway, but it is not practical to use in known electrical appliances because of the ease and economy of making very powerful and efficient magnets. Synthetic magnets produce magnetic materials such as iron, nickel, and cobalt in various shapes and sizes.

To obtain very strong magnets, special alloys are used for iron or electrically magnetized steel. This is done by placing the magnetized iron piece inside a coil of insulated wire through which an electric current or pack of electrons pass through as shown in Figure 1.1, and the force that turns cutting iron into a magnet is a line of magnetic forces quite like the lines of the electric field. Synthetic magnets are classified into both temporary and permanent depending on their ability to retain magnetic properties after the magnetization is removed.

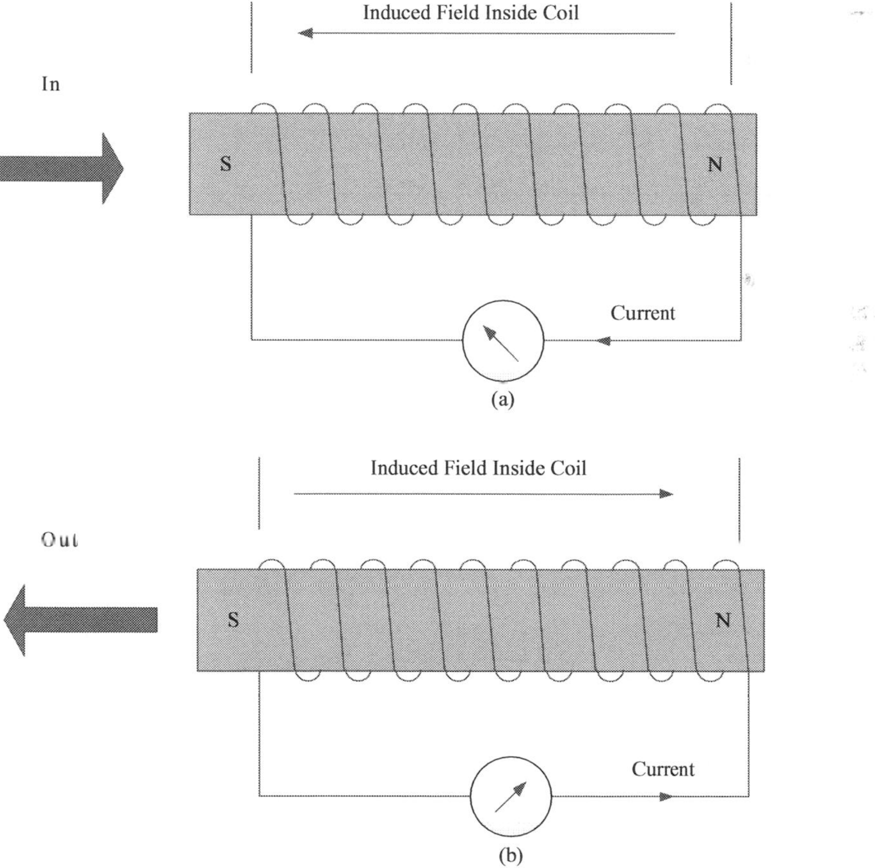

FIGURE 1.1
Magnetized piece of iron (a) When Magnetized piece of iron In direction case. (b) When Magnetized piece of iron Out direction case.

Magnetic materials such as iron, iron-silicon alloys, and nickel-iron alloys, used in the manufacture of transformers, motors, and generators have highly reluctance, which are easy to magnetize and give temporary magnets. However, these materials keep a small amount of magnetism after removing the external magnetization force. This magnetism is called *residual*, which is essential in the work of electric machines. Another criterion for measuring the susceptibility of a material to the formation of a permanent or temporary magnet is the ease with which these materials allow the lines of magnetic forces to be distributed within it. This is called *the permeability* of the material, so the permanent magnets have high impedance and low permeability while the temporary magnets low impedance and high permeability. The advantages of permanent magnets are: There is no operating cost, no maintenance, and high efficiency also doesn't need electrical power to operate, which excludes the possibility of failure of the feeding system, and they can be used in a hazardous environment, do not require electrical connections, and are not affected by shocks and vibrations. The disadvantage of permanent magnets is difficulty controlling their power. While the advantages of temporary magnets are ease of control over their ability, loadable, and withstand higher temperatures than permanent magnets, and can cut the feed and the field distance and edit the load easily. The disadvantages of temporary magnets, it needs a power source and exposure to the problems of potential disruption of this source, and that it can operate in a specific environment because it does not resist water, heat, and shocks.

1.4 Applications of Magnet

Magnets are widely used in power generation, transmission, and transmission systems. They are the basis for the work of the electrical stations that convert mechanical energy (in the form of heat to generate steam or waterfalls) to electrical power. In addition, it is the basis for the work of electrical transformers that convert the voltage from the amount supplied to other quantities needed by electrical and electronic devices. Magnets are found in electric motors which convert electrical energy into kinetic energy to move household, office, or industrial equipment.

Permanent magnets are used in sensitive and durable devices in sensitive and accurate devices such as loudspeakers, earphones, and measuring devices, in addition to vital and important uses in the field of magnetic recording and computer storage units and in audio-visual systems, see Figure 1.2.

1.5 Magnetic Materials

Magnetic materials are classified according to the properties they possess.

Ferromagnetism: These material exchange electronic forces in ferromagnetism are very large, thermal energy eventually overcomes the exchange and produces a randomizing effect. This occurs at a particular temperature called *Curie temperature*. In this process, magnetic fields are automatically directed in one direction at the Curie temperature to generate a magnetic field. At the Curie temperature, these magnetic fields are directed back to their random direction of

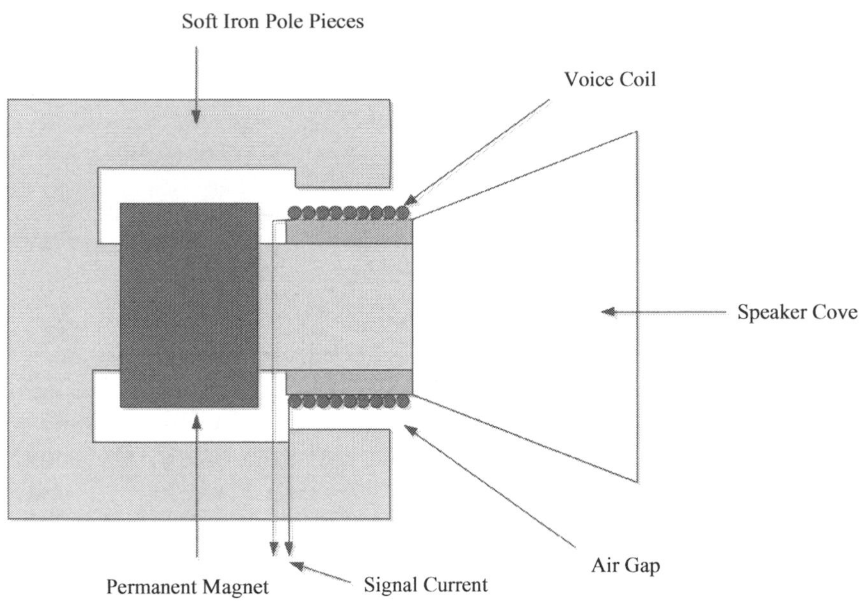

FIGURE 1.2
Magnet applications in the loudspeakers.

high temperature. The intensity of one field is equal in size and is the direction of one field parallel to some in the same direction. Also, a total torque of each field does not have to be in the direction of the other field's intensity itself. Thus, the piece of the electromagnetic material may not have a total magnetic torque, but when placed within an external magnetic field, the inhibition of the individual fields will go together in the same direction forming a total determination. So, the notable magnetic property of the ferromagnetic material is to turn into another kind of magnetic material at Curie temperature. The elements Fe, Ni, and Co are typical ferromagnetic materials which are characterized by the ability to attract and repulse with other magnetic objects.

Paramagnets: This substance that shows a positive response to the magnetic field, but it's a weak response. The amount of response is determined by a standard called *magnetic susceptibility*, which represents the ratio of the magnetic force of the material to the magnetic field strength, which is a quantity without units. The paramagnetic phenomenon is observed in substances whose atoms contain single electrons (while these electrons are usually conjugal and opposite), so its magnetic torque can't be zero. Paramagnetic materials include transient and rare substances in nature or generally substances whose atoms contain non-conjugate electrons. The magnetic properties of these substances depend on the temperature. Examples of these substances are sodium, potassium, and liquefied oxygen.

Diamagnetism: Is the material that is generated when placed within an external magnetic field, an opposition torque that contradicts the direction of the outer field and explains that as a result of the currents that are excited in atoms of the material with non-bilateral electrons are generated according to the law of Ampere of an opposite that contrasts with the field caused by this note that these materials contradict with magnets close to them. The relative permeability of these materials is slightly less than one, i.e., their magnetic intensity is negative. Examples of diamagnetism materials are copper, gold, carbon, diamonds, nitrogen gas, and carbon dioxide.

1.6 Lines of Magnetic Forces

A simple experiment is executed to acquire the properties or the behavior of magnetic materials. For the experiment, a permanent magnetic rod is used to put it on paper and scattered iron filings on the rod. It is noted that the distribution of these filings would be in such a way that remarks to the direction of the magnetic force lines generated by the magnetic bar as shown in Figure 1.3. The shape represents the nature of the magnetic field represented by the distribution of lines of forces that exit from the one terminals of the magnet and enter the other end of the so-called terminals. In this case of poles, the lines of forces involve the three-dimensional space surrounding the magnetic rod, while the experiment can show only two dimensions of them and poles include the entire length of the bar. But the intensity of the impact is concentrated at the terminals strongly and the party which the lines of force outcome is called *the north pole* and the party that enters the lines of forces called *the south pole* and symbolized with the letters N, and S, respectively. The magnetic field is called a dipole. This means that the magnetism of the bar (or any other magnet) is concentrated in two poles that are strongly equal to the opposite field in the direction of effect when cutting the bar magnetic to two parts each has a bipolar magnet, so that any number of bipolar magnets can be generated until one atom is reached, confirming that the atom is the primary source of the magnet. The magnetic field strength is measured by the number of lines of forces that leave a pole to interfere with the other, depending on the magnetic material and the magnetism and magnetic form. The magnets are usually in different shapes and sizes and are used in the form of discs, rods, or the horseshoe.

The latter is a rod-shaped like a letter (U) as shown in Figure 1.4. In this case, the magnet contains two poles, the lines of the forces of the pole, and interfere in the other exactly

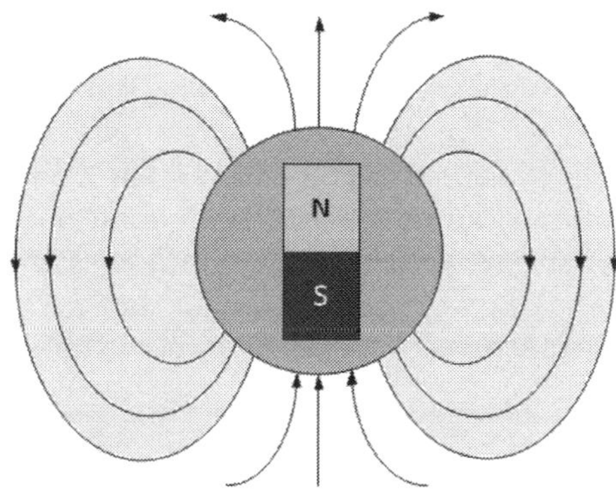

FIGURE 1.3
The distribution of lines of magnetic forces.

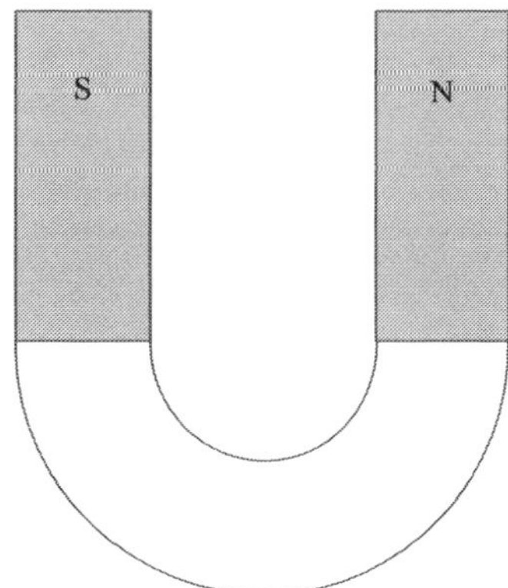

FIGURE 1.4
Magnets in the shape of a horse's suit.

how much in the case of the bar except that the magnetic field is more focused between the poles because they converge and direct lines between them in the air. When the magnets are placed on a magnetic rod or when the magnet is in the form of a loop, the circle of the path of the lines is closed and does not need to pass through the air, but will pass the majority in the metal body in the form of closed rings.

Some specialized devices are used to detect the lines of magnetic forces that can be sensed or influenced. These lines are surrounded through closed rings that emerge from the north pole to enter to mediate as shown in Figure 1.5. The speed of metal materials is useful to estimate the magnet forces such as iron when rounding from it. These lines belong to many properties:

1. Never intersect with each other.
2. All are identical and equal in intensity.
3. Take the shortest path with the least resistance between the poles.

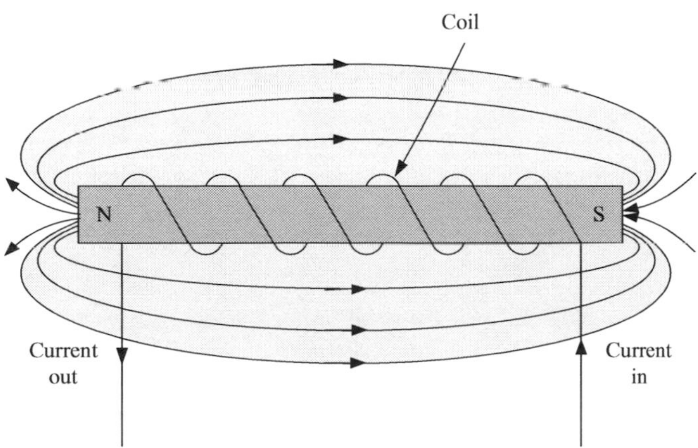

FIGURE 1.5
The lines of magnetic forces.

4. The intensity of the lines decreases as the distance between poles increases.
5. Heading from the south pole to the north pole inside the magnet and vice versa outside.
6. The static lines do not move and impose a trend as if they are moving.

1.7 Magnetic Force

There is a similarity between the magnetic and electric fields, and there are certain differences between them, and there is generally a very strong reciprocal relationship between them, and we know in our study of the field of electricity that the two different charges and the two similar charges are released because of the force affecting them according to the law of Coulomb. In the magnetic field, magnetism behaves in the same context as the conduction of charges in the electric field.

The principle of magnetic effect is to attract another object and convert them into another magnet when it is placed nearby. When two magnetic objects are close by, a certain force will affect them, and this force acts as proportional to the magnitude of the field and inversely proportional with the square of the distance between them. Either this force may be an attraction of the opposite poles or a repulsion force when the polarity of the poles is similar. If only one of the two bodies is magnetic, then the magnetic body strongly attract the other non-magnetic body to be a ferromagnetic material.

If it is a magnetic material, dissipation occurs weakly in terms of its principle. The magnetic force affects within a limited distance that is described as the field of force. Each magnetic pole generates an area around where each magnetic pole generates, turn sheds the force of the objects in this field. This field is conceived as the set of force lines such as in the case of the magnetic field. These lines converge near the poles and spread outward as they become more distant from the poles as they diverge from some, and their number in the unit area corresponds to the intensity of the field in the designated area.

1.8 The Direction of Magnetic Field Lines

A magnetic field is induced when an electric current pass through a conductor, and the generated magnetic field line will be and directed to the direction of current passing through the conductor. This phenomenon can be observed by taking an isolated straight conductor passing through the center of a carton plate as in Figure 1.6. When a current is passed in this conductor, the iron filings are scattered on the carton plate. Due to passing current, the generated magnetic field makes the iron bar take concentric rings surrounding the conductor. Compasses are placed to know the direction of these rings. All of which affect the same direction around the loop. When the direction of the current passing through the conductor changes, the direction of the compasses will change by 180°.

In general, the direction of the force lines generated by the current passing through the conductor can be determined by using the right-hand rule, which is to capture the conductor with the right hand so that the thumb is parallel to the axis of the conductor and pointing to the current. The four fingers holding on the conductor will indicate the direction of the force lines as shown in Figure 1.7.

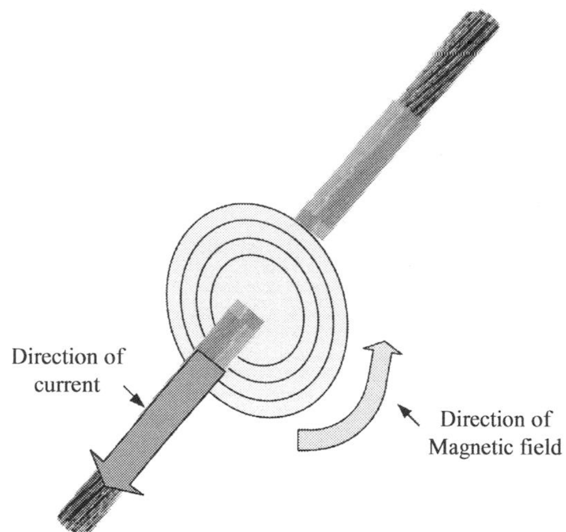

FIGURE 1.6
The direction of the generated field lines in the conductor.

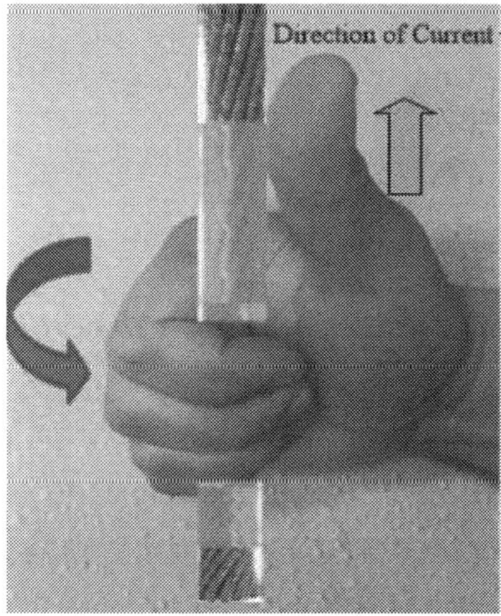

FIGURE 1.7
Directions of field lines.

The head of the arrow represents the direction of the current toward us, and the bottom of the arrow represents the direction of the current toward the page. Screwdriver method is used to detect the direction of the lines of the field. On Figure 1.8, screwdriver method is shown as well as the direction of currents and field. The spiral is in the direction of clockwise indicates the force lines, and the direction of the spiral is the direction of the current. When the coil is not plugged, the direction of the force lines is in the opposite direction of the anticlockwise. The characteristics of these lines are summarized as follows:

1. The magnetic field consists of regular concentric circles whose center corresponds to the axis of the conductor.
2. The direction of the lines depends on the direction of the current in the connector and the density of these lines on the current intensity.

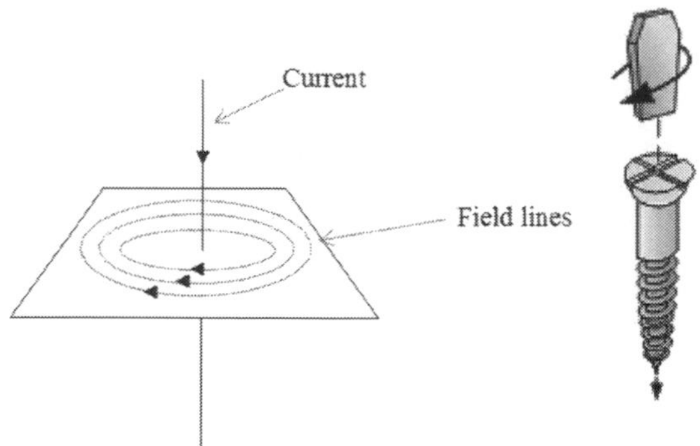

FIGURE 1.8
Apply the screwdriver method.

3. Field strength equal to zero in the axis of the connector, the maximum at the surface and less as move away from the surface of the connector.
4. All the rings are symmetrical relative to the axis of the connector, and if not, otherwise it is evidence of the presence of an influential external force that leads to it.
5. The lines of magnetic forces in two conductors carry a current remove each other, which constitutes the strength of attraction if the direction of the currents is similar, as in Figure 1.9, and support each other of the strength of repulsion if the direction of the two currents in contrast.

The magnetic field describes the most major functions is the lines of forces, that gives their number and total called *magnetic flux*, it symbolizes the letter (ϕ) and the unit is (Weber), and the flux is the flow of magnetic force represented by lines of force that move between the poles of the north and south of the part magnet. The properties of the magnetic field are given the density of the lines at a certain point in the field, and this is called *flux density*, it symbolizes the letter (B), and its unit is the Weber per the unit area (Wb/m^2) or (Tesla).

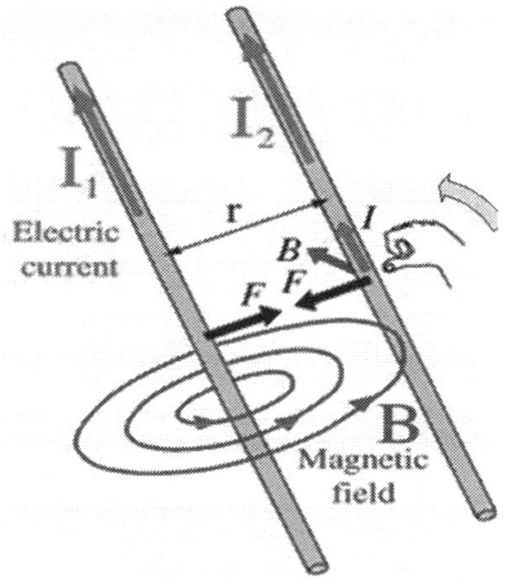

FIGURE 1.9
Two conductors carrying current. (http://WWW.phys4arab.net).

The one Weber is equal to one hundred million field lines (1 Wb = 10^8 field line) and given by thc formula:

$$B = \phi/A, \tag{1.1}$$

where A is the area of the cross-section (m^2) where the density of the lines is to be found.

Example 1.1

What is the density of the flux in the area of the number of lines passing through a rectangle of 40 × 250 cm, which are 12 million lines?

Solution

We convert the number of lines to Weber (the unit of the flux), where the Weber equals one hundred million lines:

$$\Phi = 12 \times 10^6 / 10^8 = 0.12\ Wb.$$

The area of the passage through which the lines of forces are:

$$A = 250 \times 40 \times 10^{-4} = 1\ m^2$$

So, the density of the flux is equal:

$$B = 0.12/1 = 0.12\ Wb/m^2.$$

1.9 Magnetic Field and Its Polarity

In 1820, Oersted discovered the relationship between magnetism and electrical when he showed that the current carrying wire generates its own magnetic field. When the wire forms a solenoid and a current is passed, the sphere of single coil rings accumulates to generate a strong magnetic field within the axis of the coil, as shown in Figure 1.10. In the case of putting a rod of iron or plastic material of ferromagnetic within this coil as shown in Figure 1.11, the magnetic field will increase several times. This arrangement is very

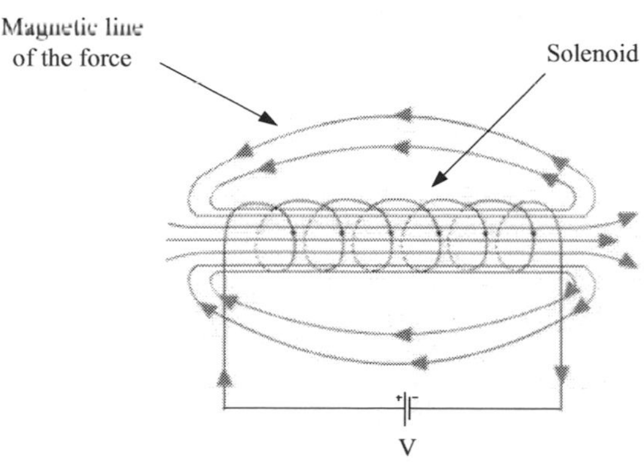

FIGURE 1.10
Field lines of the magnetic field through and around a current carrying solenoid.

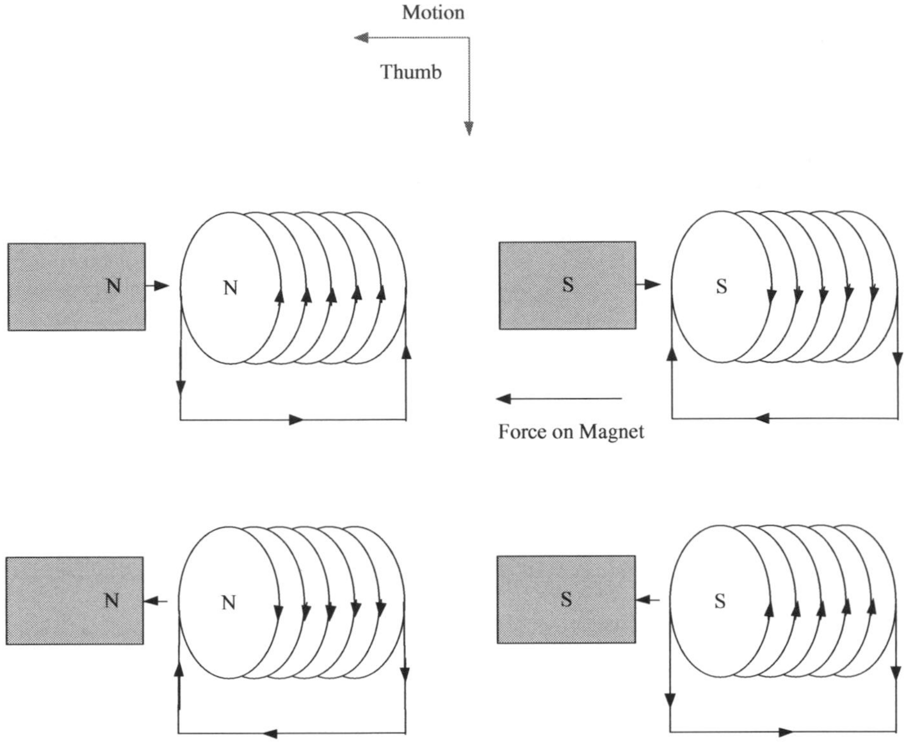

FIGURE 1.11
Change the direction of winding.

important and has many uses and is called *electromagnetic* that the force lines come out of the coil carrying the current on the north (N) pole and the enter in the other side of the south (S) pole.

The polarity of the solenoid is determined by using the right-hand rule, which states: Hold the coil with the right hand so that the four fingers represent the direction of the current passing through the coil. The parallel thumb of the coil axis, in this case, will affect the north pole of the magnetic field formed. When both the direction of the current and the direction of the winding of the coil as in (A), (B), or (C) and (D) in Figure 1.11, the polarity of the coil remains unchanged when the direction of the current is changed only, the polarity will change, and when the direction of the coil is changed only the polarity will change as shown in Figure 1.11, where there are four possibilities for these polar accordingly.

1.10 Magnetism of Magnetic Materials

Conversion of an ordinary piece of metal into a magnet is basically placed inside a spiral coil. This conversion takes place by regaining the magnetic properties and generating a magnetic moment in it. When the current passes through the solenoid, the magnetic field will pass its lines inside the iron, the magnetization of the iron piece by placing it within the magnetic field called indirect. The iron can be magnetized directly

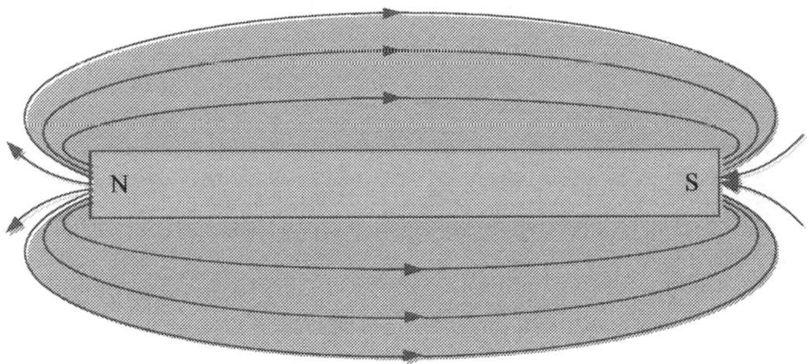

FIGURE 1.12
Field lines around a magnet bar.

by making it part of an electric circuit. The electric current passing through it is called *the magnetic current*, whose value determines the field strength of the object. The magnet can be directly cut off when it is rounded by permanent magnets. When the external magnetic field of the solenoid or permanent magnets is removed from the iron piece, its magnetic susceptibility will decay except for a small fraction called residual. When the length of the iron piece inside the solenoid is several times greater than the diameter, the lines of forces that pass through the iron piece are straight and parallel to the axis and denser within them much more than in the air surrounding this piece. This is because the air resistance to the passage of force lines is much more than that of the iron piece. When the lines of forces from the north pole of the magnetized piece are released, they are spread in the air in a large area because it can't carry this number of lines in the limited volume unit, as shown in Figure 1.12. The number of lines of magnetic forces or the flux (ϕ) generated by the solenoid is directly proportional to the current (I) and the number of coils (N) it consists of. The result of multiplying these two amounts is called *the magnetic motive force* (MMF) and is symbolized by letter (F) and its unit is ampere-turn (AT) is expressed mathematically as:

$$F = I * N. \tag{1.2}$$

1.11 Force Generated in the Field

When the coil is placed within an external magnetic field, its field will interact with the external field, and the nature of the reaction depends on the current of the current passing through the coil and on the polarity of the outer field or the direction of the force lines in it. To simplify the situation, we take one conductor carrying a current toward the page and put it within the field of external magnetic lines from the north pole to the south pole as shown in Figure 1.13. Note that the external field will emit the force of the conductor within this area because the lines of the outer areas, and to the connector to the right of the axis of the poles are in the same direction increasing density, while to the left of the axis of the poles and the opposite of each other, and become a few in this area.

The direction of force generated by the interaction between two fields can be determined using the rule of the left hand or using the filming rule as in Figure 1.14. The rule of the left-hand states that the left hand is extended within the magnetic field to the palm

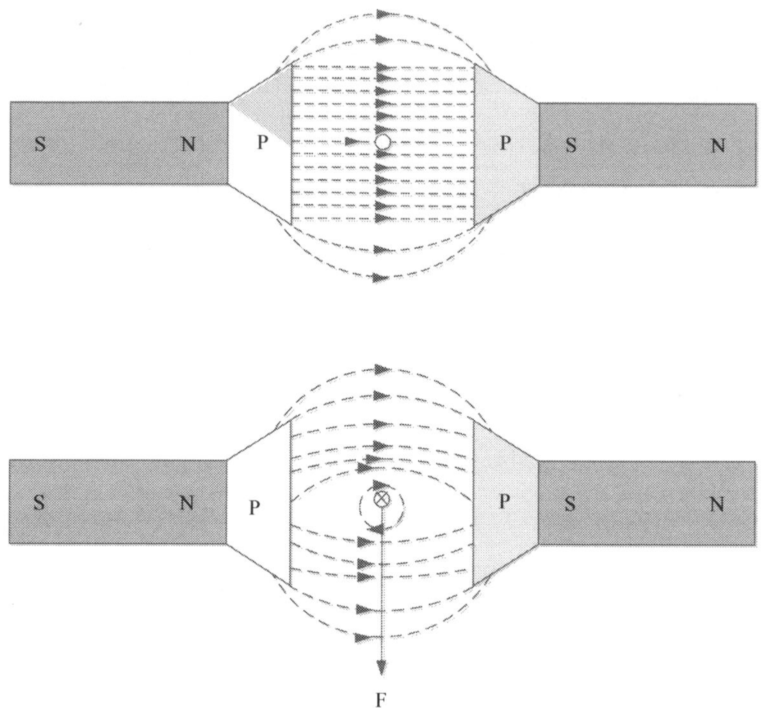

FIGURE 1.13
The departure of lines from the Arctic to the South Pole.

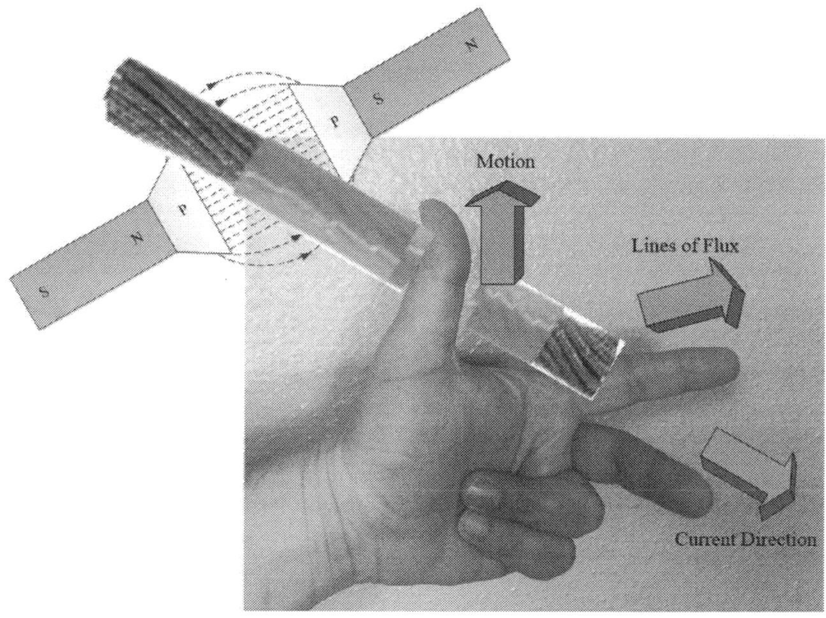

FIGURE 1.14
The rule of the left hand.

of the hand. The four fingers point to the direction of the current passing through the conductor. The orthogonal thumb with the four fingers indicates the direction of the expelling force (f). There are four possibilities for the direction of this force and depends on the direction of the lines of the outside field or its polarity and the direction of current passing through the conductor. The amount of force generated depends on the current of the current (I), the intensity of the force lines of field (B), and the effective length of the conductor within the field (L), which is usually equal to the length of the electrodes. This force is expressed as follows:

$$f = B * I * L. \tag{1.3}$$

The value of the force increases when the value of any amount on the right side of the equation increases. The length is also increased when making the conductor in the shape of a coil inside the magnetic field and often design the coil in the form of a frame that has two sides and when placed within the external magnetic field and make it moves freely around the axis, this carrier of the current will start circling around the axis. Because the direction of the upper and lower forces is one, the direction of the frame is identical to the clockwise movement. When the current is changed in the coil or the polarity of the external magnetic field changes, the direction of the frame rotation will be reversed. This phenomenon is one of the most important principles of the utilization of the transformation of electric energy into the kinetic energy of electric motors.

When the direction of the conductor movement within the magnetic field at a certain angle (θ), with the direction of the lines of the magnetic forces of this field, the expression of force becomes:

$$f = B * i * L * \sin\theta, \tag{1.4}$$

where: L distance of moving current i. The magnetic field affects the moving charge (Q) and the charge generates a moving current for a distance (L) in time (T) within the field, and the direction of the movement of the charge is at an angle (θ), the expression of force becomes:

$$f = B * Q * V * \sin\theta. \tag{1.5}$$

Example 1.2

When a 20 cm conductor is passed in which a current of 20 A is within a magnetic field, the force exerted on the conductor is 24 N. What is the density of the lines of this field?

Solution

The density of the field lines or the density of the magnetic flux is:

$$B = f/IL = 2.4/(20 \times 20 \times 10^{-2})$$

$$= 0.6 \text{ Wb/m}^2.$$

1.12 Hysteresis Loop

The (B-H) curve basically emphasizes the flux intensity and field strength of the material with additionally expresses the relationship between magnetic and non-magnetic materials and describes the former condition of it. The curve mentioned the relationship of a magnetic equation of the magnetizing and demagnetizing experiences which is drawn by a closed loop. In fact, it is also known as the most essential characteristics that represent the relationship of the field-density (B-H) of the material. It experiences a condition of material before the method of heat treatment and mechanical tension.

The magnetization curve can be drawn when a magnetic force is cast on a sample like steel or iron. When the magnetic force is applied, the curve starts to increase from zero to maximum, and it again goes to zero after removing the magnetization as shown in Figure 1.15. From the figure below, it can be seen that the field strength value goes higher when the magnetization field is lower which indicates the value of the flux intensity (B) is less than the field strength (H). After continuing to the value increase of magnetic force toward the negative direction, it goes again zero to maximum and again to zero as previous.

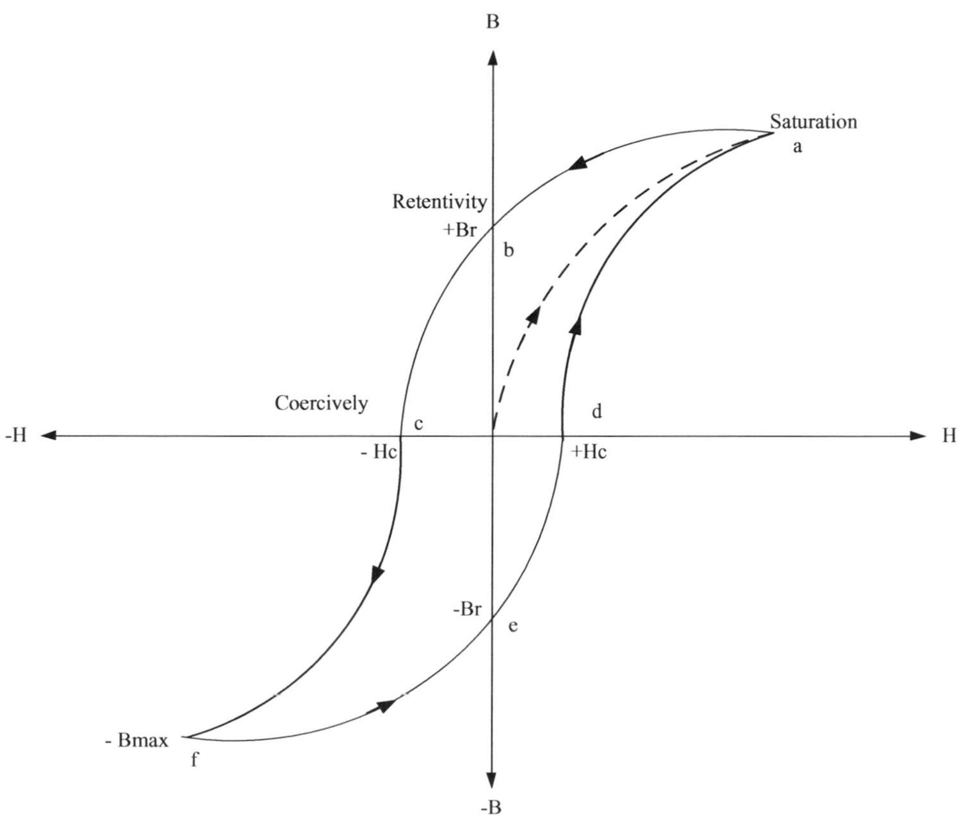

FIGURE 1.15
B-H curve.

From the figure, two very important values can be defined as below:

Residual Magnetism (B_r): is the amount of the intensity of the flux when the field strength is zero. It is symbolized as a residual.

Coercive Force (H_c): is the amount of field strength that makes the magnetism of (B) zero.

It is symbolized as a coercive force.

The hysteresis loop represents the properties of the magnetic material and the number of its basic quantities. The hinged ring may be wide. That is, the amount of (H_c) is roughly equal to (B_r) so that the magnetic material is low permeability and high impedance, and both the residual magnetism and coercive force are of high value. The loop may be tight, in other words, the amount of (B_r) is much higher than the amount of (H_c). The material is highly permeable and low impedance, with high residual magnetism and low coercive force.

Problems

1.1 What is the practical rule for passing the direction of the electric motive generated in a conductor moving within a magnetic field?

1.2 When the electric current passes in a coil, a magnetic field is formed. What are the polar potentials of this area and what do they depend on?

1.3 What are the factors in which the wide loop of hysteresis differs from those in the narrow hysteresis loop?

1.4 What are the common characteristics of all the ferromagnetic materials?

1.5 A conductor 0.32 m long with 0.025 Ω resistance is located within and normal to a uniform magnetic field of 1.3 T. Determine (a) the voltage drop across the conductor to cause a force of 120 N to be exerted on the conductor; (b) repeat part (s) assuming an angle of 65 degrees between the conductor and the magnetic field.

1.6 When a 25 cm conductor is passed in which a current of 30 A is within a magnetic field, the force exerted on the conductor is 20 N. What is the density of the lines of this field?

1.7 Find the linear velocity of a 0.54 m conductor that will generate 30.6 V when cutting flux in 0.86 T magnetic field.

1.8 What is the density of the flux in the area of the number of lines passing through a rectangle of 40 * 250 cm, which is 12 * 10^6 lines?

1.9 A coil of wire with 80 turns has a cross-sectional area of 0.04 m^2. A magnetic field of 0.6 T passes through the coil. Calculate the total magnetic flux passing through the coil.

2

Magnetic Circuit

Education is what remains after one has forgotten everything he learned in school.

Albert Einstein

Magnetic circuits are those parts of devices that employ magnetic flux due to inducing a voltage. Such devices include generators, transformers, motors, and other actuators as solenoid actuators and loudspeakers. In such devices, it is necessary to produce magnetic flit. This is usually done with pieces of ferromagnetic. In this sense, the magnetic circuits are like the electric circuits in which conductive material such as aluminum or copper has high electric conductivity and are used to guide electric current. The analogies between electric and magnetic circuits are two: the electric circuit quantity of current is analogous to magnetic circuit quantity flth. The electric circuit quantity of voltage or electromotive force (EMF) is analogous to the magnetic circuit quantity of magnetomotive force (MMF). EMF is the integral of electric field E, and MMF is the integral of magnetic field H.

2.1 Magnetic Quantities

The previous chapter discussed the various topics of many quantities describing the magnetic field. This section will define these quantities and give the mathematical expression and its units used to measure them as follows.

2.1.1 Flux Density

It represents the magnetic field near the poles of the magnet, where they emerge from the north pole and enter the south pole, and the total of these lines called *the magnetic flux* (ϕ), which is known as the total number of lines of forces in the magnetic field and the number of lines passing in the vertical area unit on the direction of the flux (B) and the flux unit is (Wb/m^2) or Tesla.

$$B = \phi/A \tag{2.1}$$

2.1.2 Permeability

It's characterized by the properties of magnetic materials, and the permeability is divided into absolute magnetic permeability, which is characterized by magnetic and non-magnetic materials and symbolized by the letter (μ), and relative permeability and symbolized by the letter (μ_r), which is the ratio of the absolute permeability of the material to the absolute permeability of air ($\mu_o = 4\ \pi * 10^{-7}$), and it's unit given in the Henry/meter of length, so the relative permeability, which is without unit, is equal to:

$$\mu_r = \mu/\mu_o \tag{2.2}$$

The relative permeability of air and non-magnetic materials ($\mu_r = 1$) and alloys of iron, nickel, and cobalt to tens of thousands.

Also, the absolute permeability of the material is known as changing the intensity of the flux (Δ B) relative to the corresponding change. The magnetic field intensity (Δ H):

$$\mu = \Delta B/\Delta H \tag{2.3}$$

Field intensity is the magnetic force of magnetic material or the amount of magnetic motive force in the unit of length necessary for the flow of magnetic flux:

$$H = F/L = I.N/L \tag{2.4}$$

Its unit is Amp. turn/meter or (A.T/m), and the density of the flux is approximately proportional to the field strength, and the relatively constant here is the permeability of the medium in which these lines of force pass:

$$B = \mu.H \tag{2.5}$$

The permeability of magnetic materials depends on the technique of manufacturer and arrangement of this material and gives the casting plants tables showing the permeability of their materials or the relationship between the density of the flux and the intensity of the field or may give the curves of the relationship (B-H), it's called *magnetization curves.*

2.1.3 Magnetic Reluctance

Is the resistance that faces the lines of forces passing in the middle and called *magnetic resistance* or *reluctance* and symbolized by the letter (S). The reluctance is opposing to the generation of magnetic flux and depends on the amount of the distance of the medium through which this flux and permeability of this medium can be derived expression through the flux on what follows:

$$\phi = B.A = F/S \tag{2.6}$$

This expression is like the Ohm's law (I = V/R) for this reluctance is equal:

$$S = \frac{L}{\mu A}, \left(\frac{\mu}{H/\mu \times \mu^2} = \frac{1}{H} \right) \tag{2.7}$$

Also:

$$S = \frac{F}{\phi} \tag{2.8}$$

So, the reluctance unit is a reciprocal Henry or amp-turn/Weber.

The magnetic motive force (F = I.N) is the source that generates the motive and drives it in the magnetic circuit. The passivity of this circuit is called *the permeance*, which is like the conductivity in the circuit. Which is equal to the reciprocal resistance where the equalization of the reciprocal reluctance in Henry:

$$\frac{1}{S}=\frac{\mu}{L}.A=\frac{\phi}{F} \tag{2.9}$$

The magnetic circuit is not very different from the electrical circuit, consisting of three main components, as in Figure 2.1:

- A voltage source (E)
- Connection wires as a conductor for the current
- Load (Lamp) consumes electricity and converts it to another form
- Resistance (R).

The magnetic circuit as shown in Figure 2.2 consisting of:

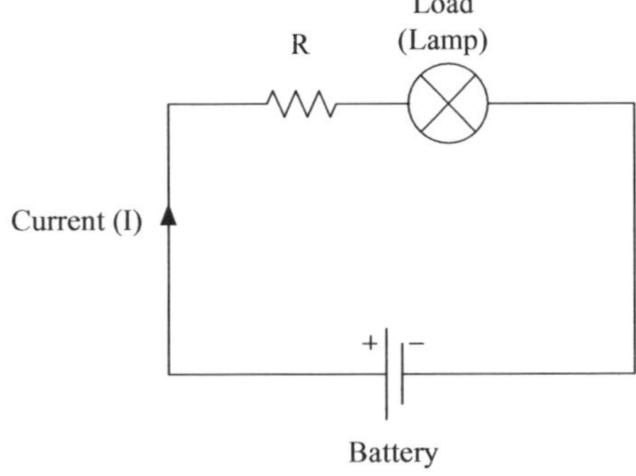

FIGURE 2.1
Simple electrical circuit.

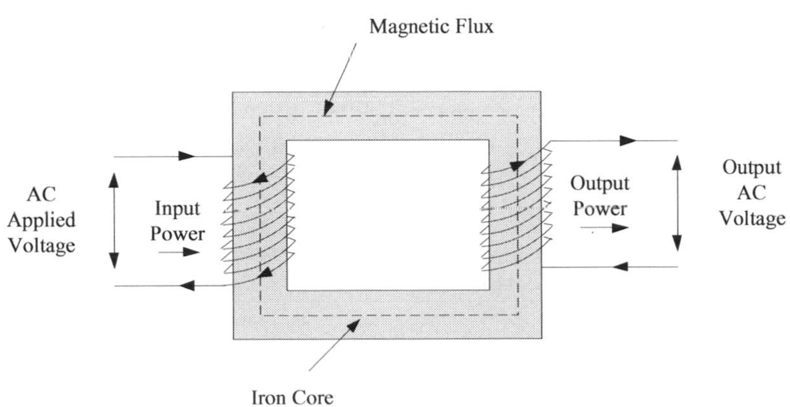

FIGURE 2.2
The magnetic circuit.

- A source of magnetic motive force (F) is simply a coil consisting of a number of turns (N_1) and passes by the current (I_1) because of the voltage applied (V_1).
- An arrangement of a magnetic material that is a path to the passage of the magnetic flux (ϕ).
- Load receives magnetic energy and transfers it into another form. It is here in the form of a second coil consisting of the winding (N_2), and where the ratio of the two voltages (V_2/V_1) as the ratio of the number of turns in the coils (N_2/N_1).

Magnetic path is one of the most important parts of the magnetic circuit and is usually made of a highly suitable ferromagnetic material for magnetization in the form of a closed loop, often different, for example, a circular ring or square or rectangular frame. The path may be simply consisting of a single series or complex consisting of several rings formed of more than two parallel branches. The magnetic path may be made without cutting. There may be parts in the closed, if empty, called *the air gap*, when the conduct of a single material with a specific reluctance and without an air gap, it is uniform, and the calculation of its coefficients is simple. When the path consists of two or more different objects with different reluctances or parts of which have a different sectioned area or when there is an air gap, the behavior is not uniform, and some complexity is complicated by the method of calculation. When calculating the magnetic circuit, you must know the dimensions and nature of the material of each part.

There are two ways to calculate the parameters of the magnetic circuit, depending on the information known and what is needed to find in them:

First: From the knowledge of finding the path and the current that passes through the coil required to find the magnetic flux and to achieve this follow the following steps:

1. Calculate the length of the wire (L) and the segment area (A).
2. Assume several values of the flux density (B) in the circuit and extract the corresponding flux ($\phi = B.A$).
3. We use the B–H of the path material to find the appropriate field strength (H) for each density of flux (B), and then according to the magnetic motive force of each amount ($F = H.l$).
4. Plot the relationship curve between ϕ and F and for the coil defined by its data, we find the amount ($F = I.N$).

Second: From the knowledge of the dimensions of the path, the coil, and the flux that passes through the path to be found and the amount of current that achieves it.

1. From the knowledge of magnetic flux, we find the density of the flux ($B = \phi/A$), and using the magnetization curve (B–H), we find the field strength (H).
2. From the knowledge of (L) and (H), we find the amount of magnetic motive force ($F = HL$), and from the knowledge of the number of turns, we find the amount of current passing through the coil ($I = F/N$).

As in the case of a continuous circuit, the same flux passes through all its parts. There is no branch or node that leads to its distribution.

Example 2.1

The coil has 200 turns and the flux of magnetization passing through it 1 A is placed on magnetic conduction equal to the same permeability 0.2. Calculate the value of the flux passing through the path with an average diameter of 10 cm and the diameter of the ring segment 4 cm.

Solution

First, the length of the ring:

$$L = \pi.D_a = \pi \times 10 \times 10^{-2} = 0.314 \text{ m}.$$

The cross section Area:

$$A = \pi.\left(\frac{D}{2}\right)^2 = \pi \times 2^2 \times 10^{-4} = 0.001256 \text{ m}^2.$$

The magnetic reluctance:

$$S = L/\mu.A = 0314/0.2 \times 0.001256 = 1250 \text{ AT/Wb}.$$

Magnetic motive force is:

$$F = I.N = 1 \times 200 = 200 \text{ AT}.$$

The magnetic flux passing through the path is:

$$\phi = F/S = 200/1250 = 0.16 \text{ Wb} = 160 \text{ mWb}.$$

Example 2.2

A medium diameter iron ring 20 cm is made of a metal rod with a diameter of 2 cm, with a coil number of turns 100 turns. Which is the necessary magnetic current to pass in the coil to generate a magnetic flux of 0.5 mWb? If the relative permeability of material 400 and for the air $4\,\pi \times 10^{-7}$.

Solution

The length of the path of force lines:

$$L = \pi.D_A = \pi \times 20 \times 10^{-2} = 0.628 \text{ m}.$$

Wire cross section area:

$$A = \pi\,(2/2)^2 \times 10^{-4} = 3.14 \times 10^{-4} \text{ m}^2.$$

The magnetic flux density is equal to:

$$B = \phi/A = (0.5 \times 10^{-3})/(3.14 \times 10^{-4}) = 1.56\ \text{Wb/m}^2.$$

Field strength is equal to:

$$H = B/\mu = B/\mu_o = 1.56/4 \times 10^{-7} \times 400 = 310\ \text{A/m}.$$

The magnetization current is then equal (Tables 2.1 and 2.2):

$$I = H/N = 310/100 = 3.1\ \text{A}.$$

TABLE 2.1

Comparison of Electrical and Magnetic Quantities

Electrical Components		Magnetic Components	
The Amount	**Symbol**	**The Amount**	**Symbol**
Voltage or electric motive force	V or E	Magnetic motive force	$F = NI = HL$
Current	I	Flux	ϕ
Resistance	$R = \rho L/A = L/GA$	Reluctance	$S = L/\mu A$
Conductivity	$G = 1/\rho$	Permeability	μ

TABLE 2.2

Comparison between Electric and Magnetic Circuit

Electrical Circuit	Magnetic Circuit
The path traced by the current is known as the electric current.	The path traced by the magnetic flux is called as a magnetic circuit.
The current is actually flows, i.e., there is a movement of electrons.	Due to MMF, flux gets established and does not flow in the sense in which current flows.
There are many materials which can be used as insulators (air, PVC, synthetic resins, etc.) which current cannot pass.	There is no magnetic insulator as flux can pass through all the materials, even through the air as well.
Energy must be supplied to the electric circuit to maintain the flow of current.	Energy is required to create the magnetic flux, but is not required to maintain it.
The resistance and conductivity are independent of current density under constant temperature. But may change due to the temperature.	The reluctance, permanence, and permeability are dependent on the flux density.
Electric lines of flux are not closed. They start from a positive charge and end on a negative charge.	Magnetic lines of flux are closed lines. They flow from N pole to S pole externally while S pole to N pole internally.
There is continuous consumption of electrical energy.	Energy is required to create the magnetic flux and not to maintain it.
Kirchhoff current law and voltage law is applicable to the electric circuit.	Kirchhoff MMF law and flux law is applicable to the magnetic flux.
The current density.	The flux density.
EMF is the driving force in the electric circuit. The unit is volts.	MMF is the driving force in the magnetic circuit. The unit is ampere-turns.
There is a current I in the electric circuit which is measured in amperes.	There is flux Φ in the magnetic circuit which is measured in the Weber.
The flow of electrons decides the current in the conductor.	The number of magnetic lines of force decides the flux.

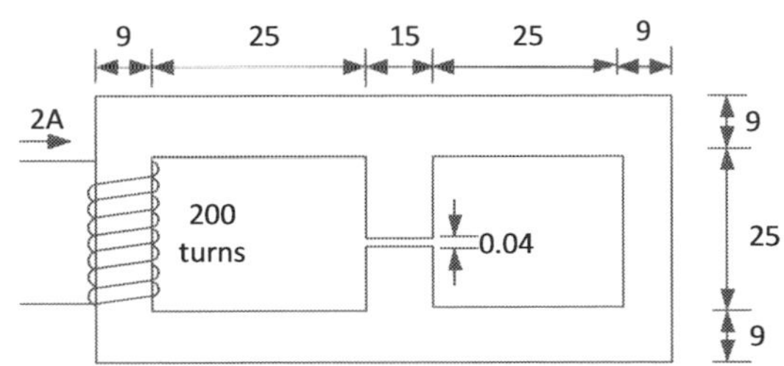

FIGURE 2.3
The magnetic circuit of Example 2.3.

Example 2.3

A magnetic core with three legs is shown in Figure 2.3. Its depth is 0.04 m, and there are 200 turns on the left most leg. The relative permeability of the core is 1500 with core lengths as shown. For the transformer circuit derive the following:

1. Equivalent circuit diagram with calculated values of magneto motive forces and reluctances. Neglect fringing at air gap
2. Flux Φg
3. Flux density in the left leg
4. Write MATLAB code to verify the answer in a, b, and c above.

Solution

Length right = length left = ($L_r = L_l$) = 2(0.075 + 0.25 + 0.045) + 0.09 + 0.25 = 1.08 m

Length of center path = L_C = 0.34 m

Air gap length (L_{ag}) = 0.0004 m

a) $$R = \frac{L}{\mu_o \mu_r A}$$

$$R_{right} = R_{left} = \frac{1.08}{4\pi \times 10^{-7} \times 1500 \times 0.09 \times 0.04} = 159154.94 \ AT/wb$$

$$R_{center} = \frac{0.34}{4\pi \times 10^{-7} \times 1500 \times 0.15 \times 0.04} = 30062.6 \ AT/wb$$

$$R_{air} = \frac{0.0004}{4\pi \times 10^{-7} \times 0.15 \times 0.04} = 53051.65 \ AT/wb$$

$$R_{total} = \frac{(53051.65 + 30062.6) \times 159154.94}{53051.65 + 30062.6 + 159154.94} + 159154.94 = 213755.54 \ AT/wb$$

b) $$\varphi_{total} = \varphi_{left} = \frac{200 \times 2}{213755.54} = 1.8713 \ mWb$$

$$\varphi_{right} = 1.8713 \times 10^{-3} \times \frac{53051.65 + 30062.6}{53051.65 + 30062.6 + 159154.94} = 0.6419 \;\; mWb$$

c) $Flux\;\; Density_{left} = \dfrac{1.8713 \times 10^{-3}}{0.09 \times 0.04} = 0.5198\; Wb/m^2.$

d) MATLAB Code:

```
clc;
clear all;
N=200; I= 2;
ur= 1500; u0= (4*pi*10^-7);
Lr= 1.08; Ll= 1.08; Lc= 0.34; Lair= 0.0004;
Ar=(0.09 * 0.04)
Ac=(0.15 * 0.04)
Aair=(0.15 * 0.04)
Rright= ((Lr)/(ur*u0*Ar))
Rleft= ((Ll)/(ur*u0*Ar))
Rcenter= ((Lc)/(ur*u0*Ac))
Rair= ((Lair)/(u0*Ac))
Rtotal= ((Rcenter+Rair)*(Rleft)/(Rcenter+Rair+Rleft))+Rright
Fluxtotal= (N*I)/Rtotal
FluxLeft= Fluxtotal
Fluxright= Fluxtotal*((Rcenter+Rair)/(Rcenter+Rair+Rright))
Fluxdensityleft= Fluxtotal/(Ar)
```

MATLAB Output:

Ar = 0.0036

Ac = 0.0060

Aair = 0.0060

Rright = 1.5915e+05

Rleft = 1.5915e+05

Rcenter = 3.0063e+04

Rair = 5.3052e+04

Rtotal = 2.1376e+05

Fluxtotal = 0.0019

FluxLeft = 0.0019

Fluxright = 6.4198e–04

Fluxdensityleft = 0.5198

2.2 Electromagnetic Induction

The English scientist Faraday first developed the rules of the relationship between electricity and magnetism when he discovered the magnetic effect of the electric current, followed by Lenz several years later, which is one of the most important of the progress of human civilization is the basis of the design and work of transformers, motors, generators, and a lot of electrical equipment and appliances. The Faraday simple experiment is shown in Figure 2.4.

When moving a piece of a magnet near the coil, the electric motive force will shrink in this coil to verify the movement of the index galvanometer bound with him and that the placement of a conductor within the magnetic field fixed as in Figure 2.5 and move the conductor to intersect the lines of the field of electrodes, an electric motive force will be induced in this conductor. From these experiments, the following facts can be inferred:

1. The direction of the magnet movement in the first experiment or the conductor in the second experiment has a direct effect on the direction of the displacement of the galvanometer index or on the direction of the electric motive force being produced
2. The magnetism of the coil close to the coil has a direct effect on the direction of displacement of the galvanometer index or the direction of the electric motive force being generated
3. Do not induce electrical motive force and the galvanometer remains stationary when both the coil and magnets are stationary, that is, there must be a relative movement between them to induce the force
4. When the speed of the magnet movement is constant and constant, the amount of motion in the coil will be alternately alternating depending on the speed of movement.

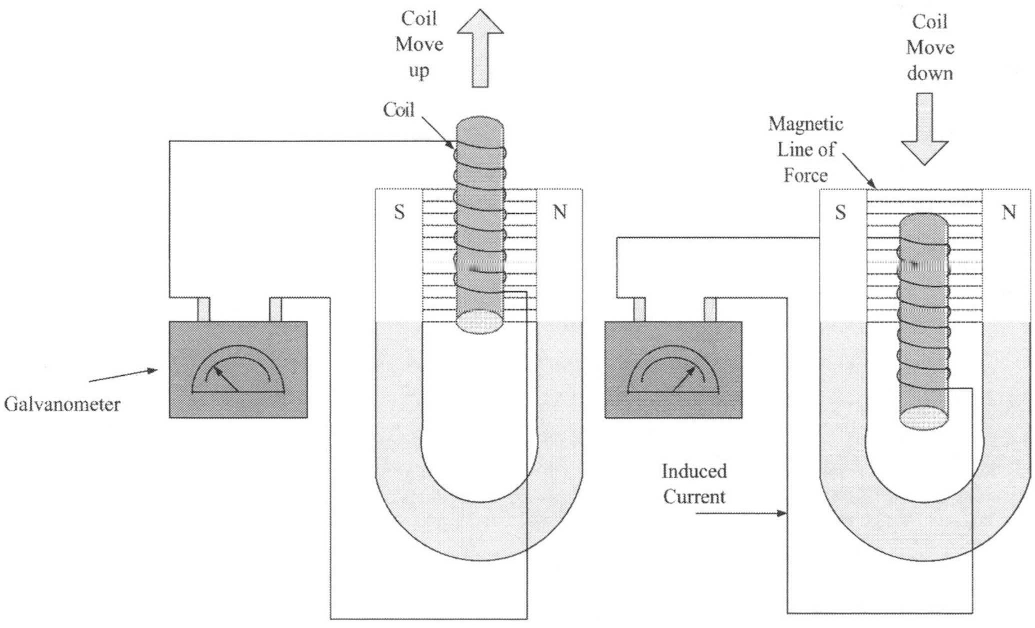

FIGURE 2.4
The Faraday experiment.

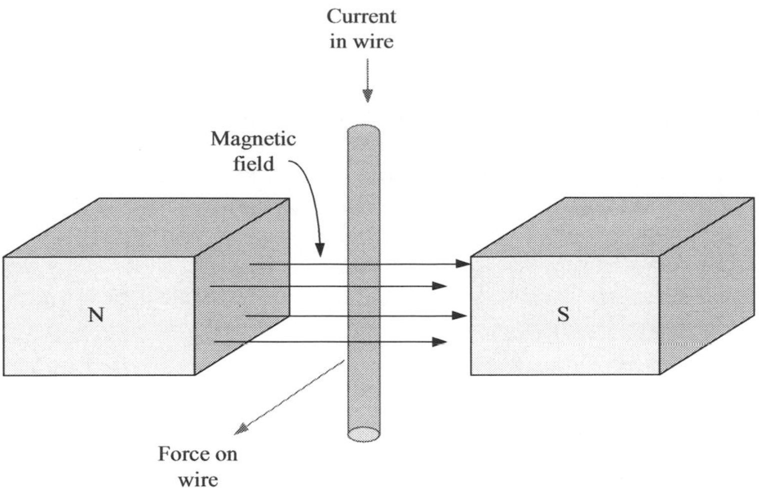

FIGURE 2.5
Place a conductor within a fixed magnetic field.

The amount of electrical motive force is dependent on the following:

1. The number of force lines that intersect the coil turns or the amount of the intensity of the flux (B) of the magnetic field
2. The number of coil turns because the total amount of electrical motive force generated in the coil is equal to the sum of the amount of electrical motive force produced in a single turn in the number of turns
3. The length of the coil (or conductor) that falls within the magnetic field and intersects its lines and is usually equal to the length of the electrodes (L)
4. The speed of movement of the magnet (or conductor) within the sphere (v), (the rate of the intersection of the force lines with the coil)
5. The direction of the motion of the magnet for the coil, or the conductor for the direction of the force lines, which is expressed in angle (θ) where equal zero when the direction of motion is parallel to the direction of force lines, and equal 90° when it is vertical.

In general, the amount of electrical motive force (e) induced by the following expression can be determined from:

$$e = B.L.N.\ sin\ \theta \tag{2.10}$$

Its direction is determined by using the right-hand rule, which provides the right-hand rest within the magnetic field so that the force lines enter it. If the orthogonal thumb with the rest of the four fingers indicates the direction of the movement of the conductor, the direction of the four fingers indicates the direction of the electric motive force, generally.

The direction in which a conductor is in a magnetic field is contrary to the direction of movement that caused it. This text is defined by the Lenz's law.

When the electric motive is driven in a loop-enclosed conductor, an electric current will pass through it, so that the magnetic effect of this current is counter to the change in the amount of flux.

This is consistent with the Lenz's law, which asserts that the direction of the current must be in such a way as to make the density of the field lines high in the path of the conductor to obstruct this movement. To illustrate this, we take Figure 2.6 where the current of the current in the conductor is within the magnetic field toward the page, and then the force F is generated to block the conductor movement within the field, the effect of the force generated in this case (F) helps in the movement of the conductor, which is not true because it is contrary to the Lenz's law. It is this conclusion that according to the right-hand rule referred to above that the direction of the current must match the direction which caused it to apply the Lenz's law to find the passing current in a spiraling coil due to the approximation of a magnet piece of it as in Figure 2.7a and b, the effect of the current must be obstructing the movement of the magnetic segment. This is done with the magnetic pole of the coil, like the magnet pole nearby. To determine the poles of the coil, hold it with the right hand so that the thumb is parallel to the axis of the coil and an indicator toward the north pole, the direction of the current toward the four fingers.

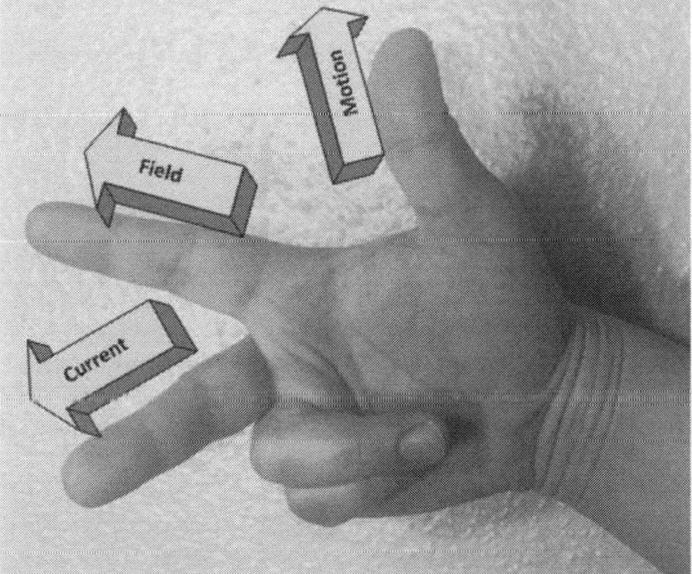

FIGURE 2.6
Direction of the current in the conductor.

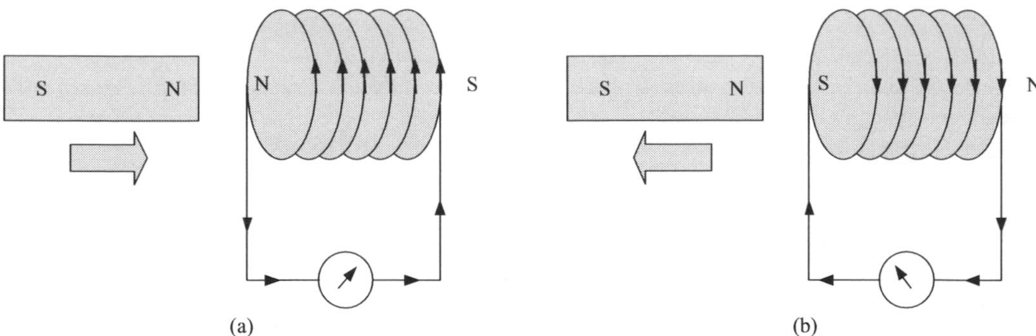

FIGURE 2.7
Direction of current passing in a spiraling coil. (a) Direction from S to N. (b) Direction from N to S.

2.3 Induced Electric Motive Force (EMF)

In summary of what has been mentioned, the conductor in the conductor moves within a magnetic field equal to the law of Faraday interchange rate of the flux with this conductor, or the coil of the N turn, this electric motive force will cause a current in the conductor to reverse the force of the conductor under the Lenz's law, if this force is in support of the conductor movement, we will have an acceleration and an increase in potential energy. This is not possible under the energy conservation law, that is, we can not get energy from anything. This is the rate of change of time over time. This is expressed mathematically as follows:

$$\mathrm{E} = -\frac{\Delta\phi}{\Delta T} \tag{2.11}$$

The presence of a negative sign on the right side of the equation is an explanation of Lenz's law, which indicates that the induction of the object is to counteract the change in the magnetic flux. When a coil of N is placed within the magnetic field, Equation 2.11 becomes:

$$\mathrm{E} = -\mathrm{N}\frac{\Delta\phi}{\Delta T} \tag{2.12}$$

This can be explained by a conductor moving within a magnetic field in another way. A force will affect the positive and negative conductor of the nucleus and the electrons of the conductor atoms so that the direction of the effect of the two charges is reversed. Since the conductor is metallic, the free mobility of the electrons will lead to negative charges increase in the negative sign charges of the conductor and decrease in the other end, this leads to the formation of the voltage difference between the two ends of the conductor, and the voltage difference is the electric motive force.

Example 2.4

A conductor curved single-turn that moves vertically on the lines of an area of a density of a flux 0.04 T that decreases at a constant rate to zero during 20 seconds. Find the value of electrical motive force in the conductor if the cross-section area is 8 cm^2.

Solution

The field flux:

$$\phi = BA = 0.04 \times 8 \times 10^{-4} = 3.2 \times 10^{-5}\ \text{Wb}.$$

The electrical motive force induced in a coil whose number of turns is N=1 is:

$$E = -N\frac{\Delta\phi}{\Delta T} = -\left(3.2 \times 10^{-5}/20\right) = -0.16 \times 10^{-5}\ \text{V}.$$

The placing of the coil within a variable magnetic field, or moving the coil within a static magnetic field leads to the generation of electric motive force, and this is the principle of all generators. Also, when placing a metal object within the range of magnetic variable, the electric motive force will urge in the example of what happens in the coil or connector. This electric motive force will be stimulated in the metal body when moving within the static magnetic field. Considering the metal body as a closed circuit, a current will pass through it.

As shown in Figure 2.8, this current is called *the eddy current*, a normal harmful current because it converts part of the electric energy into useless heat energy. To reduce

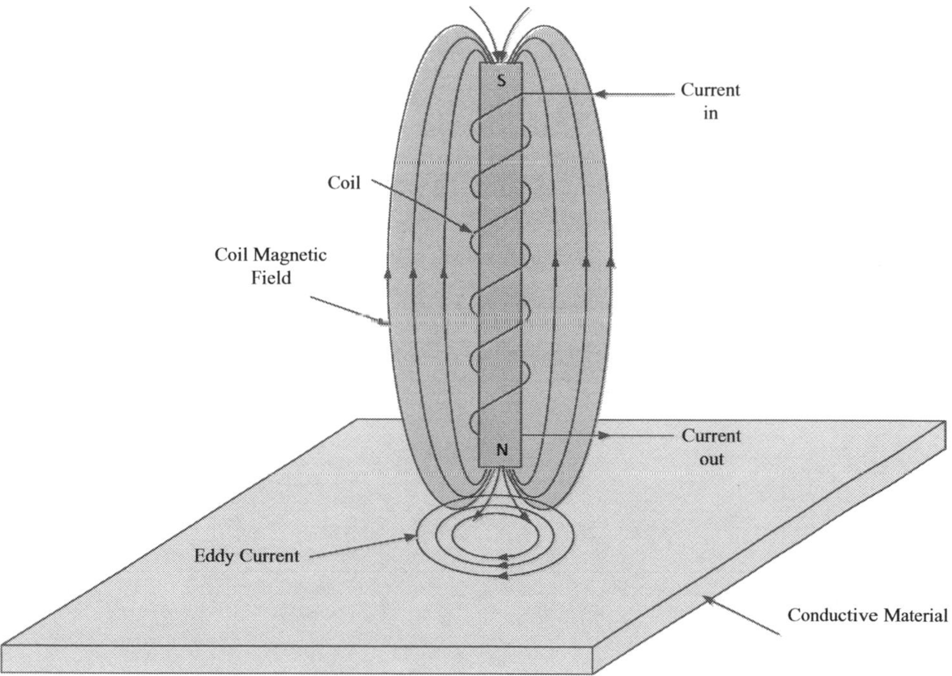

FIGURE 2.8
The eddy current.

the amount of this current increases the resistance of the body by adding silicon in metal casting, as well as to make the body a narrow segment isolated to reduce some of the total currents. It is worth mentioning that the currents are sometimes useful when used for industrial purposes such as modern furnaces.

2.4 Types of Inductance

There are two types of inductance, self and mutual.

2.4.1 Self-Inductance

When an electric current pass through a coil, it will generate a magnetic flux that intertwines with the coil turns and corresponds to the amount of this current. The proportional ratio between the flux and the current is called *inductance* and is symbolized by the letter L. The proportion is linear when the coil is wrapped alone in the air or on a non-magnetic object such as wood and paper and is not linear when the body is ferromagnetic. Inductance is expressed as:

$$L=\frac{N\phi}{I}=\frac{N}{I}\times\frac{NI}{S}=\frac{N^2}{S} \tag{2.13}$$

The inductance unit is Henry (H), and the only one is the inductance of the coil, which evokes an electric impulse of 1 volt when the current passing through the coil changes at a rate of 1 ampere per second.

Example 2.5

An iron ring with a coil on it of 200 turns with a current of 20 A. When the current decreases to 12 A, the flux changes from 1 mWb to 0.8 mWb. Find the amount of inductance of the coil.

Solution

Change the power supply:

$$\Delta I=20-12=8\,A.$$

And magnetic flux change:

$$\Delta\phi=1\times10^{-3}-0.8\times10^{-3}=0.2\times10^{-3}\,Wb.$$

The inductance is equal:

$$L=N.\phi/I=N.\frac{\Delta\phi}{\Delta I}=200\ (0.2\times10^{-3}/8)$$

$$=5\times10^{-3}H=5mH$$

In Equation 1.18, the inductance is directly proportional to the number of coil turns and vice versa with reluctance (S = L/NA), the self-inductance is directly proportional to the permeability of the magnetic medium and, conversely, with the length of the coil. The change of the flux with the coil leads to the induction of electric motive force, which is expressed as follows:

$$E=-N.\Delta\phi/\Delta t \tag{2.14}$$

From the knowledge that the flux of linkage is directly proportional to the current, the self-sustaining (e_L) in the coil is expressed as:

$$e_L=-\frac{\Delta(Li)}{\Delta t}=-L\frac{\Delta i}{\Delta t} \tag{2.15}$$

This expression also indicates that the inductance is the property of the magnetic circuit in its induction when the current in its coil changes. When the coil is wrapped around a ferromagnetic body, the permeability of the magnetic medium changes with the current change.

Example 2.6

A resistance of the coil is 50 Ω and the number of turns 200 turns, connecting in a row with a resistive galvanometer 300 Ω, and placed in the magnetic field amount of 2 mWb. Find the value of current in the coil when moving during 0.2 sec to 0.6 mWb.

Solution

The change the magnetic flux:

$$\Delta\phi = 2\times10^{-3}-0.6\times10^{-3}=1.4\times10^{-3}\text{ Wb.}$$

The time when the flux changes:

$$\Delta t = 0.2-0 = 0.2\text{ sec.}$$

The self-sustaining amount is:

$$E = N\frac{\Delta\psi}{\nabla t} = 200\left(1.4\times10^{-3}\right)/0.2 = 1.4\,\text{V}$$

The amount of current in the coil is equal:

$$I = E/R = 1.4/(300+50) = 4\text{ mA.}$$

2.4.2 Mutual Inductance

When placing two adjacent coils coupled on a metal rod as shown in Figure 2.9, you will be prompted to close (or open) the first key (S) in the first coil (e_1). This coil is self-stimulated and is proportional to the current (e_1) in the first circuit, and the factor of proportionality is self-inductor (L). This is expressed as follows:

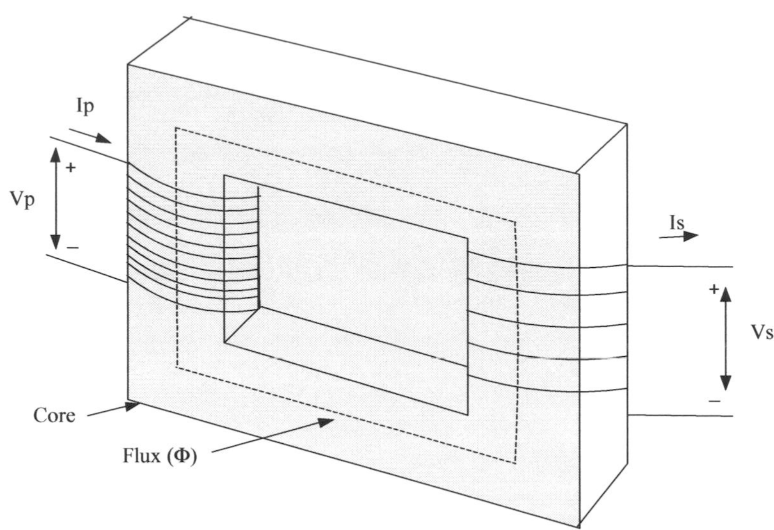

FIGURE 2.9
Mutual induction effect.

$$e_1 = -L1\frac{\Delta i1}{\Delta t} \tag{2.16}$$

The second one, which is generated in the second coil (e_2), is mutually induced by the change of current and therefore the magnetic flux in the first circuit. The expression of this unit is as:

$$E_{21} = -M_{21}\frac{\Delta i1}{\Delta t} \tag{2.17}$$

The self-sustaining factor is directly proportional to the rate of current change over time, and the proportionality factor is called mutual inductance and is denoted as letter *M*. In Equation 2.17, the mutual inductance (M_{21}) is the proportionality factor between the amounts of the object as reflected in the second coil due to the change in the amount of current in the first coil. The expression of the updated volume is also the rate of change of the first syllable of the coil with the second coil turns and the following:

$$E_{21} = -N_2\frac{\Delta \phi 1}{\Delta t} \tag{2.18}$$

The negative sign in this expression according to Lenz's law indicates that mutual induction seeks to create an opposite effect of the current change in the first coil when the second coil is closed. Thus, a current and therefore variable flux will consist of the intersection of the lines of the two coils. Two forces motives are energized: the first is self-contained in the second coil (e_2), the other is mutually induced in the first coil (e_{12}). In this case, M12 is the first coil inductance relative to the second coil. The mutual inductance unit is also Henry, and this is equal to 1 Henry when the current change in the rate of 1 ampere per second leads to a mutual exchange of 1 volt in the other coil. The amount of mutual inductance depends on the permeability of the magnetic medium and on the shape, dimensions, and position of the exchanged coils.

We conclude that the expression of the coupling factor between the two coils is:

$$K_c = \frac{M}{\sqrt{L_1.L_2}} \tag{2.19}$$

In the case where the coupling factor is ideal ($K_c = 1$), the mutual inductance is equal:

$$M = \sqrt{L_1.L_2} \tag{2.20}$$

Example 2.7

Two coils that are adjacent to the first current 0.5 A, and change at the rate of 0.01 sec. If the coefficient of the mutual inductance of both coils 0.1 H, calculate the electric motive force that arises in the second coil.

Solution

$$E_2 = M_{12}\frac{\Delta i}{\Delta t}$$

$$E_2 = 0.15 \times (0.5 / 0.01) = 7.5\,V$$

Example 2.8

A coil contains 200 turns with an electric current of 0.5 A, causing a magnetic flux of 1.5 mWb at 0.01 sec, place adjacent to a second coil with 50 turns and changing the current by 0.2 A causing a change in the flux by 2 mWb at a rate of 0.015 seconds. Calculate the mutual inductance and the electric motive force that arise in each of the two coils due to mutual induction if k = 0.9.

Solution

Induction coefficient in the first coil:

$$L_1 = N_1\frac{\Delta\phi 1}{\Delta I1} = 200 \times \left(1.5 \times 10^{-3}\right)/0.5 = 0.6\,H$$

Induction coefficient in the second coil:

$$L_2 = N_2\frac{\Delta\phi 2}{\Delta I2} = 50 \times \left(2 \times 10^{-3}\right)/0.2 = 0.5\,H$$

Mutual inductance between the two coils:

$$M = K\sqrt{L_1.L_2} = 0.9 \times \sqrt{0.6 \times 0.5} = 0.9 \times \sqrt{0.3} = 0.493\,H$$

The induced EMF in the first coil:

$$E_1 = M_{12}\frac{\Delta i1}{\Delta t1} = 0.493 \times (0.5/0.01) = 24.65\,\text{volt}$$

The induced EMF in second coil:

$$E_2 = M_{12}\frac{\Delta i2}{\Delta t2} = 0.493 \times (0.2/0.015) = 6.57\,\text{volt}$$

2.5 Stored Energy

When placing a coil number (N) on a metal rod, the opening of the switch (S) or the circuit breaker, the magnetic field will be gradually reduced. This means the completion of work goes to convert the magnetic energy into electrical energy, all of which are dissipated to heat in the coil turns, and this work is accomplished if the magnetic field disappears. When the S switch is closed again, the current (I) will pass in the coil (L), and the electric energy will come back again, and the magnetic field is storage. This stored energy is expressed as follows:

$$W = \frac{1}{2}LI^2 \tag{2.21}$$

Example 2.9

A coil has 500 turns, connect to a constant current source. The amount of generated magnetic flux 0.1 mWb, when passing through the current 1 A and 0.001 seconds, calculate the energy stored in the coil.

Solution

$$W = \frac{1}{2}LI^2$$

$$L = \frac{N\phi}{I}\left(500 \times 0.1 \times 10^{-3}\right)/1 = 0.05\,\text{H}$$

$$W = \frac{1}{2} \times 0.05 \times 1^2 = 0.025\,\text{Joule}$$

Problems

2.1 What are the areas of comparison between the electrical and magnetic circuits?

2.2 How do you determine the total inductance of two components, (a) in series (b) in parallel, in the cases of mutually reinforcing and antagonistic?

2.3 Two conductors carrying two currents in the same direction 200 A and 300 A, respectively, and distance from some distance 180 cm. Find the amount of force affecting each meter of length for the two conductors.

2.4 A new ring with the relative permeability of 800, and its circumference rate 120 cm. It is made of iron bar diameter 4 cm and a coil with 400 turns. What is the magnetization current required generating flux with an excess of 0.5 Wb?

2.5 A metal ring with a relative permeability of 800 and a diameter of 50 cm. A 20 mm diameter rod is formed, which a piece of length 2 mm is cut, the coil is placed on it the number of turns is 200 turns. How much current is needed to pass the coil to generate a magnetic flux 0.2 mWb in the air gap?

2.6 What is the relative permeability of the magnetic circuit length 20 cm, and the area of its regular section 4 cm^2, and its coil consist of 200 turns, and when the passing of 2 A in this coil generates a flux of 6.3 mWb.

2.7 The coil has 300 turns and the flux of magnetization passing through it 2 A is placed on magnetic conduction equal to the same permeability 0.2. Calculate the value of the flux passing through the path with an average diameter of 10 cm and diameter of the ring segment 4 cm.

2.8 A medium diameter iron ring 25 cm is made of a metal rod with a diameter of 3 cm, with a coil number of turns 120 turns. Which is the necessary magnetic current to pass in the coil to generate a magnetic flux of 0.25 mWb? If the relative permeability of material 400 and for the air $4\pi \times 10^{-7}$.

2.9 A conductor curved single-turn that moves vertically on the lines of an area of a density of a flux 0.05 T that decreases at a constant rate to zero during 25 seconds. Find the amount of electrical motive force in the conductor if the cross-section area is 10 cm^2.

2.10 An iron ring with a coil on it of 250 turns with a current of 25 A. When the current decreases to 18 A, the flux changes from 1 mWb to 0.8 mWb. Find the amount of inductance of the coil.

2.11 A resistance of the coil is 100 Ω and the number of turns 250 turns, connecting in a series with a resistive galvanometer 300 Ω and placed in the magnetic field amount of 2 mWb. Find the value of current in the coil when moving during 0.5 sec to 0.8 mWb.

2.12 A coil contains 250 turns with an electric current of 1.5 A, causing a magnetic flux of 1.5 mWb at 0.01 sec, place adjacent to a second coil with 50 turns and changing the current by 0.2 A, causing a change in the flux by 2.5 mWb at a rate of 0.005 seconds, calculate the mutual inductance and the electric motive force that arise in each of the two coils due to mutual induction if $k = 0.8$.

3

Alternating Current Power

Setting goals is the first step in turning the invisible into the visible.

Anthony Robbins

Alternating Current (AC) is used in large areas and in all different life facilities, for ease of generating. AC generators are electrical machines that operate on the principle of electromagnetic induction to generate electric power. AC is defined as the current whose value and direction change continuously over a period on a wave called *the sinusoidal wave* and as shown in Figure 3.1.

The advantages of alternating current are as follows:

1. Changes in value and direction and the polarity are not fixed
2. Generates mechanical methods by cutting magnetic fields such as generators
3. Widely used
4. Low cost of production, especially when generated with large power
5. It can be changed to constant current using electrical components
6. It can be converted from low voltage to high voltage and vice versa using electric transformers
7. It can transfer to long distances using high voltage towers.

3.1 Sinusoidal Wave Cycle and Frequency

The complete cycle of the sinusoidal, or the sine wave, is called *the period*, and the number of oscillations per second is called *the frequency*, and it is denoted by the letter F. From Figure 3.2, the portion between the two points (A–B) is called *the full wave*, the part bound between (A–C) is called *half the positive wave*, and (C–B) is called *half the negative wave*. The relationship between the time of the wave T and the frequency F for one cycle is:

$$F = \frac{1}{T} \tag{3.1}$$

$$T = \frac{1}{F} \tag{3.2}$$

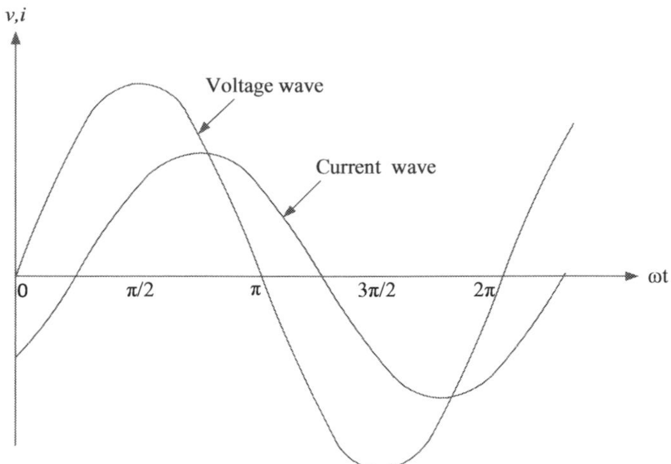

FIGURE 3.1
Voltage and current sine wave.

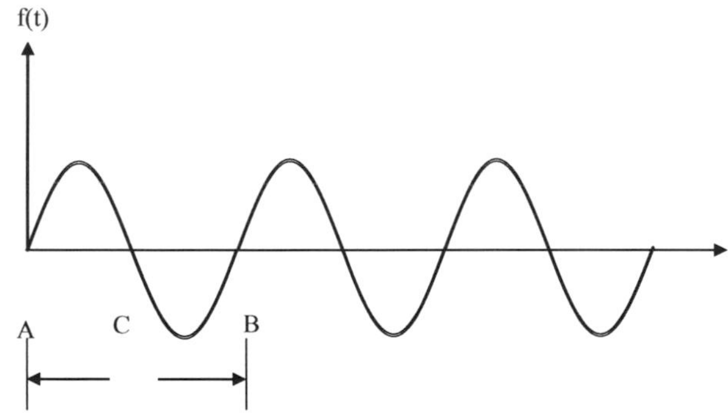

FIGURE 3.2
Sinusoidal waveform.

Example 3.1

If the one cycle is for a sine wave 10 msec, what is the frequency?

Solution

$$F = 1/T = 1/(10 \times 10^{-3}) = 1000/10 = 100 \text{ Hz}.$$

Example 3.2

Find the time of the one cycle if the frequency is 50 Hz.

Solution

$$T = 1/F = 1/50 = 0.02 \text{ sec}.$$

3.2 Electric Power Generation

Figure 3.3 shows the model of a generator that generates alternating current.

The above generator consists of a coil with copper wire in the form of a rectangular frame that connects its ends to isolated copper conductors installed on a rotational axis moving at a constant speed within two magnetic poles north and south.

Because of the conductor rotation and a side intersection with the lines of the magnetic forces in which an electric motive force is produced as in Figure 3.4, the value depends on the magnetic flux density (B), the length of the conductor within the magnetic field (L), and the velocity of the conductor (V). The angle formed by the direction of rotation with the direction of the lines of the magnetic field and the following expression for any moment is:

The instantaneous value of electric motive force.

$$e = B.\,L.\,V.\,sin\,\theta \tag{3.3}$$

As the coil revolves (n) of the cycles per minute, the frequency calculates by the following equation:

$$F = \frac{n}{60} \tag{3.4}$$

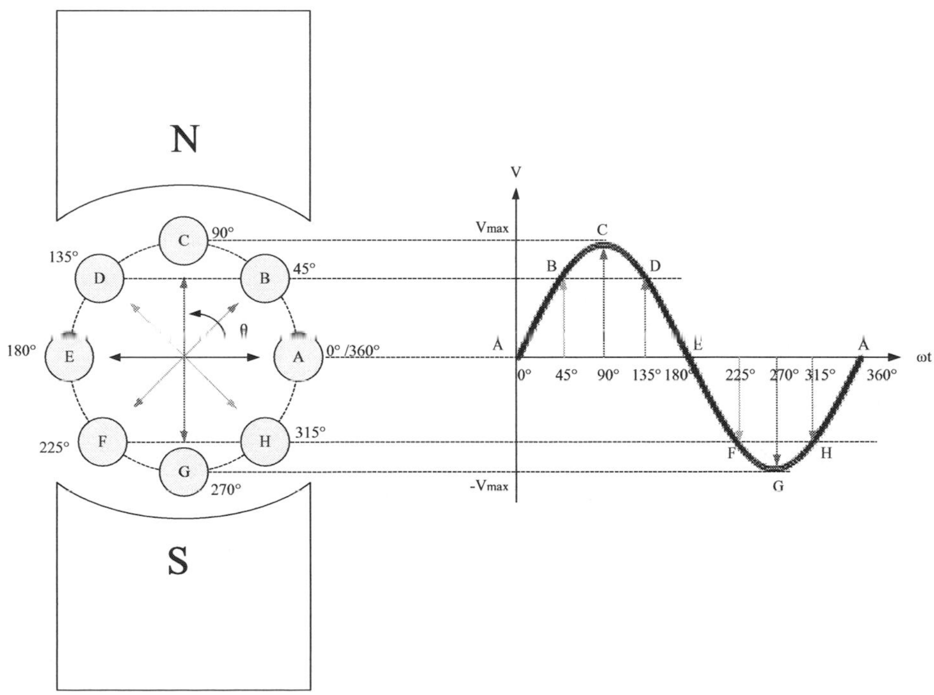

FIGURE 3.3
The model of a generator that produces alternating current.

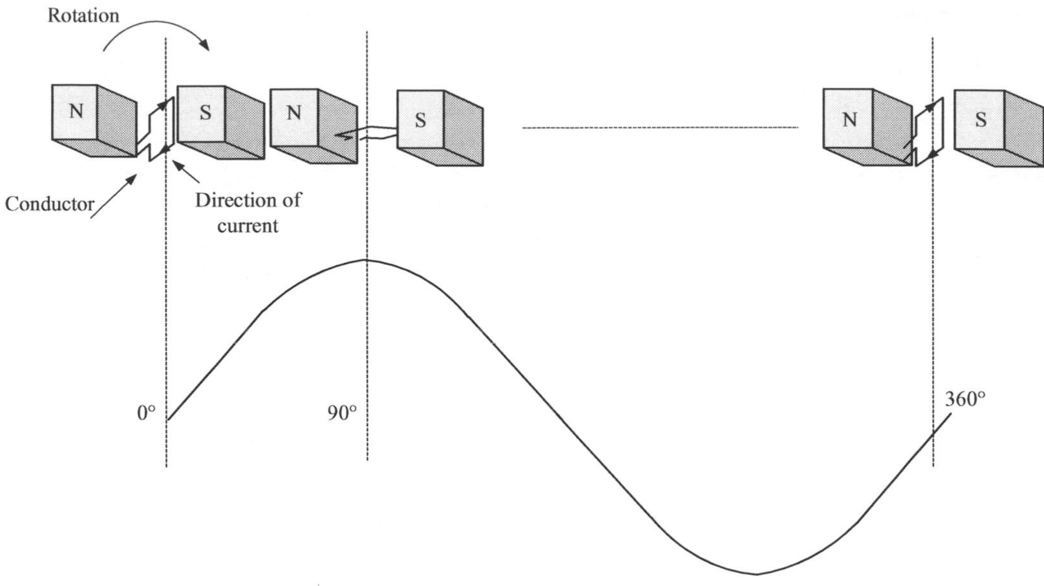

FIGURE 3.4
The process of generating (EMF) in the conductor.

3.3 Terms and Concepts

Each cycle generates one period of the Electric Motive Force (EMF) within two poles of the magnet for any moment, called *the instantaneous value* of the electric motive force as in the following equation:

$$E=2\pi.\phi.f.N.\sin\theta \tag{3.5}$$

where N is the number of coil turns.

When the direction of the movement of the coil is vertical on the lines of the magnetic field, i.e., the angle is equal to 90°, i.e., ($\theta = 90°$), so ($\sin 90° = 1$). On this basis, the maximum value of the voltage is:

$$E_{max}=2\pi\,\phi\,f\,N \tag{3.6}$$

Therefore, the expression of the real value of the voltage is:

$$E = E_{max}\sin\theta \tag{3.7}$$

The actual value of the electric voltage (V_{eff}) is the value indicated by the measuring instruments is equal to:

$$V_{eff} = (0.707)\,V_{max} \tag{3.8}$$

Example 3.3

A coil containing 100 turns rotates at a speed of 1500 rpm within the two magnetic poles, calculate: (i) Frequency, (ii) Time, and (iii) Immediate electric motive force if you know that the angle between the movement of the conductor and the lines of the field is 30°, and the maximum value of voltage 3.14 V.

Solution

i) $f = n/60$

$= 1500/60 = 25$ Hz.

ii) $T = 1/f$

$= 1/25 = 0.04$ sec.

iii) $e = E_{Max} \sin\theta$

$= 3.14 \times \sin 30°$

$= 3.14 \times 0.5 = 1.57\,V$

Example 3.4

A coil containing 50 turns rotates at a regular velocity inside a regular magnetic field. Calculate:

(a) The speed at which the coil must be rotated, by the frequency of the electric motive force 25 Hz
(b) A number of poles
(c) The maximum value of electric motive force if the current value 15 V and the amount of angle 30°
(d) The actual value of electric motive force.

Solution

a) $F = n/60$

$N = 60 \times F = 60 \times 35 = 1500$ rpm.

b) $p = (120F)/n = (120 \times 25)/1500 = 2$ poles

c) $e = E_{max} \cdot \sin\theta$

$$E_{max} = \frac{e}{\sin\theta} = \frac{e}{\sin 30°} = \frac{15}{0.5} = 30V.$$

d) $E_{eff} = 0.707\,E_{max}$

$= 0.707 \times 30 = 12.21$ V.

3.4 AC Current Values

The alternating current changes in value and direction on a sine wave. Some important values of the AC current can be calculated as follows.

3.4.1 The Maximum Value of the Alternating Current

It is the highest value of the current direction, and arises in the case where the direction of the movement of the terminals of the coil is vertical on the lines of the magnetic field, and symbolizes it (I_{max}) and shall be calculated from the following formula:

$$I_{max} = \sqrt{2} \cdot I_{eff}$$
$$I_{max} = 1.414 \cdot I_{eff} \tag{3.9}$$

3.4.2 Average Value of Alternating Current (Mean Value)

The AC value of the sinusoidal wave can be calculated according to the following equation and symbolized by the symbol (I_{av}):

$$I_{av} = 0.637\ I_{max} \tag{3.10}$$

3.4.3 Actual AC Value

It is the reading that is indicated by the measuring instruments that are read on these devices, symbolized by the symbol (I_{eff}) from the following equation:

$$I_{eff} = 0.707\ I_{max} \tag{3.11}$$

3.4.4 The Instantaneous Value of Alternating Current

Which is the value we get at any moment of the current, and symbolized by the letter (i) and calculated from the following equation:

$$I = I_{max} \cdot \sin\theta \tag{3.12}$$

Example 3.5

What is the maximum value voltage of AC source has a voltage 220 V?

Solution

$$V_{eff} = 0.707\ V_{max}.$$

$$V_{max} = V_{eff}/0.707.$$

$$V_{max} = 220/0.707 = 311\text{ V}.$$

Example 3.6

An ammeter is connected to a circuit through which an alternating current passes. If the reading recorded by this device is 2 A, find the maximum value of the current.

Solution

$$I_{eff} = 0.707\ I_{max}.$$

$$I_{max} = I_{eff}/0.707.$$

$$I_{max} = 2/0.707.$$

$$I_{max} = 2.82\ A.$$

Example 3.7

A wire rotates at a speed of 300 rpm in a uniform magnetic field. If the current of the circuit is at 50 Hz. Calculate the number of poles.

Solution

$$N = \frac{60\ f}{P}.$$

$$P = \frac{60\ f}{n} = \frac{60 \times 50}{3000}.$$

P = 3000/3000 = 1 pole pair or 2 poles.

3.5 AC Circuits

3.5.1 AC Circuit Containing Pure Resistance

Figure 3.5a shows a pure resistance connected to an AC source, where the current wave is in phase with the voltage wave as shown in Figure 3.5b, both waves start from zero and reach the greatest value at 90° angle, then fall to zero at 180°, and so the waves are completed simultaneously, as in Figure 3.5b represents of the two waves with directional lines.

The resistance in the circuit shall be calculated as follows:

$$R = \frac{V}{I} \tag{3.13}$$

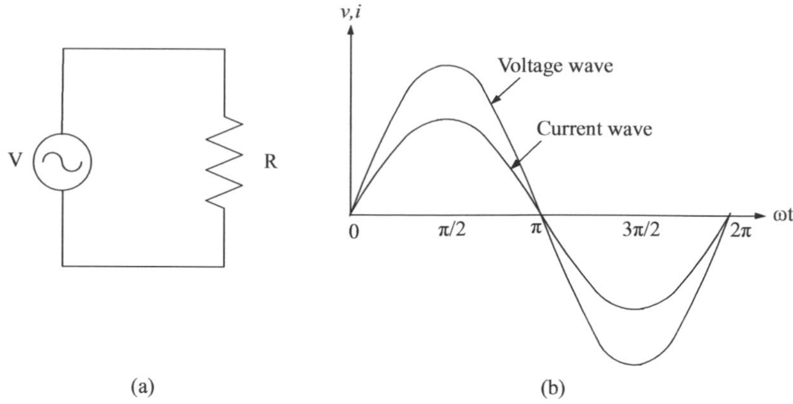

FIGURE 3.5
AC circuit containing only pure resistance. (a) Circuit diagram. (b) Voltage and current waveforms.

or of the following formula:

$$R = \frac{\rho L}{A} \tag{3.14}$$

where:
ρ = resistivity and its unit Ω.mm^2/m
L = Conductor length in meters
A = cross section area in mm^2.

3.5.2 AC Circuit with Inductive Reactance

Figure 3.6 represents the connection of inductive in the circuit of the alternating current, as the current wave is delayed from the voltage wave at an angle of 90° because of the phenomenon of self-induction. As the back EMF at the moment of zero, the current wave passes until the voltage wave reaches the upper value at an angle of 90°. As the current wave starts to appear because the back EMF at this moment is equal to zero. Figure 3.7 represent the voltage and current waves in the circuit contain inductive load connected to the AC source.

The inductive reactance is represented by the symbol "X_L", and the unit is measured by the Ω (Ω). It is calculated from the following formula:

$$X_L = 2\pi \cdot f \cdot L \tag{3.15}$$

where:
f = frequency
L = Self inductive factor

$$\omega = 2\pi \cdot f \tag{3.16}$$

ω = Angler speed (rad/sec)

$$X_L = \omega \cdot L \tag{3.17}$$

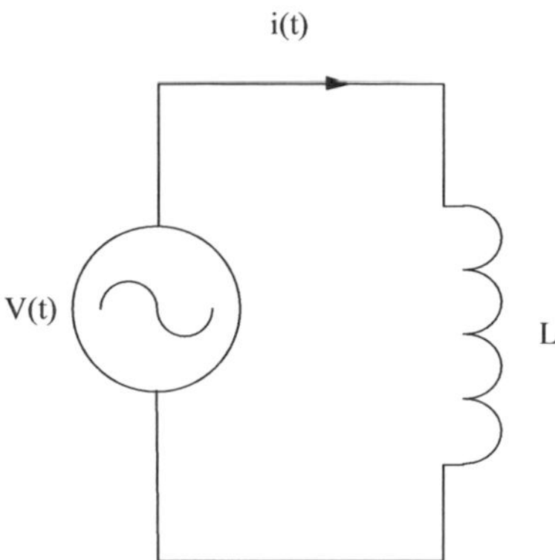

FIGURE 3.6
AC circuit containing inductive reactance.

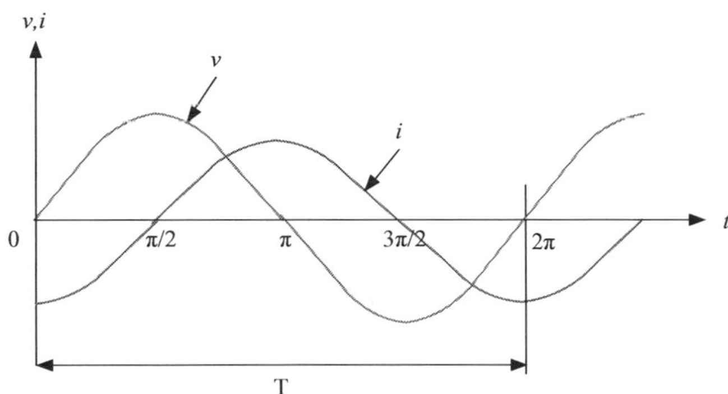

FIGURE 3.7
The voltage and current currents of the inductor.

From the above, scientifically, each coil has a natural resistivity (R = ρ.L/A), and magnetic resistance ($X_L = 2\pi.f.L$), and these two resistors are always connected to the series.

3.5.3 AC Circuit with Capacitive Reactance

Figure 3.8 represents a capacitor in the circuit of the alternating current, as the current wave in this case, ahead of the voltage wave at an angle 90°, for the passage of current in the capacitor to charge, and then show the voltage on the ends gradually, as shown in Figure 3.9a. Figure 3.9b represents the voltage and current waves with directional lines.

The capacitive reactance (Xc) and its unit of measurement (Ω), and its value depends on frequency (f) and capacitor (C) and is calculated from the following formula:

$$Xc = \frac{1}{2\pi \cdot F \cdot C} \tag{3.18}$$

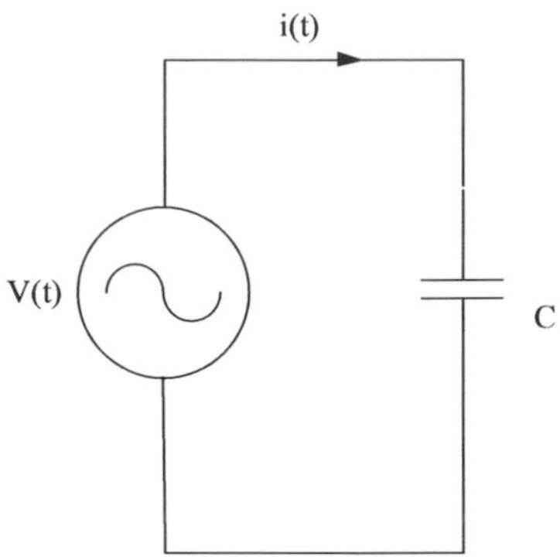

FIGURE 3.8
Capacitor in the circuit of the alternating current.

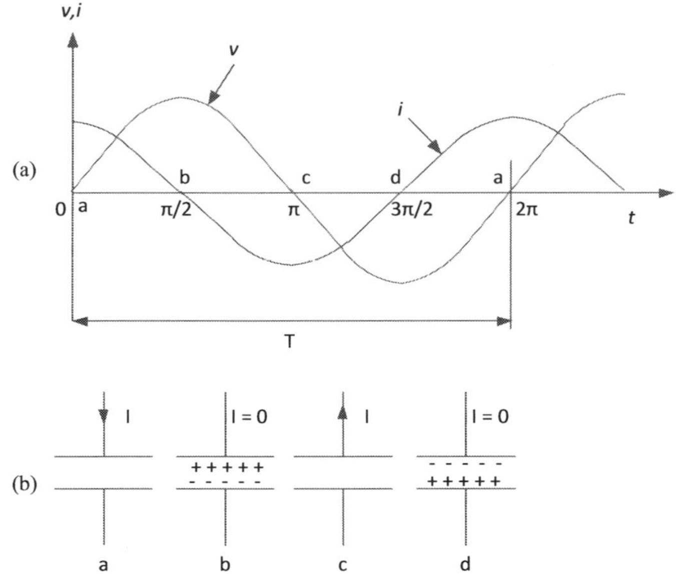

FIGURE 3.9
Voltage and current waveforms and the direction of the charges and current follow. (a) Voltage and current waveforms (b) Direction of the current.

where C = Capacitor capacity and calculated by F (Farad):

$$\omega = 2\pi \cdot f \tag{3.19}$$

$$X_C = \frac{1}{\omega C} \tag{3.20}$$

ω = angular speed in (rad/sec).

From the above, scientifically, where both resistance (R) and capacitive reactance $\left(Xc = \frac{1}{2\pi fc}\right)$ are always connect to each other.

3.6 Series Impedance Connection to the AC Circuit

3.6.1 R-L Series Circuit

Figure 3.10a represents a series connection of resistance and inductor with an AC voltage source. From Figure 3.10b, note that the voltage (V_R) on the resistance (R) is in phase with the angle of the current in the circuit, and the voltage across inductor (V_L) ahead of the current at an angle 90°, so the total voltage of the circuit (V) is the sum of both voltages.

From the right-angled triangle in Figure 3.10b, according to Pythagoras' theory, the total voltage is equal to:

$$V=\sqrt{V_R{}^2+V_L{}^2}$$

Thus, the power factor can be calculated from this triangle:

$$\cos\theta=\frac{VR}{VL} \tag{3.21}$$

where cos θ = power factor of the circuit.

The power factor can also be calculated from the impedance triangle Figure 3.11.

$$\cos\theta=\frac{R}{Z} \tag{3.22}$$

From Figure 3.10a, the current is equal in all parameters of the circuit, so the total impedance of the circuit is equal to:

$$Z=\frac{V}{I} \tag{3.23}$$

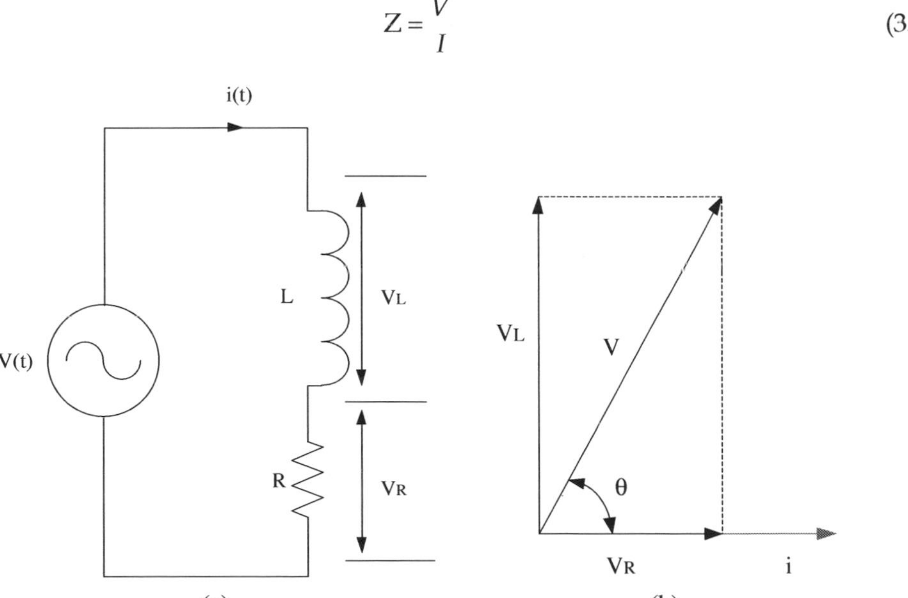

FIGURE 3.10
Circuit and phasor diagram of R-L series circuit.

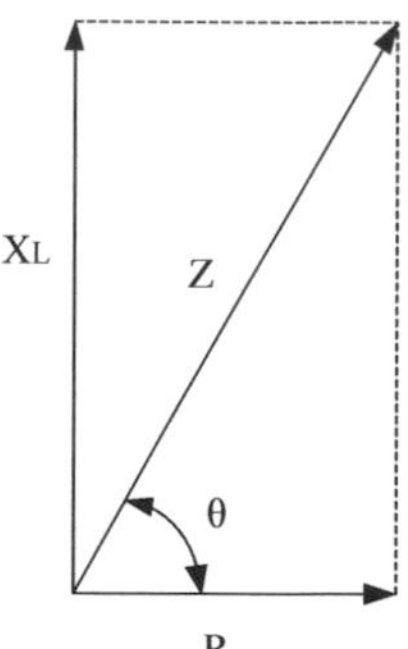

FIGURE 3.11
The impedance triangle.

where:

Z = total impedance of the circuit and calculated by Ω
V = the total voltage of the circuit and calculated by volts
I = total circuit current and calculated in amperes,

and

$$R = V_R/I \tag{3.24}$$

$$X = V_L/I \tag{3.25}$$

Example 3.6

A self-inductance of 10 mH and its resistance 5 Ω connect to AC source of voltage 200 V, and frequency of 60 Hz, calculate the current in the circuit and voltage on both ends of resistance and reactance.

Solution

$$X_L = 2\pi \cdot f \cdot L$$

$$X_L = 2 \times 3.14 \times 60 \times 0.01$$

$$X_L = 3.77\ \Omega$$

$$Z = \sqrt{R^2 + X_L^2} = \sqrt{5^2 + 3.77^2} = 6.262\ \Omega$$

$$I = V/Z = 200/6.262 = 31.93\ \text{A}$$

$$V_R = I \times R = 31.93 \times 5 = 159.7\ \text{V}$$

$$V_L = I \times X_L = 33.9 \times 3.77 = 120.41\ \text{V}.$$

3.6.2 R-C Series Circuit

Figure 3.12a represents a series connection of resistance and capacitor with an AC voltage source. From Figure 3.12b, the voltage on the resistance is aligned with the current, with

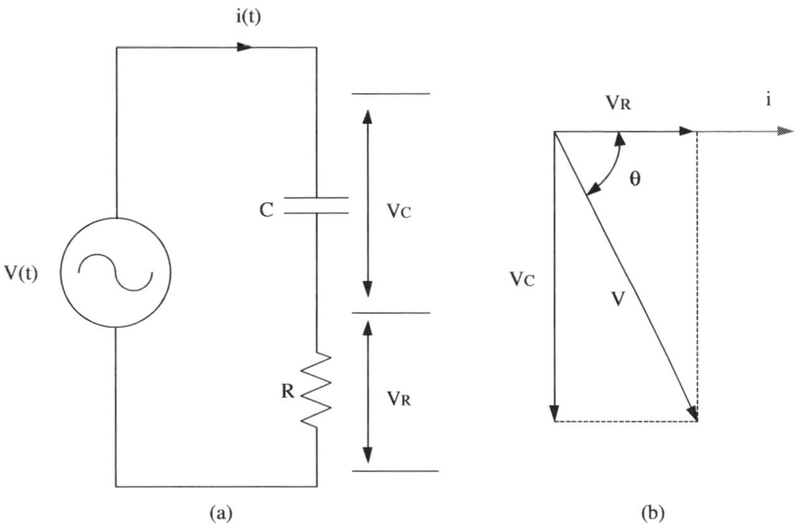

FIGURE 3.12
Resistors and capacitor in series. (a) Circuit diagram (b) Voltage and current directions.

the current of the circuit, and the voltage on the capacitive reactance is 90° below the current. Therefore, the total voltage of the circuit is the sum of both voltages.

Figure 3.12b represents the voltage triangle and can be found in the power factor formula:

$$\cos\theta = \frac{VR}{V} \tag{3.26}$$

As for the impedance triangle as in Figure 3.13, the power factor formula can be calculated:

$$\cos\theta = \frac{R}{Z} \tag{3.27}$$

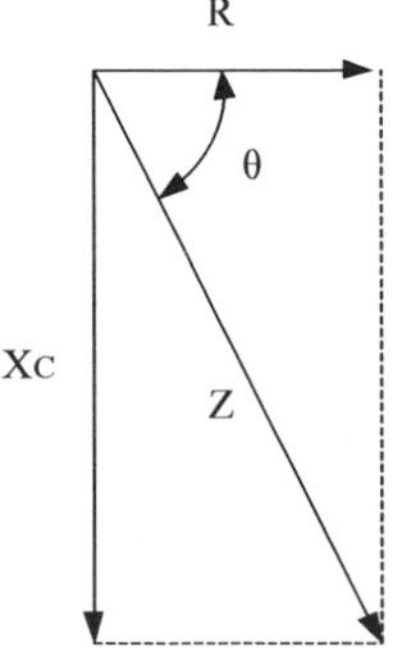

FIGURE 3.13
RC impedance triangle.

Example 3.7

A resistance of 30 Ω and a capacitor of 80 μF are connected in series to AC source of voltage 200 V, 50 Hz. Calculate the total current of the circuit and the voltage on the capacitor.

Solution

$$Xc = \frac{1}{2\pi fC} = \frac{1}{2\times 3.14\times 50\times 80}$$

$$= 40\ \Omega$$

$$Z = \sqrt{R^2 + X_C{}^2}$$

$$= \sqrt{30^2 + 40^2}$$

$$= 50\ \Omega$$

$$I = \frac{V}{Z} = \frac{200}{50}$$

$$= 4\,A$$

$$V_C = I\times X_C = 4\times 40$$

$$= 160\ V.$$

3.6.3 R-L-C Series Circuit

Figure 3.14a represents a series connection of resistance, inductor, and capacitor with an AC voltage source. From the phasor diagram shown in Figure 3.14b, we note that the voltage on the resistance is in phase with the current. The phase angle between the voltage on the inductance and the current is 90°, while the phase angle between the voltage on the capacitance and current is 90° as shown in Figure 3.14b. The voltage on the inductive reactance, the voltage on the capacitance reactance on one straightness, and at an angle of 180° is reversed in the direction.

From Figure 3.14b, we note:

$$V_{LC} = V_L - V_C$$

$$V = \sqrt{V_R^2 + V_{LC}^2}$$

$$V = \sqrt{V_R^2 + (V_L - V_C)^2} \qquad (3.28)$$

$$I.Z = \sqrt{I^2.R^2 + (I.X_L - I.X_C)^2}$$

$$I.Z = I\sqrt{R^2 + (X_L - X_C)^2}$$

$$Z = \sqrt{R^2 + (X_L - X_C)^2} \quad \text{if } X_L > X_C$$

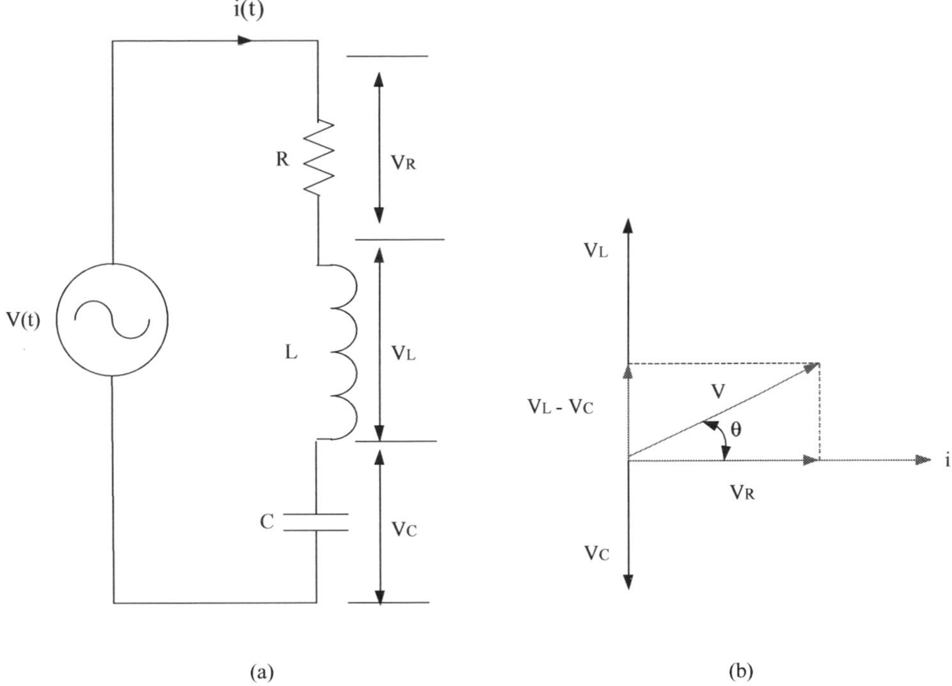

FIGURE 3.14
Circuit diagram and voltage vectors for the (R-C-L) circuit. (a) Circuit diagram (b) phasor diagram.

$$Z=\sqrt{R^2+(X_C-X_L)^2} \quad \text{if } X_L < X_C \tag{3.29}$$

$$\cos\theta=\frac{R}{Z} \tag{3.30}$$

Example 3.8

An R-L-C circuit containing a resistance of 20 Ω, inductance of 0.16 H, and capacitor of 91 μF, has connected an alternating current source of 200 V, and its frequency 50 Hz. Calculate the total impedance and the total current in the circuit.

Solution

$$X_L=2\pi.f.L$$

$$X_L=2\times3.14\times50\times0.16=50\,\Omega$$

$$X_C=1/2\pi fC$$

$$X_C=1/\left(2\times3.14\times50\times91\times10^{-6}\right)$$

$$X_C = 10^6/(91 \times 314) = 35\Omega$$

$$Z = \sqrt{R^2 + (X_L - X_C)^2}$$

$$Z = \sqrt{20^2 + (50-35)^2} = \sqrt{20^2 + 15^2} = \sqrt{400^2 + 225^2}$$

$$Z = 25\ \Omega$$

$$I = 200/25 = 8\ A.$$

3.7 Parallel Connection

3.7.1 Parallel R-L Circuit

Figure 3.15a represents a parallel connection of resistance and inductor with an AC voltage source. In Figure 3.15b, we note that the voltage (V) of the circuit is equally on terminals of both the resistance and inductance. The drawing of the directional lines shows the horizontal line representing the voltage of the circuit. With the voltage, the current passing through the resistance in phase with voltage while the current following in the reactance is late with the voltage by 90°.

From Figure 3.15b and according to the Pythagoras theory:

$$I^2_T = I^2_R + I^2_L$$

$$I_T = \sqrt{I_R{}^2 + I_L{}^2} \tag{3.31}$$

$$\sin\theta = I_L/I_T \tag{3.32}$$

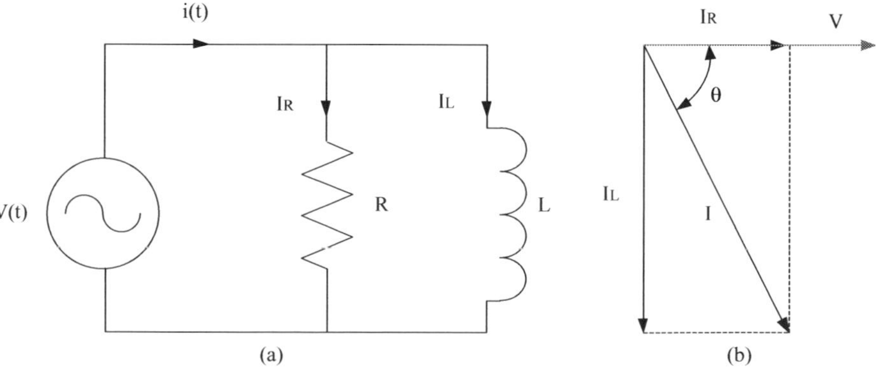

FIGURE 3.15
The R-L parallel circuit. (a) Circuit diagram. (b) Phasor diagram.

$$I_L = I_T . \sin\theta \quad (3.33)$$

$$\cos\theta = I_R / I_T \quad (3.34)$$

$$I_R = I_T . \cos\theta, \quad (3.35)$$

where:
I_R = resistive current
I_L = inactive current
I_T = total current.

Example 3.9

A circuit with 50 Ω of resistance, parallel to the self-induction coefficient 89 mH, connected to the source of the alternating voltage 200 V and its frequency 50 Hz. Calculate the total current of the circuit.

Solution

$$I_R = V/R$$

$$I_R = 200/50 = 4\text{ A}$$

$$X_L = 2\pi f L$$

$$X_L = 2 \times \pi \times 50 \times 0.089$$

$$X_L = 28\ \Omega$$

$$I_L = 200/28 = 7.12\text{ A}$$

$$I_T = \sqrt{I_R^2 + I_L^2}$$

$$I = \sqrt{4^2 + 7.12^2} = 8.18\text{ A}.$$

3.7.2 Parallel R-C Circuit

The electric circuit in Figure 3.16 represents R-C circuit connected in parallel to the voltage supply. The voltage of the circuit (V) is the same at terminals of both resistance and capacitance. The current in the resistance is in phase with the voltage and current in the capacitance ahead of the voltage by an angle 90°, and the sum of both currents is the total current (I) as in Figure 3.16b.

From Figure 3.16b.

$$I^2_T = I^2_R + I^2_C$$

$$I_T = \sqrt{I_R^2 + I_C^2}, \quad (3.36)$$

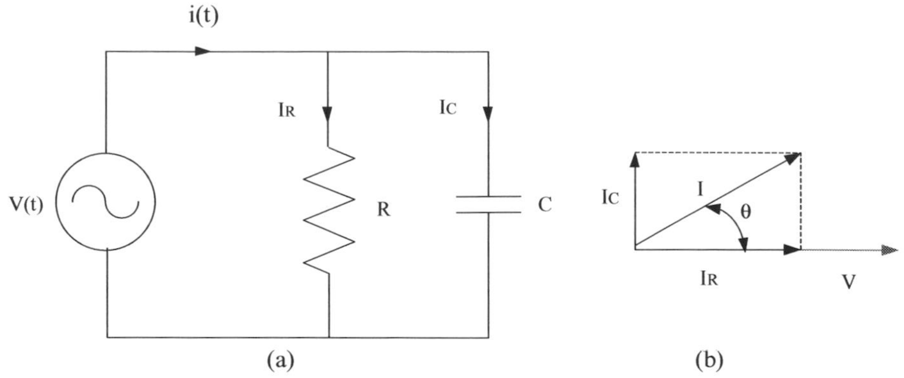

FIGURE 3.16
The voltage and current vectors of the parallel circuit R-C. (a) Circuit diagram (b) phasor diagram.

where:

I_T = the total current
I_R = the current in the resistance
I_C = the current in the capacitance.

Example 3.10

A parallel R-C circuit containing 25 Ω of resistance, and a capacitor of 95.5 μF, connected to a voltage source of 200 V and a frequency of 50. Calculate the total current of the circuit.

Solution

$$X_C = \frac{1}{2\pi.F.C}$$

$$= 1/\left(2\times 3.14\times 50\times 95.5\times 10^{-6}\right)$$

$$= 10^6/29987$$

$$= 33.34\ \Omega$$

$$IC = V/X_C$$

$$= 200/33.34 = 6\ \text{A}$$

$$I_R = V/R$$

$$= 200/25 = 8\ \text{A}$$

$$I_T = \sqrt{I_R{}^2 + I_C{}^2}$$

$$= \sqrt{8^2 + 6^2}$$

$$= \sqrt{64 + 36}$$

$$= 10\ \text{A}$$

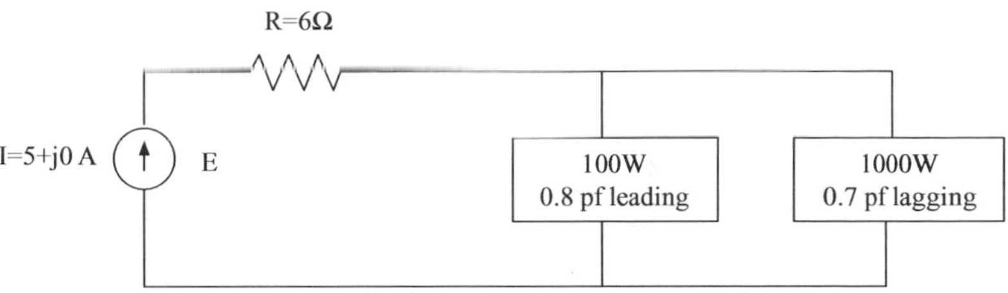

FIGURE 3.17
Circuit of Example 3.11.

Example 3.11

The following Figure 3.17, power system has two loads attached to a 6 Ω line with a supply current of values of $5\angle 0°A$, if the loads data are given as:

Load 1: P = 100 W at 0.8 pf leading
Load 2: P = 1000 W at 0.7 pf lagging. Determine:

a) The value of the total watts, VAR, and VA for the circuit
b) Value of the supply voltage E
c) The power factor for the circuit
d) Type of element and their impedance in each box
e) Write a MATLAB program to verify the answers.

Solution

Given Load 1: $P_1 = 100$ W, $\cos\varphi_1 = 0.8$ pf leading.
Reactive power is:

$$Q_1 = P_1 \tan \varphi_1 = 100 \times \tan(\cos^{-1}0.8) = 75 \text{ VAR (Volt Amper Reactive).}$$

The total complex power for Load 1 is:

$$S_1 = P_1 - jQ_1 = 100 - j75 \text{ VA (Volt Amper).}$$

For Load 2: $P_2 = 1000$ W, $\cos\varphi_2 = 0.7$ pf lagging.
Reactive power is:

$$Q_2 = P_2 \tan \varphi_2 - 100 \times \tan(\cos^{-1}0.7) = 1020.2 \text{ VAR.}$$

The total complex power for Load 2 is:

$$S_2 = P_2 - jQ_2 = 1000 + j1020.2 \text{ VA.}$$

The P and Q of 6 ohm line is:

$$P_r = I^2 \times R = 52 \times 6 = 150 \text{ W.}$$

The reactive power of resistor is zero i.e., $Q_r = 0$ VAR.

a) The total watts are:

$$P = P_1 + P_2 + P_3 = 150 + 100 + 1000 = 1250 \text{ W.}$$

The total VAR is:

$$Q = Q_2 - Q_1 = 1020.2 - 75 = 945.2 \text{ VAR}.$$

The total VA is:

$$S = P + jQ = 1250 + j945.2\,\text{VA} = 1567.13\angle 37.1^\circ VA$$

b) The supply voltage E:

$$E = \frac{S}{I^*} = \frac{1567.13\angle 37.1^\circ}{5\angle 0^\circ} = 313.43\angle 37.1\ V.$$

c) The power factor for the circuit:

$$\text{pf}= \cos\left(\tan^{-1}\frac{945.2}{1250}\right) = 0.798\,\text{lagging}.$$

d) The voltage across loads is:

$$V = E - I \times R = 313.43\angle 37.1^\circ - 5\angle 0^\circ \times 6 = 290.06\angle 40.67^\circ V.$$

The current through Load 1 is:

$$I_1 = \left(\frac{S_1}{V}\right)^* = \left(\frac{100 - j75}{290.06\angle 40.67^\circ}\right)^* = 0.431\angle 77.54\ A.$$

The impedance of Load 1 is:

$$Z_1 = \frac{V}{I_1} = \frac{290.06\angle 40.67^\circ}{0.431\angle 77.54^\circ} = 538.4 - j403.8\ \Omega$$

Resistor and capacitor as load as pf is leading:
Current through Load 2 is:

$$I_2 = I - I_1 = 5\angle 0^\circ - 0.431\angle 77.54^\circ = 4.925\angle -4.9^\circ A.$$

The impedance of Load 2 is:

$$Z_2 = \frac{V}{I_2} = \frac{290.06\angle 40.67^\circ}{4.925\angle -4.9^\circ} = 41.23 - j42.06\,\Omega$$

Resistor and inductor as load as pf is lagging.

e) MATLAB CODE

```
P1 = 100;
pf1 = 0.8;
Q1=P1*tan(acos(pf1));
P2 = 1000;
pf2 = 0.7;
Q2=P2*tan(acos(pf2));
I=5;
```

```
Pl=I^2 * 6;
P=Pl+P1+P2
Q=Q2-Q1
S=P+Q*i;
VA=abs(S)
E=(S/conj(I));
magE=abs(E)
angleE=angle(E)*180/pi
powerFactor=cos(atan(Q/P))
V=E-(I*6);
I1=conj((P1-Q1*i)/V);
Z_1=V/I1
I2=I-I1;
Z_2=V/I2
P1 = 100;
pf1 = 0.8;
Q1=P1*tan(acos(pf1));
P2 = 1000;
pf2 = 0.7;
Q2=P2*tan(acos(pf2));
I=5;
Pl=I^2 * 6;
P=Pl+P1+P2
Q=Q2-Q1
S=P+Q*i;
VA=abs(S)
E=(S/conj(I));
magE=abs(E)
angleE=angle(E)*180/pi
powerFactor=cos(atan(Q/P))
V=E-(I*6);
I1=conj((P1-Q1*i)/V);
Z_1=V/I1
I2=I-I1;
Z_2=V/I2
```

Ans:

P = 1250

Q = 945.2041

VA = 1.5671e + 03

magE = 313.4269

angleE = 37.0952

powerFactor = 0.7976

Z_1 = 5.3847e + 02 – 4.0385e + 02i

Z_2 = 41.2269 + 42.0598i.

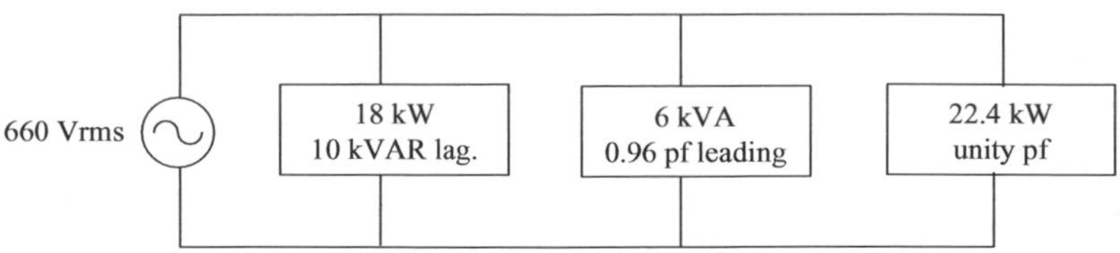

FIGURE 3.18
Circuit of Example 3.12.

Example 3.12

Three loads are connected in parallel across 660 V(rms), 60 Hz line as shown in Figure 3.18.

Load 1: absorbs 18 kW and 10 kVAR with lagging power factor
Load 2: absorbs 6 kVA at 0.96 leading power factor
Load 3: absorbs 22.4 kW at unity power factor. Determine:

a) The value of the total watts, VAR, and VA for the circuit
b) The power factor for the circuit.

Solution

Given data:

Load 1: absorbs 18 kW and 10 kVAR with lagging power factor
Load 2: absorbs 6 kVA at 0.96 leading power factor
Load 3: absorbs 22.4 kW at unity power factor

For Load 2, the reactive power is:

$$Q_2 = S_2 \sin \varphi_2 = 6 \times \sin (\cos^{-1} 0.96) = 5.76 \text{ kVAR.}$$

The total complex power for Load 2 is:

$$S_2 = P_2 + jQ_2 = 6 + j5.76 \text{ kVA.}$$

For Load 3: $P_3 = 22.4$ kW, $\cos\varphi_3 = 1$.
Reactive power is $Q_3 = 0$ VAR (unity power factor).
The total complex power for Load 3 is:

$$S_3 = P_3 - jQ_3 = 22.4 + j0 \text{ kVA.}$$

a) The total watts are:

$$P = P_1 + P_2 + P_3 = 18 + 5.76 + 22.4 = 46.16 \text{ kW.}$$

The total VAR is:

$$Q = Q_1 + Q_2 + Q_3 = 10 + (-1.68) = 8.32 \text{ kVAR.}$$

The total VA is:

$$S = P + jQ = 46.16 + j8.32 \text{ kVA} = 46.9\angle 10.21° \text{ kVA.}$$

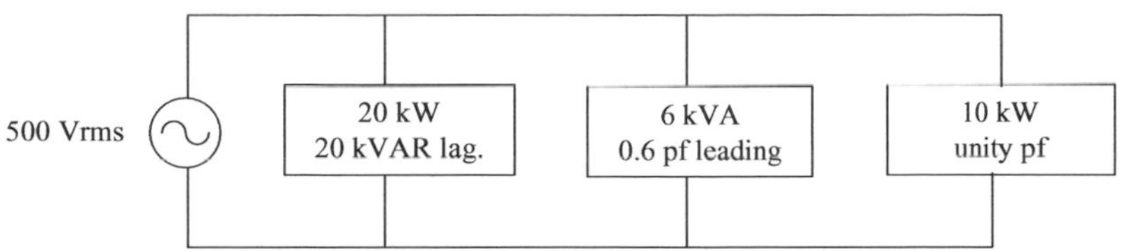

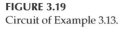

FIGURE 3.19
Circuit of Example 3.13.

b) The power factor for the circuit:

$$\text{p.f} = \cos\left(\frac{46.16}{46.9}\right) = 0.798 \text{ lagging.}$$

Example 3.13

Three loads are connected in parallel across 500 V(rms), 60 Hz line as shown in Figure 3.19.

Load 1: absorbs 20 kW and 10 kVAR with lagging power factor
Load 2: absorbs 6 kVA at 0.6 leading power factor
Load 3: absorbs 10 kW at unity power factor. Determine:

a) The value of the total watts, VAR, and VA for the circuit
b) The power factor for the circuit
c) The value of capacitance C connected in parallel across the loads that will raise the power factor to unity.

Solution

Given data:

Load 1: absorbs 20 kW and 20 kVAR with lagging power factor
Load 2: absorbs 6 kVA at 0.6 leading power factor
Load 3: absorbs 10 kW at unity power factor.

For Load 2, the reactive power is:

$$Q_2 = S_2 \sin\varphi_2 = 6 \times \sin(\cos^{-1}0.6) = 4.8 \text{ kVAR}.$$

The total complex power for Load 2 is:

$$S_2 = P_2 + jQ_2 = 6 + j4.8 \text{ kVA}.$$

For Load 3: $P_3 = 10$ kW, $\cos\varphi_3 = 1$.
Reactive power is $Q_3 = 0$ VAR (unity power factor).
The total complex power for Load 3 is:

$$S_3 = P_3 - jQ_3 = 10 + j0 \text{ kVA}.$$

a) The total watts are:

$$P = P_1 + P_2 + P_3 = 20 + 6 + 10 = 36 \text{ kW}.$$

The total VAR is:

$$Q = Q_1 + Q_2 + Q_3 = 20 + (-4.8) = 15.2 \text{ kVAR}.$$

The total VA is:

$$S = P + jQ = 36 + j15.2\,\text{kVA} = 39\angle 22.89°\,kVA.$$

b) The power factor for the circuit:

$$\text{p.f} = \cos(22.89) = 0.921 \text{ lagging.}$$

c) With required unity power factor $S = P = 36$ kVA, and $Q = 0$,so $Q_c = 15.2$ kVAR.

$$X_c = \frac{V}{Q_c} = \frac{500}{15.2 \times 10^3} = 0.033\,\Omega$$

$$C = \frac{1}{2\pi f X_c}$$

$$C = \frac{1}{2\pi \times 60 \times 0.03} = 88.42\ mF.$$

Problems

3.1 What are the advantages of alternating current?

3.2 What are the phase shift angles between voltage and current for series R-L, R-C, and R-L-C circuits?

3.3 Explain your answer to the diagram about the current and voltage relationship in the AC circuit containing:
a) pure resistance, b) inductance, and c) capacitance.

3.4 Define the following: alternating current, instantaneous current, wave time, and frequency.

3.5 Give the reason: the current wave is delayed by the voltage wave in the AC circuit containing only inductance.

3.6 Give the reason: the voltage wave is delayed by the current wave in the AC circuit containing only capacitance.

3.7 An electrical generator, four poles, and the speed of 1800 rpm. Calculate the value of frequency.

3.8 A copper wire length of 400 m, and a cross-section area 2 mm^2. Calculate the resistance if its resistivity of $\rho = 1.68 \times 10^{-8}\ \Omega$.m.

3.9 Calculate the speed of the coil revolving within the magnetic field arising from the 12 poles, and the frequency is 60 Hz.

3.10 Calculate the value of the reactance of the self-inductor of 0.1 H if it is to be connected to a circuit of frequency 60 Hz.

3.11 Calculate the capacitive reactance of a capacitor 31.8 μF if it is known to be connected to a 50 Hz voltage source.

3.12 Calculate the value of the reactance of a coil, connected with a resistance of 8 Ω, the circuit fed by a voltage source of 110 V, and the current passing through the circuit of 11 A.

3.13 A series circuit of 20 Ω resistance, and a capacitor of 100 μF. This circuit is connected to an AC source of 200 V at 50 Hz. Calculate:

a. The total current of the circuit

b. The voltage on both terminals of the resistance

c. The voltage on both terminals of the capacitor

d. The total impedance of the circuit

e. Power factor.

3.14 Find the maximum value of a current 10 A.

3.15 Find power factor for an electric circuit with a resistance of 6 Ω, and connected in series, with a reactance of 8 Ω.

3.16 A reactance of 10 Ω, and this self-inductance of 150 mH calculated this frequency.

3.17 A reactance of 62.25 Ω, calculate its self-inductive factor if this coil connected with an AC voltage source of 240 V and the frequency of 25 Hz. The total resistance of the circuit is equal to 80 Ω.

3.18 A series circuit contains resistance of 10 Ω, and a self-inductance coefficient of 0.2 H, connected with an alternating voltage source of 64 V and frequency of 50 Hz., calculate:

a. The total current of the circuit

b. The current that passes through both resistance and the coil

c. The voltage on the terminals of the coil

d. The power factor of the circuit.

4

Transformers

Successful people are always looking for opportunities to help others. Unsuccessful people are always asking, "What's in it for me?"

Brian Tracey

An electric transformer is a device which converts electrical energy of a specific voltage into another voltage through a pair of electrical windings. It can also be defined as a static electromagnetic device which converts AC power from certain components to other components.

Transformers are efficient electrical machines that can convert 99.75% of the input power of the transformer to the output of the transformer. The manufacturing of electric transformers is done in different sizes ranging from thumb-sized transformers (where the microphone can consist of a pair of these transformers) to very large sizes transformers weighing hundreds of tons and used in power grids. The operation of all types of transformers is based on the same basic principles despite difference sizes. Different types and sizes of electrical transformers are given in Figure 4.1. In this chapter, the focus is on single-phase electrical transformers.

4.1 Installation of the Transformer

The transformer consists of two windings wires wound around an iron-core. The terminal connected to a source called *the primary*, while the associated end of the load is called *the secondary winding*, as shown in Figure 4.2a, i.e., the transformer can be considered to be composed of two circuits: a magnetic circuit consisting of the core, and the other is an electric circuit consisting of primary and secondary windings. The components of the transformer are placed inside a container to maintain the components of dust, moisture, and mechanical shocks.

The transformer is used to change the value of the voltage in the AC power transmission system where the transformer cannot operate in DC systems. If the secondary voltage is less than the primary voltage, the transformer is step-down voltage, and if the secondary voltage is higher than the primary voltage, the voltage step-up is shown in Figure 4.2.

The transformer windings must contain the following:

- High mechanical strength, enough to protect it from distortions which may result from short or excess currents
- Sufficient thermal strength so that the high temperature does not lead to thermal breakdown of the insulation material

FIGURE 4.1
Different sizes and shapes of electrical transformers. (https://www.udvavisk.com/cfd-analysis-transformer-room/; http://www.procontransformers.com/; http://www.mpja.com/24V-10A-Center-Tapped-12-0-12-Transformer/productinfo./)

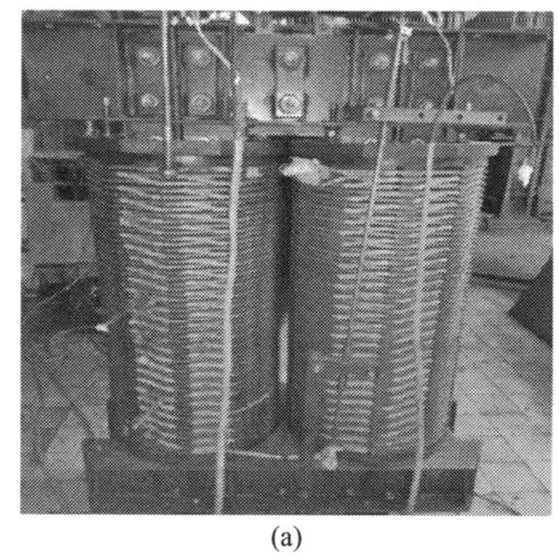

(a)

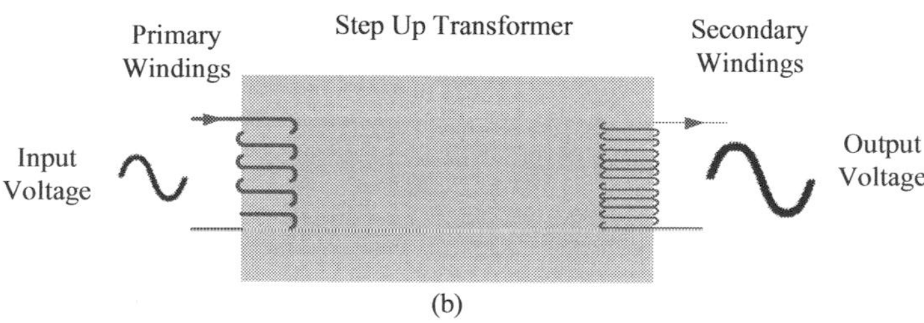

(b)

FIGURE 4.2
Two windings step-up transformer. (a) Two windings transformer. (b) Step-up voltage transformer.

- Enough electrical strength so that the insulation materials and insulation distances are sufficient to prevent electric breakdown or electric arc.

The transformer windings vary according to their currents and nominal voltages and are made of copper or aluminum wires with circular or rectangular sections.

4.2 Core Shape

The core's installation of the transformer depends on the following factors: voltages, current, and frequency. The size and cost of installation of the core are also considered, and the air-core is usually used for soft-core or steel-core. The selection of a particular type of core depends on the area of the transformer's use. Generally, air-cored transformers are used when the source voltage is high frequency, and the iron-core switches are usually used when the source voltage is low frequency, transmitting a better power than the air-core transformer. Soft iron-core transformers are very useful when the size of the transformer is small. The steel-cored transformer loses heat easily and thus is used to efficiently convert power. There are two types of slides that make up the steel-core of the transformers: the first is called *the core-shaped* transformer, as shown in Figure 4.3. The core is made of several steel strips, as shown (Figure 4.4):

The second shape of the core is the shell-core frame, which is the most common and efficient, as shown in Figure 4.5. Each layer or slice of the core consists of a part of letter E and the other a letter I, when combined produces a single slide shape, isolates these segments by isolation, and presses to be the core.

In this type of transformer, the windings are placed around the center column of the iron-core, as shown in Figure 4.6.

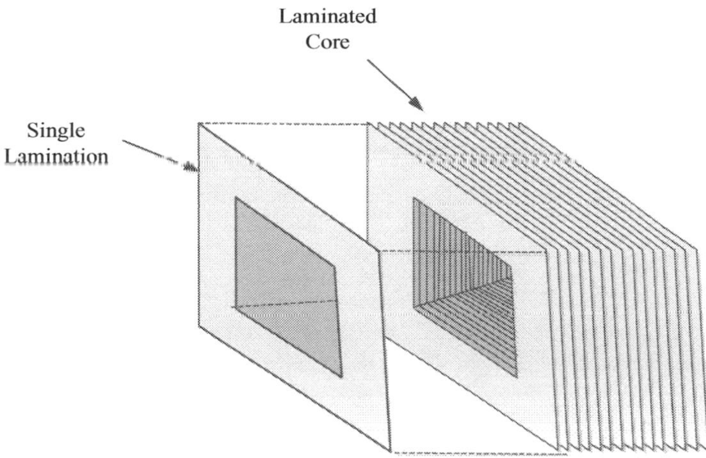

FIGURE 4.3
The core-shaped transformer.

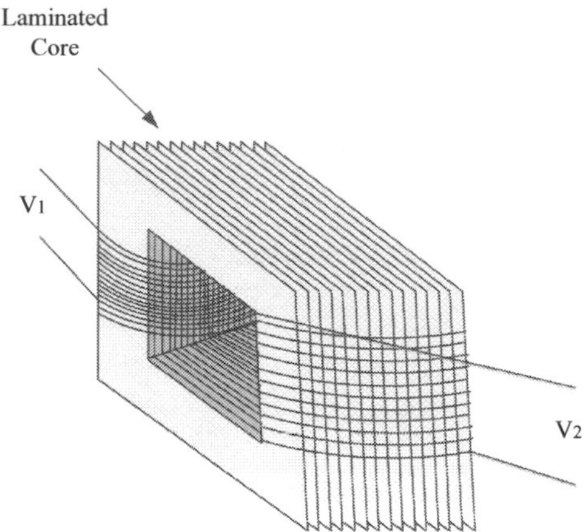

FIGURE 4.4
The location of converted coils around the core.

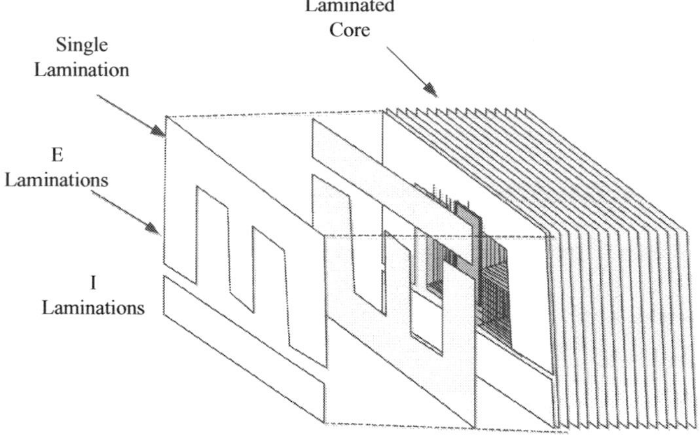

FIGURE 4.5
The shape of the frame-frame slides.

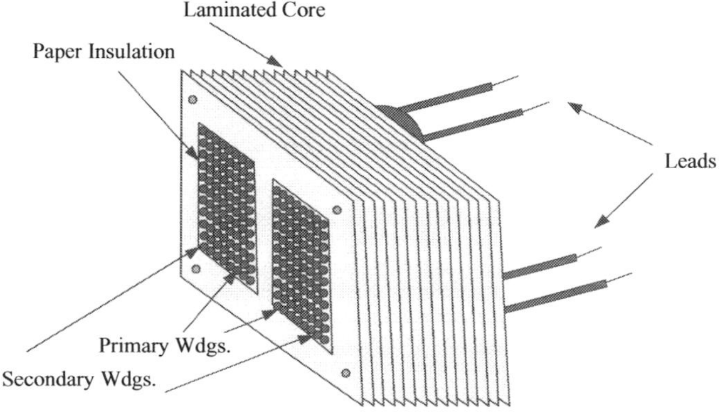

FIGURE 4.6
The location of converted coils with the frame.

4.3 Principle of Operation

The principle work of the electric transformer is based on Faraday's law of electromagnetic induction, which states that the value of the electric force (electric voltage) is directly proportional to the rate of change of the magnetic flux. For this reason, the transformer does not work in DC systems because the DC creates a static magnetic field. The amount of its changed is zero, so it is not possible to create an electrical voltage in a way of induction, the main reasons for why the preference of the current AC at the time when there is not even a practical and economical method to adjust the value of the voltage.

When an electric current I_1 is applied in the primary windings, a magnetic flux ϕ_{12} can be determined by the right-hand rule. When the right-hand fingers point to the direction of the coils, the thumb indicates the direction of the magnetic flux and, as there isn't contact between the primary and secondary terminals, the magnetic flux is applied in a magnetic circuit between the two ends. When the flux of the secondary party coils reaches the flux of the current in these coils I_2, the direction can be determined in the above manner. But this time by making the direction of the thumb first correspond to the direction of magnetic flux ϕ_{21}, pointing to the direction of current flux in the coils, as shown in Figure 4.7.

4.4 Ideal Transformer

The ideal transformer is a theoretical assumption only and is used to understand the real transformer. The ideal transformer assumes that there is no loss of energy where the energy moves from the primary coil circuit to the secondary coil circuit. The ideal transformer also assumes that the coils have no resistance to the flow of the current and there is no leakage in the magnetic flux. These hypotheses help to infer the different relationships. The ideal transformer is composed of two coils that have only an inductive impedance and are wound around an iron-core, as shown in Figure 4.8. If the primary coil is connected by a variable voltage source, it produces an alternating magnetic flux and frequency as well as the number of primary coil turns. This alternating flux is intertwined with the secondary coil, generating an alternating voltage dependent on the number of secondary coil turns.

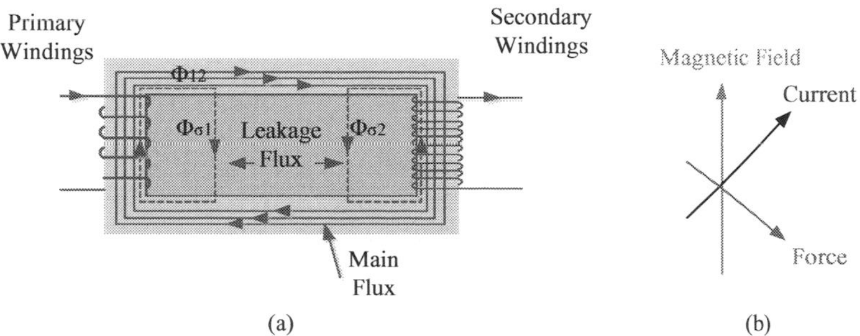

FIGURE 4.7
The principle of the work of the electric transformer. (a) Direction of current (b) Direction of current, force and magnetic field.

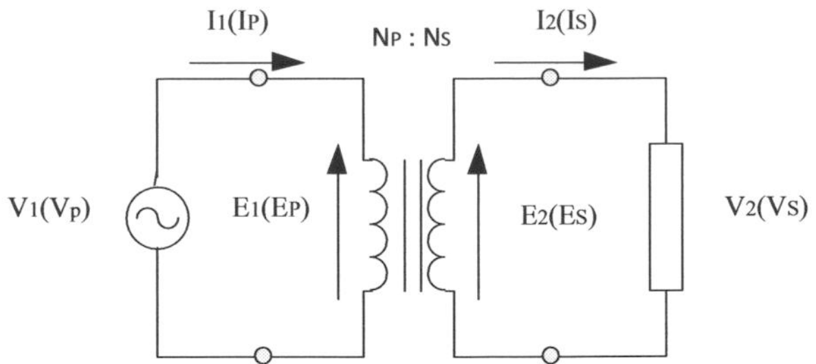

FIGURE 4.8
The composition of the ideal transformer and its components.

If we assume that the initial voltage is V_1, and the resulting magnetic flux is an inverse electric force (E_1) is generated in the primary coil given by:

$$E_1 = V_1 = N_1 \times \frac{d\Phi}{dt} \tag{4.1}$$

where:
- E_1 = the induce electric motive force in the primary windings (volt)
- N_1 = the number of primary windings turns
- Φ = the amount of magnetic flux (Wb)
- t = time (second).

The following of a variable electric current in the secondary windings causes the generation of an induced voltage according to Faraday's law equal to:

$$E_2 = V_2 = N_2 \times \frac{d\Phi}{dt} \tag{4.2}$$

where:
- E_2 = the induce electric motive force in the secondary windings (volt)
- N_2 = the number of secondary windings turns.

By dividing Equation 4.2 into Equation 4.1, it results in:

$$E_2/E_1 = V_2/V_1 = N_2/N_1. \tag{4.3}$$

The ideal transformer shown in Figure 4.8, in the case of connecting its secondary coil to the load, will follow a current through which the electric power will be transferred from the primary circuit to the secondary circuit. In the ideal case (by neglecting the losses), all incoming energy will be transferred from the primary circuit to the magnetic field and to the secondary circuit, so that the input power can be equal to the output power as given in the following equation:

$$P_{in} = I_1 . V_1 = P_{out} = I_2 . V_2. \tag{4.4}$$

That is, the ideal transformer equation is equal to:

$$V_2/V_1 = N_2/N_1 = I_1/I_2. \quad (4.5)$$

It is possible to calculate voltages on both ends of the primary windings in terms of voltages on both ends of the secondary windings by the following equation:

$$V_1 = (N_1/N_2) * V_2. \quad (4.6)$$

It is also possible to calculate the current following through the primary windings in terms of current following through the secondary windings by the following equation:

$$I_1 = (N_2/N_1) * I_2, \quad (4.7)$$

where:

V_P = the primary windings voltages, V_S: secondary windings voltages
N_P = the number of primary windings turns, N_S: the number of secondary windings turns
I_P = the primary windings current, and
I_S = the secondary windings current.

So, the ideal transformer equation is as follows:

$$V_S/V_P = N_S/N_P = I_P/I_S.$$

If the transformer is ($V_S > V_P$), then the current will be lower ($I_S < I_P$) by the same ratio, this ratio is called *transformation ratio* and symbolized by the symbol (α). The induced electrical motive force of the primary windings is calculated by the following equation:

$$E_1 = 4.44\, \phi f N_1. \quad (4.8)$$

And the induced electrical motive force generated in the secondary windings is calculated by:

$$E_2 = 4.44\, \phi f N_2. \quad (4.9)$$

The following assumptions are made in the analysis of an ideal transformer:

1. The transformer windings are perfect conductors, meaning that there is zero winding resistance.
2. The core permeability is infinite, meaning that the reluctance of the core is zero.
3. All magnetic flux is confined to the transformer core, meaning that no leakage flux has occurred.
4. The core losses are hypothetically assumed to be zero.

4.5 Transformer Rating

Transformers carry ratings related to the primary and secondary windings. The ratings refer to the power in kVA and primary/secondary voltages. A rating of 10 kVA, 1100/110 V means that the primary is rated for 1100 V, while the secondary is rated for 110 V ($\alpha = 10$).

The kVA rating gives the power information, with a kVA rating of 10 kVA and a voltage rating of 1100 V, the rated current for the primary is 10,000/1100 = 9.09 A, while the secondary rated current is 10,000/110 = 90.9 A.

Example 4.1

A transformer gives 500 A at 24 V when fed with a source voltage of 120 V. How many turns are required in the secondary side? When the number of turns of the primary side is 3000 turns, and how much is the primary current?

Solution

$$N_2/N_1 = V_2/V_1$$

$$N_2 = V_2/V_1 \times N_1$$

$$N_2 = (24/120) * 3000$$

$$N_2 = 600 \text{ Turns}$$

$$N_2/N_1 = I_1/I_2$$

$$I_1 = N_2/N_1 * I_2$$

$$I_1 = (600/3000) * 500 = 100 \text{ A.}$$

Example 4.2

In a step-down voltage transformer, if the number of high-voltage side is equal to 500 turns, while the number of turns of the low voltages 100 turns, the load current is 12 A, calculate:

i. The transformation ratio (α)
ii. The primary windings current.

Solution

i. The transformation ratio $\alpha = N_1/N_2$

$$\alpha = 500/100$$

$$\alpha = 5$$

ii. $N_2/N_1 = I_1/I_2$

The primary windings current $I_1 = 100/500 * 12$

$$I_1 = 2.4 \text{ A.}$$

Example 4.3

A single-phase transformer operates at a frequency of 50 Hz. If the iron-core is a square shape, the length of its side 20 cm and the maximum magnetic flux density allowed follow in the iron-core 10000 line/cm², calculate the number of coils to be placed for both the primary and secondary windings to be a voltage transformation ratio 220/3000 V.

Solution

$$\phi = A \times B = 20 \times 20 \times 10^{-8} \times 10000 = 0.04 \text{ Wb}$$

$$E_1 = 4.44\ \phi\ f\ N_1$$

$$N_1 = E_1/4.44\ \phi\ f\ N_1 = 3000/4.44 \times 0.04 \times 50 = 338 \text{ turns}$$

$$E_2 = 4.44\ \phi.\ f.\ N_2$$

$$N_2 = E_2/4.44\ \phi\ f\ N_2 = 220/4.44 \times 0.04 \times 50 = 25 \text{ turns.}$$

4.6 Transformer Operation

A transformer is said to be on "no-load" when its secondary side winding is open circuited, in other words, nothing is attached and the transformer loading is zero. When an electrical load is connected to the secondary winding of a transformer and the transformer loading is, therefore, greater than zero, a current flows in the secondary winding and out to the load.

4.6.1 The Transformer Operation at No-Load

It is known that the theory of transformer operation depends on electromagnetic induction. When the transformer is connected to an alternating current source, a current follows through the primary winding called *no-load current* I_o. The flux of this current results in a variable magnetic flux followed by the current. This flux cut the primary and the secondary windings and generate in each of them an opposite electric motive force proportional to the number of turns and the rate of change in the flux to time as mentioned above.

The flow of the no-load current is the result of two currents:

1. The vertical component is called *the effective current*, which is the current responsible for the iron losses in the iron-core of the transformer (the heat produced in the iron-core when the transformer is working) and is symbolized by the symbol (I_e).
2. The horizontal component is called *the magnetism current*, which is the current responsible for the magnetic field that occurs in the iron-core and is symbolized by the symbol (I_m), as shown in Figure 4.9.

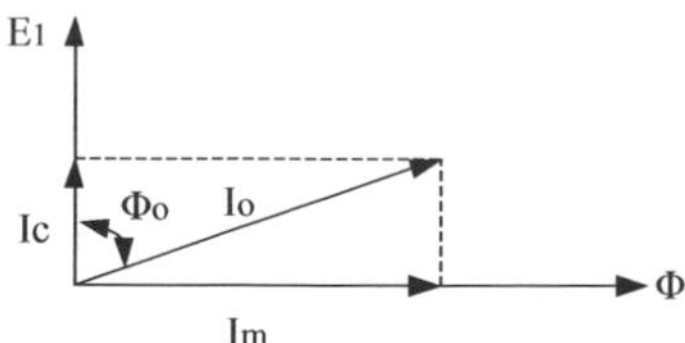

FIGURE 4.9
The flowing current and its two components.

The power dissipated to be carried out is the iron ΔP_{Fe}, which is calculated from the relationship:

$$\Delta P_{Fe} = V\, I_o \cos \phi_o \tag{4.10}$$

$$\Delta P_{Fe} = V\, i_e \tag{4.11}$$

$$I_e = \Delta P_{Fe}/V\,. \tag{4.12}$$

In applying Pythagoras' theorem to the right-angled triangle, we find that:

$$I_o^2 = I_e^2 + I_m^2$$

$$\mathrm{I_m} = \sqrt{I_o^2 - I_e^2} \tag{4.13}$$

Example 4.4

A 250 V transformer takes a current of 0.5 mA with a power factor at no-load of 0.3 lagging. Calculate the magnetic current.

Solution

$$\mathrm{I_e = I_o \cos \theta_o}$$

$$\mathrm{I_e} = 0.5 \times 0.3$$

$$\mathrm{I_e} = 0.15\ \mathrm{A}$$

$$\mathrm{I_m} = \sqrt{0.5^2 - 0.15^2}$$

$$\mathrm{I_m} = 0.477\ \mathrm{A}.$$

4.6.2 The Operation Transformer at Load

If the load is connected at the terminals of the secondary windings, a current followed called *the secondary windings current I_2*, which causes a magnetic flux and an induced electric motive force $-E_2$ to counteract the electric motive generated in the primary windings

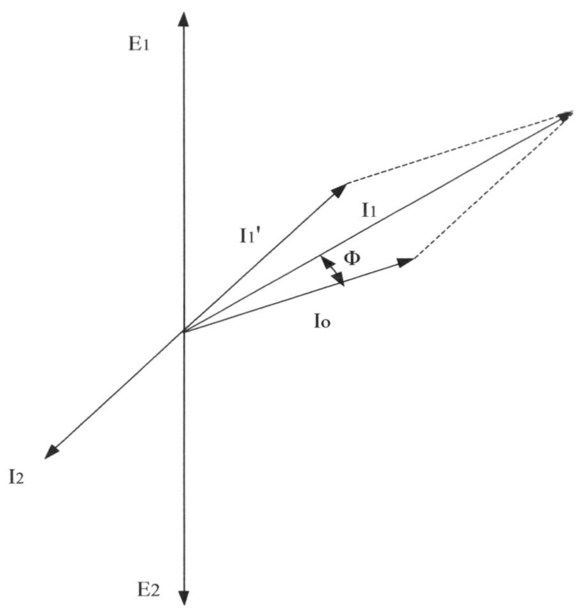

FIGURE 4.10
The current of the load current.

E_1. Thus, the current in the primary windings I_1' is equal to the secondary current I_2, which is equivalent to the amount and reverses it in the direction, as well as the current of the no-load I_o, so the current following through the primary windings with the load is large because it is the sum of two currents, I_0 and the load current of I_2 or I_1', as shown in Figure 4.10.

Transformers can be divided according to use into:

1. Power Transformers: Convert the voltage from one level to another and are used for power transmission and in various manufacturing areas, as shown in Figure 4.11 and in the household uses
2. Regulation Transformers: Used to obtain different voltage values in laboratories, research centers, and automated control
3. Transformers to change the number of phase currents (m): The alternating frequency (f) and pulse shape are mainly used in electronic devices, wired communications, and automatic control that does not exceed the capacity such as these transformers have several VA
4. Measurement Transformers: Such as serial current transformer and serial voltage transformer are used in electrical measurements and distribution boards and processes like feeding.

The transformers are divided in terms of the number of phases to:

1. Single-phase Transformers
2. Three-phase Transformers
3. Polyphase Transformers.

FIGURE 4.11
Power transformers.

Also divided in terms of conversion rates, which are:

1. Step-down Transformers: The high primary windings voltage V_1 is converted to a low voltage V_2 ($V_1 > V_2$)
2. Step-up Transformers: The primary windings voltage V_1 is converted to high voltage secondary V_2 ($V_1 < V_2$).

Also divided in terms of its cooling method into ways like following:

1. Dry Transformers: Cooled by natural or forced air and are usual transformers with small and medium capacities.
2. Transformers immersed in oil: Are cooled with oil as power transformers with medium and large capacity used in different power stations, and these transformers are characterized by the dangers of explosion and therefore provide advanced control circuits.
3. Transformers refrigeration with SF_6: Recently successfully tested and are now being used in indoor environments.

Figure 4.12 shows the types of electrical transformers.

4.7 Non-ideal Transformer Equivalent Circuits

The non-ideal transformer equivalent circuit in Figure 4.13 expresses the various ways in which all the loss terms that are neglected in the ideal transformer model. The individual loss terms in the equivalent circuit are:

R_1, R_2_primary and secondary winding resistances (losses in the windings due to the resistance of the wires)

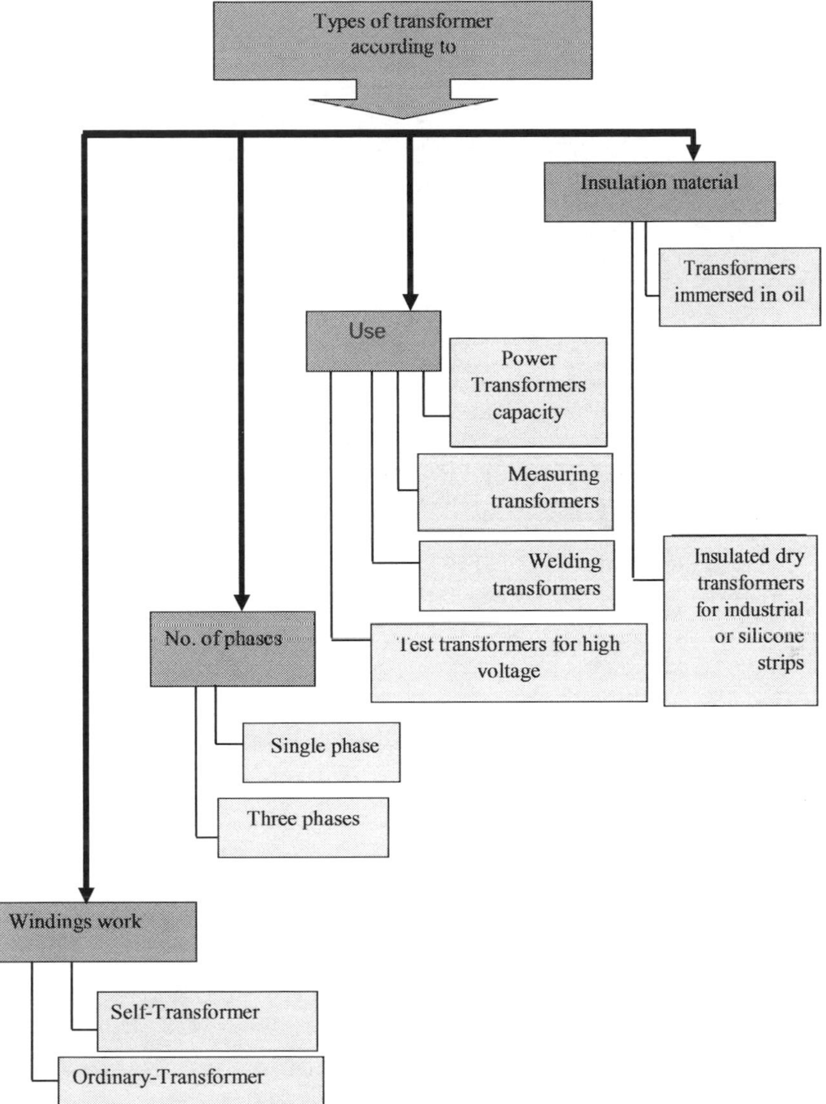

FIGURE 4.12
Types of electrical transformers.

X_1, X_2_primary and secondary leakage reactance's (losses due to flux leakage out of the transformer core)

R_c_core resistance (core losses due to hysteresis loss and eddy current loss)

X_m_magnetizing reactance (magnetizing current necessary to establish a magnetic flux in the transformer core).

Using the impedance reflection technique, all the quantities on the secondary side of the transformer can be reflected on the primary side of the circuit. The resulting equivalent

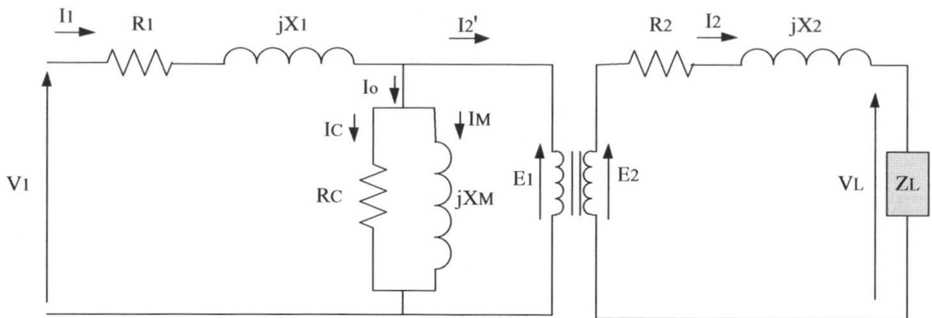

FIGURE 4.13
The non-ideal transformer equivalent circuit diagram.

circuit is shown below. The primed quantities represent those values that equal the original secondary quantity multiplied by voltages, divided by a current or multiplied by a^2 impedance components. Figure 4.14 shows the non-ideal transformer equivalent circuit diagram referred to the primary side and Figure 4.15 equivalent circuit for the secondary side of the transformer.

$$V_2' = a \cdot V_2 \tag{4.14}$$

$$I_2' = \frac{I_2}{a} \tag{4.15}$$

$$R_2' = a^2 \cdot R_2 \tag{4.16}$$

$$X_2' = a^2 \cdot X_2 \tag{4.17}$$

$$Z_2' = a^2 \cdot Z_2 \tag{4.18}$$

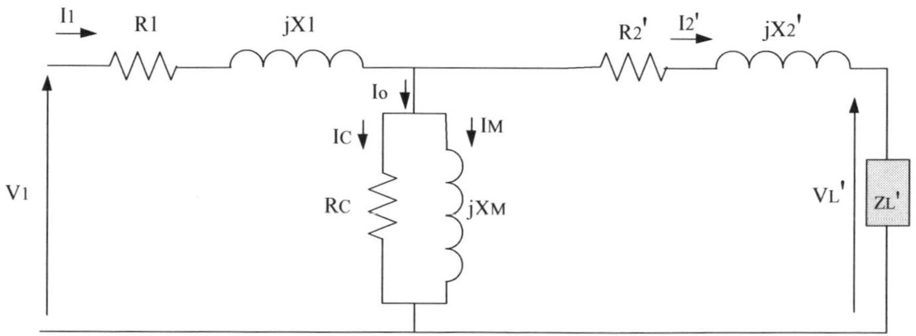

FIGURE 4.14
The non-ideal transformer equivalent circuit diagram referred to the primary side.

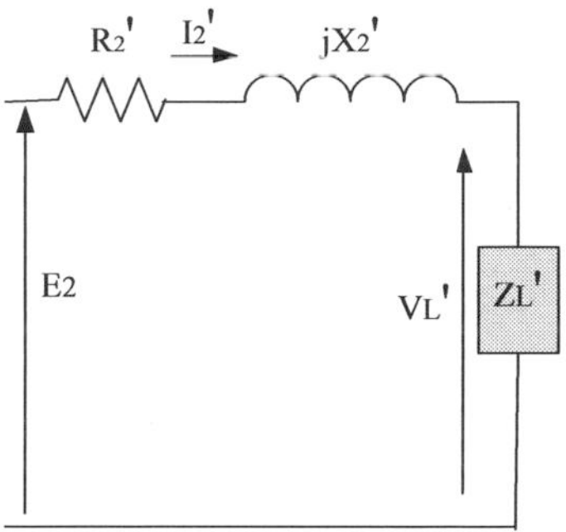

FIGURE 4.15
Equivalent circuit for the secondary side of the transformer.

4.8 Determination of Equivalent Circuit Parameters

To utilize the complete transformer equivalent circuit, the values of R_1, R_2, X_1, X_2, R_c, X_m, and *a* must be known. These values can be computed given the complete design data for the transformer including dimensions and material properties. The equivalent circuit parameters can also be determined by performing two simple test measurements. These measurements are the no-load (or open-circuit) test and the short-circuit test.

4.8.1 No-Load Test (Determine R_c and X_m)

The rated voltage at the rated frequency is applied to the high-voltage (HV) or low-voltage (LV) winding with the opposite winding open-circuits. Measurements of current, voltage, and real power are made on the input winding (most often the LV winding, for convenience). Figure 4.16 shows the equivalent circuit for no-load test.

$$P_L = \frac{V_L^2}{R_{CL}} \tag{4.19}$$

$$I_{CL} = \frac{V_L}{R_{CL}} \tag{4.20}$$

$$I_L = \sqrt{I_{CL}^2 + I_{ML}^2} \tag{4.21}$$

$$I_{ML} = \sqrt{I_L^2 - I_{CL}^2} \tag{4.22}$$

$$V_L = I_{ML} \cdot (jX_{ML}) \tag{4.23}$$

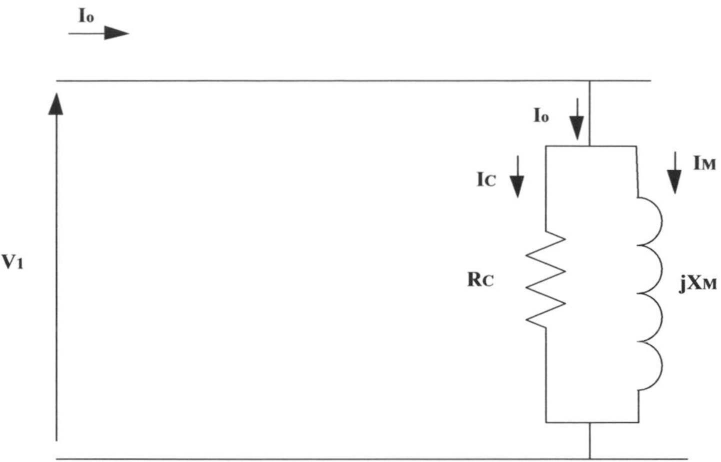

FIGURE 4.16
Equivalent circuit for no-load test.

4.8.2 Short-Circuit Test (Determine $R_{eq.H}$ and $X_{eq.H}$)

Either the LV or HV winding is short-circuited, and a voltage at the rated frequency is applied to the opposite winding, such that the rated current results. Measurements of current, voltage, and real power are made on the input winding (most often the HV winding, for convenience, since a relatively low-voltage is necessary to obtain rated current under short-circuit conditions). Figure 4.17 shows the equivalent circuit for the short-circuit test.

The values measured on the HV winding (primary) in the short-circuit test need to be referred to the LV side. Note that turns ratio is given by:

$$V_H' = \frac{V_H}{a} \tag{4.24}$$

$$I_H' = a \cdot I_H \tag{4.25}$$

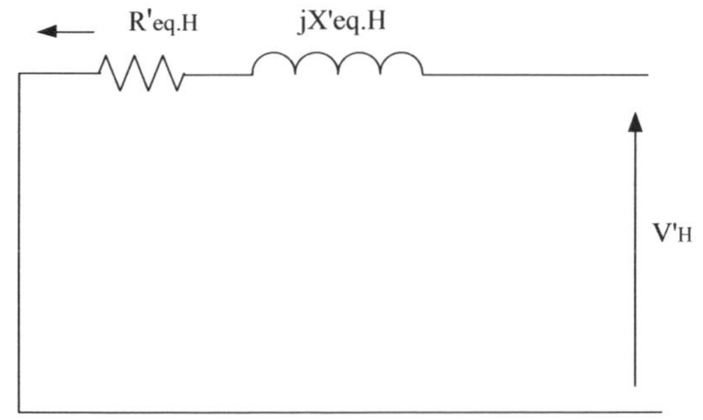

FIGURE 4.17
Equivalent circuit for short-circuit test.

$$R_{eq.L} = \frac{P_H}{I'^2_H} \quad (4.26)$$

$$V'_H = \left|R_{eq.H} + jX_{eq.H}\right| \cdot I'_H = Z_{eq.H} \cdot I'_H \quad (4.27)$$

and

$$X_{eq.H} = \sqrt{Z^2_{eq.H} - R^2_{eq.H}} \quad (4.28)$$

4.9 Transformer Voltage Regulation

For a given input (primary) voltage, the output (secondary) voltage of an ideal transformer is independent of the load attached to the secondary. As seen in the transformer equivalent circuit, the output voltage of a realistic transformer depends on the load current. If the current through the excitation branch of the transformer equivalent circuit is small in comparison to the current that flows through the winding loss and leakage reactance components, the transformer approximate equivalent circuit referred to the primary is shown below. Note that the load on the secondary (Z_2) and the resulting load current (I_2) have been reflected the primary (Z_{N2}, I_{N2}).

The percentage voltage regulation (V_R) is defined as the percentage change in the magnitude of the secondary voltage as the load current changes from the no-load to the loaded condition.

$$V_R = \frac{|V_2|_{NL} - |V_2|_L}{|V_2|_L} \quad (4.29)$$

Example 4.5

A 15 kVA, 2400:240/120 V, 60 Hz, the two-winding transformer is to be reconnected as a 2400:2520 V step-up transformer. From test work on the two-winding transformer, it is known that its rated voltage core losses and coil losses are 280 and 300 W, respectively. For this autotransformer, (*a*) determine the apparent power rating and (*b*) the full-load efficiency if supplying 2520 V to a 0.8 PF lagging load.

Solution

a)

$$I_H = I_2 = \frac{15{,}000}{120} = 125 \text{ A}$$

$$V_H = V_1 + V_2 = 2400 + 120 = 2520$$

$$S_X = S_H = V_H I_H = (2520)(125) = 315 \text{ kVA}$$

b) The core and copper losses are unchanged from the two-winding transformer.

$$P_o = S_H PF = (315{,}000)(0.8) = 252 \text{ kW}$$

$$\eta = \frac{P_o(100)}{P_o + \text{losses}} = \frac{(252{,}000)(100)}{252{,}000 + 280 + 300} = 99.77\%.$$

Example 4.6

Determine the value of the coefficient of coupling (*k*) for the transformer of Example 4.5.

Solution

The turns ratio is:

$$a = \frac{V_1}{V_2} = \frac{240}{120} = 2$$

$$M = \frac{X_m}{\omega a} = \frac{400}{2\pi(60)(2)} = 0.5305 \text{ H}$$

$$L_1 = \frac{X_1}{\omega} + aM = \frac{0.18}{2\pi(60)} + 2(0.5305) = 1.0615 \text{ H}$$

$$L_2 = \frac{X_2}{\omega} + \frac{M}{a} = \frac{0.045}{2\pi(60)} + \frac{0.5305}{2} = 0.2654 \text{ H}$$

$$k = \frac{M}{\sqrt{L_1 L_2}} = \frac{0.5305}{\sqrt{(1.0615)(0.2654)}} = 0.999$$

Example 4.7

For the ideal transformer circuit of Figure 4.18, R_p = 18 Ω, R_L = 6 Ω, and X_L = 0.5 Ω. If $\vec{V}_2 = 120\angle 0° \ V$ and P_s = 5600 W, (*a*) determine the turns ratio *a*, (*b*) the source voltage $\bar{V}_S$, and (*c*) the input power factor.

(a) Equations

$$P_{R_L} = \frac{V_2^2}{R_L} = \frac{(120)^2}{6} = 2400 \text{ W}$$

$$P_{R_p} = P_S - P_{R_L} = 5600 - 2400 = 3200 \text{ W}$$

$$V_1 = \sqrt{P_{R_p} R_p} = \sqrt{(3200)(18)} = 240 \text{ V}$$

$$a = \frac{V_1}{V_2} = \frac{240}{120} = 2$$

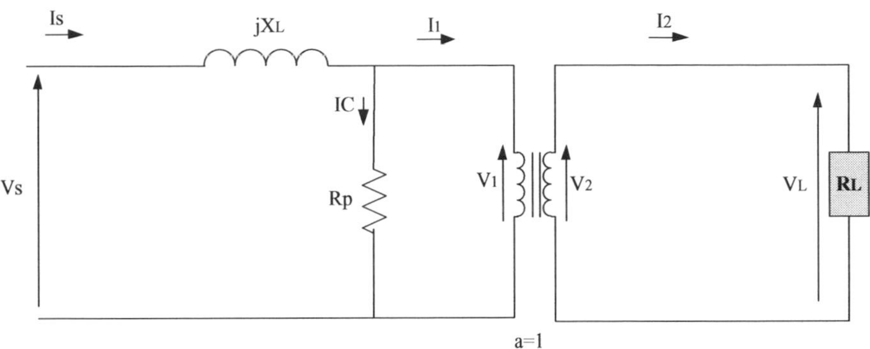

FIGURE 4.18
Circuit of Example 4.7.

(b) Equations

$$\bar{I}_2 = \frac{\bar{V}_2}{R_L} = \frac{120\angle 0^\circ}{6} = 20\angle 0^\circ \text{ A}$$

$$\bar{I}_1 = \frac{1}{a}\bar{I}_2 = \frac{1}{2}(20\angle 0^\circ) = 10\angle 0^\circ \text{ A}$$

$$\bar{I}_S = \bar{I}_1 + \frac{\bar{V}_1}{R_p} = 10\angle 0^\circ + \frac{240\angle 0^\circ}{18} = 23.33\angle 0^\circ \text{ A}$$

$$\bar{V}_S = Z_l\bar{I}_S + \bar{V}_1 = (0.5\angle 90^\circ)(23.33\angle 0^\circ) + 240\angle 0^\circ = 240.28\angle 2.78^\circ \text{ V}$$

(c) Equations

$$PF_S = \frac{P_S}{V_S I_S} = \frac{5600}{(240.28)(23.33)} = 0.999 \text{ lagging}$$

Example 4.8

Consider the circuit shown in Figure 4.19. The input to the circuit is the voltage of the voltage source $V_s(t)$. The output is the voltage across the 9 H inductor, $V_o(t)$. Determine the output voltage, $V_o(t)$. Write MATLAB program to verify the answers.

Given $V_s(t) = 75.5 \cos(4t + 26°)$ V, so $V_s = 75.5\angle 26°$ V,
where frequency $\omega = 4$ rad/sec.
The impedance offered by inductor 9H is:

$$Z_L = J\omega.$$

$$L = J(4 \times 9) = J36\ \Omega.$$

The total impedance referred to the primary side using turns ratio is given as:

$$Z = 8 + 30 \times \left(\frac{3}{2}\right)^2 + (J36) \times \left(\frac{3}{2}\right)^2$$

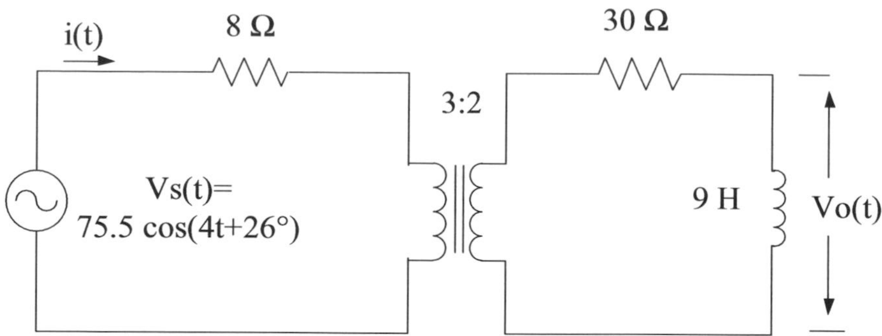

FIGURE 4.19
Circuit of Example 4.8.

$$Z = 8 + 67.5 + J81$$

$$Z = 75.5 + J81.$$

The current from the source is:

$$I_s = \frac{V_s}{Z} = \frac{75.5\angle 26^\circ}{75.5 + J81} = 0.682\angle -21.01^\circ A$$

Now current in the secondary is:

$$I_o = I_s \times \frac{3}{2} = 0.682\angle -21.01^\circ = 1.023\angle -21.01^\circ A$$

Now output voltage is:

$$V_o = I_o Z_L = 1.023\angle -21.01^\circ \times J36 = 36.82\angle 68.99 V$$

$$V_o(t) = 36.82 \cos(4t + 68.99^\circ) \text{ V}.$$

MATLAB CODE

```
W=4;
L=9;
Z_L=w*9*i;
Z=8+30*(3/2)^2+Z_L*(3/2)^2;
Vs=75.5*exp(i*26*pi/180);
Is=Vs/Z;
Io=Is*(3/2);
Vo=Io*Z_L;
magVo=abs(Vo)
phaseVo=angle(Vo)*180/pi
%The above matlab code executed in octave online compatible with matlab
below octave:5>=w;
L=9;
Z_L=w*9*i;
Z=8+30*(3/2)^2+Z_L*(3/2)^2;
Vs=75.5*exp(i*26*pi/180);
Is=Vs/Z;
Io=Is*(3/2);
Vo=Io*Z L;
magVo=36.8191
phaseVo=68.9872
```

4.10 Three-Phase Transformers

A three-phase transformer may consist of three single-phase windings on the same core inside a single tank or three single-phase transformers wired externally in wye or delta. It is also common practice to construct valid three-phase transformation using only two single-phase transformers. This configuration is called "open-delta". In three-phase systems, the default voltage designation is the line-to-line value, V_{L-L}. Commercial and light industrial systems utilizing three-phase are usually served at 208 V (with 120 V line-to-neutrals). Larger industrial systems utilize 480 V (277 V line-to-neutral). Most industrial plants may use 4160 V (2400 V line-to-neutral) or even higher ranges. A high-voltage system will transmit a given amount of power at a lower current than a lower-voltage system. This is important for many reasons, not the least of which is I^2R losses (efficiency).

There are four ways to configure a standard three-phase transformer bank: delta-delta (Δ-Δ), wye-wye (Y-Y), delta-wye (Δ-Y), and wye-delta (Y-Δ). The most commonly used configuration is delta-wye (Δ-Y). When creating a delta winding using single-phase transformers, great care must be taken to ensure proper phasing and proper polarity. If a delta is completed with the wrong polarity, it is possible to create a dead short between phases, resulting in great damage to the transformer. Never close a delta transformer connection without measuring the voltage across the open corner. The life you save may be your own.

When a three-phase load is served by two single-phase transformers, the connection is called open-delta. Open-delta transformer banks will carry less than rated capacity since both transformers carry not only their own phase current, but also a portion of the current for the third phase. For two transformers of equal size (kVA rating), wired in open-delta, and serving a balanced three-phase load, the de-rating factor is 57% of a corresponding three-phase bank. For example, two 33.3 kVA transformers that are wired in open-delta would have a three-phase capacity of 58 kVA, whereas three transformers in the delta would have a capacity of 100 kVA.

4.10.1 Three-Phase Transformer Configuration

A three-phase transformer or 3-ϕ transformer can be constructed either by connecting together three single-phase transformers, thereby forming a so-called three-phase transformer bank or by using one pre-assembled and balanced three-phase transformer which consists of three pairs of single-phase windings mounted onto one single laminated core.

The advantages of building a single three-phase transformer are that for the same kVA rating it will be smaller, cheaper, and lighter than three individual single-phase transformers connected together because the copper and iron-core are used more effectively. The methods of connecting the primary and secondary windings are the same, whether using just one three-phase transformer or three separate single-phase transformers.

4.10.2 Three-Phase Transformer Connections

The primary and secondary windings of a transformer can be connected in a different configuration as shown to meet practically any requirement. In the case of three-phase transformer windings, three forms of connection are possible: "star" (wye), "delta" (mesh), and "interconnected-star" (zig-zag).

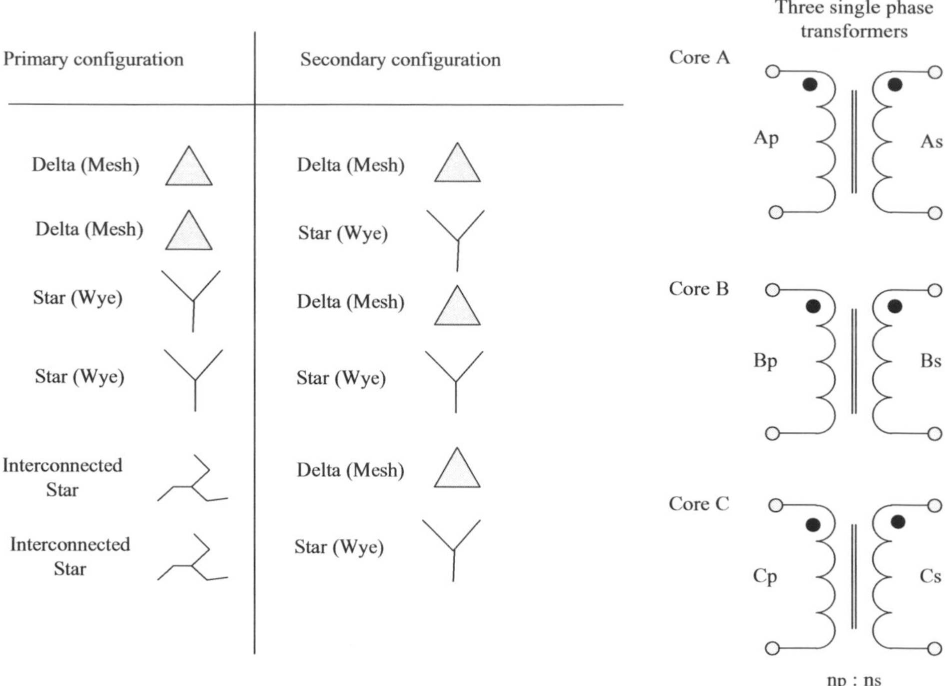

FIGURE 4.20
Three-phase transformer connections.

The combinations of the three windings may be with the primary delta-connected and the secondary star-connected, or star-delta, star-star, or delta-delta, depending on the transformers used. When transformers are used to provide three or more phases, they are generally referred to as a poly-phase transformer (Figure 4.20).

4.10.2.1 Three-Phase Transformer Star and Delta Configurations

A three-phase transformer has three sets of primary and secondary windings. How these sets of windings are interconnected determines whether the connection is a star (also known as wye) or delta (also known as mesh) configuration.

The three available voltages, which themselves are each displaced from the other by 120 electrical degrees, not only decide on the type of the electrical connections used on both the primary and secondary sides, but determine the flow of the transformers currents.

With three single-phase transformers connected together, the magnetic flux in the three transformers differs in phase by 120°. With a single three-phase transformer, there are three magnetic flux in the core differing in time-phase by 120°.

The standard method for marking three-phase transformer windings is to label the three primary windings with capital (upper case) letters A, B, and C used to represent the three individual phases of red, yellow, and blue. The secondary windings are labeled with small (lower case) letters a, b, and c. Each winding has two ends normally labeled 1 and 2 so that, for example, the second winding of the primary has ends which will be labeled B1 and B2, while the third winding of the secondary will be labeled c1 and c2, as shown (Figure 4.21).

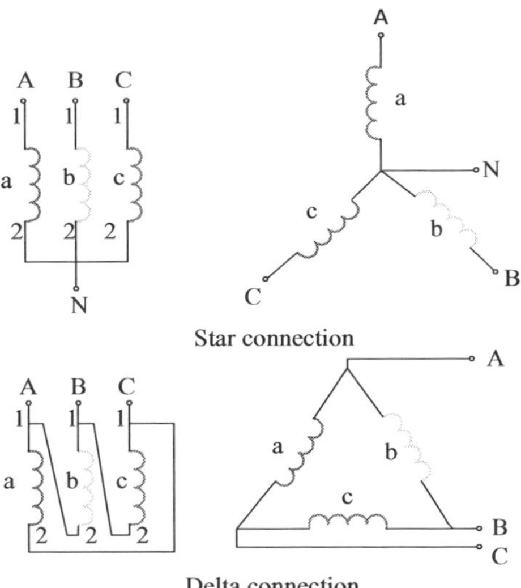

FIGURE 4.21
Transformer star and delta configurations.

4.10.2.2 Transformer Star and Delta Configurations

Symbols are generally used on a three-phase transformer to indicate the type or types of connections used with uppercase Y for star connected, Δ for delta connected, and Z for interconnected star primary windings, with lower case y, d, and z for their respective secondaries. Then, star-star would be labeled Yy, delta-delta would be labeled Dd, and interconnected star to interconnected star would be Zz for the same types of connected transformers.

Connection	Primary Winding	Secondary Winding
Delta	D	d
Star	Y	y
Interconnected	Z	z

4.10.2.3 Transformer Winding Identification

We now know that there are four different ways in which three single-phase transformers may be connected between their primary and secondary three-phase circuits. These four standard configurations are given as delta-delta (Δ-Δ), star-star (Y-Y), star-delta (Y-Δ), and delta-star (Δ-Y).

Transformers for high-voltage operation with the star connections have the advantage of reducing the voltage on an individual transformer, reducing the number of turns required, and an increase in the size of the conductors, making the coil windings easier and cheaper to insulate than delta transformers.

The delta-delta connection, nevertheless, has one big advantage over the star-delta configuration, in that, if one transformer of a group of three should become faulty or disabled, the two remaining ones will continue to deliver three-phase power with a capacity equal to approximately two-thirds of the original output from the transformer unit.

4.10.2.4 Transformer Delta and Delta Connections

In a delta connected (Δ-Δ) group of transformers, the line-voltage, V_L, is equal to the supply voltage, $V_L = V_S$. But the current in each phase winding is given as $1/\sqrt{3} \times I_L$ of the line current, where I_L is the line current.

One disadvantage of delta connected three-phase transformers is that each transformer must be wound for the full-line-voltage, for example, 120 V, as shown in Figure 4.22 and for 57.7%, line current. The greater number of turns in the winding, together with the insulation between turns, necessitates a larger and more expensive coil than the star connection. Another disadvantage with delta connected three-phase transformers is that there is no "neutral" or common connection.

4.10.2.5 Transformer Star and Star Connections

In the star-star arrangement (Y-Y), each transformer has one terminal connected to a common junction or neutral point with the three remaining ends of the primary windings connected to the three-phase mains supply. The number of turns in a transformer winding for star connection is 57.7% of that required for delta connection.

The star connection requires the use of three transformers, and if any one transformer becomes fault or disabled, the whole group might become disabled. Nevertheless, the star connected three-phase transformer is especially convenient and economical in electrical power distributing systems, in that, a fourth wire may be connected as a neutral point, (n), of the three-star connected secondary, as shown in Figure 4.23.

The voltage between any lines of the three-phase transformer is called *the line-voltage*, V_L, while the voltage between any line and the neutral point of a star connected transformer is called *the phase voltage*, V_P. This phase voltage between the neutral point and any one of the line connections is $1/\sqrt{3} \times V_L$ of the line-voltage. The primary side phase voltage, V_P, is given as:

$$V_P = \frac{1}{\sqrt{3}} \cdot V_L = \frac{1}{\sqrt{3}} \times 208 = 120\,Volts.$$

The secondary current in each phase of a star-connected group of transformers is the same as that for the line current of the supply, then $I_L = I_S$. Then the relationship between line and phase voltages and currents in a three-phase system can be summarized as follows.

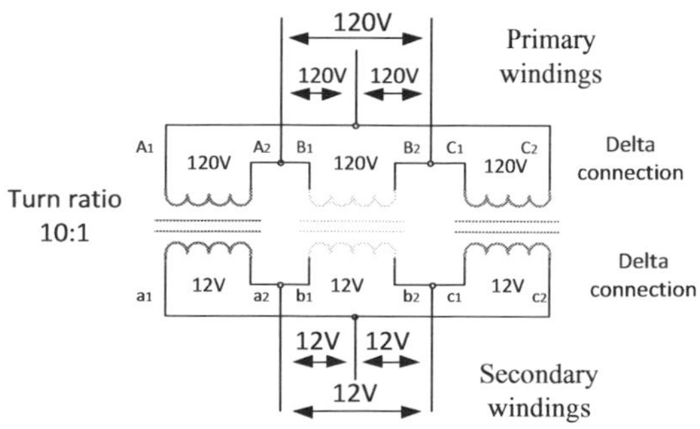

FIGURE 4.22
Transformer delta and delta connections.

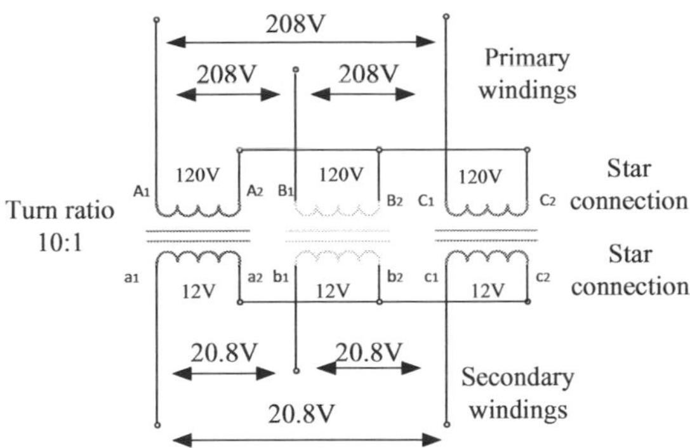

FIGURE 4.23
Transformer star and star connections.

4.10.3 Three-Phase Voltage and Current

Where again, V_L is the line-to-line-voltage and V_P is the phase-to-neutral voltage on either the primary or the secondary side. Other possible connections for three-phase transformers are star-delta Yd, where the primary winding is star-connected, and the secondary is delta-connected or delta-star (Δ-Y) with a delta-connected primary and a star-connected secondary.

Delta-star connected transformers are widely used in low power distribution with the primary windings providing a three-wire balanced load to the utility company while the secondary windings provide the required 4th-wire neutral or earth connection.

When the primary and secondary have different types of winding connections, star or delta, the overall turns ratio of the transformer becomes more complicated. If a three-phase transformer is connected as delta-delta (Δ-Δ) or star-star (Y-Y), then the transformer could potentially have a 1:1 turns ratio. That is the input and output voltages for the windings are the same.

However, if the three-phase transformer is connected in star-delta, (Y-Δ), each star-connected primary winding will receive the phase voltage, V_P, of the supply, which is equal to $1/\sqrt{3} \times V_L$. Then each corresponding secondary winding will then have this same voltage induced in it, and since these windings are delta-connected, the voltage $1/\sqrt{3} \times V_L$ will become the secondary line-voltage. Then with a 1:1 turns ratio, a star-delta connected transformer will provide a √3:1 step-down line voltage ratio.

Connection	Phase Voltage	Line-Voltage	Phase Current	Line Current
Star	$V_P = V_L/\sqrt{3}$	$V_L = \sqrt{3} \times V_P$	$I_P = I_L$	$I_L = I_P$
Delta	$V_P = V_L$	$V_L = V_P$	$I_P = I_L/\sqrt{3}$	$I_L = \sqrt{3} \times I_P$

4.10.3.1 Star-Delta Turns Ratio

For a star-delta (Y-Δ) connected transformer, the turns ratio becomes:

$$\text{Transformation ratio} = \frac{N_P}{N_S} = \frac{V_P}{\sqrt{3}V_S} \tag{4.30}$$

4.10.3.2 Delta-Star Turns Ratio

Likewise, for a delta-star (Δ-Y) connected transformer, with a 1:1 turns ratio, the transformer will provide a 1:√3 step-up line-voltage ratio. Then for a delta-star connected transformer, the turns ratio becomes:

$$\text{Transformation ratio} = \frac{N_P}{N_S} = \frac{\sqrt{3}V_P}{V_S} \tag{4.31}$$

Then for the four basic configurations of a three-phase transformer, we can list the transformers secondary voltages and currents with respect to the primary line-voltage, V_L, and its primary line current I_L, as shown in the following Table 4.1.

where n equals the transformers "turns ratio" of the number of secondary windings N_S, divided by the number of primary windings N_P, and V_L is the line-to-line-voltage with V_P being the phase-to-neutral voltage.

Example 4.8

The primary winding of a delta-star connected 50 kVA transformer is supplied with a 100 volt, 60 Hz, three-phase supply. If the transformer has 500 turns on the primary and 100 turns on the secondary winding, calculate the secondary side voltages and currents.

Solution

$$n = \frac{N_s}{N_p} = \frac{100}{500} = 0.2$$

$$V_{L(sec)} = \sqrt{3} \cdot n \cdot V_{L(pri)}$$

$$V_{L(sec)} = \sqrt{3} \times 0.2 \times 100$$

$$= 34.64\,\text{V}$$

$$V_{p(sec)} = \frac{V_{L(sec)}}{\sqrt{3}} = \frac{34.64}{\sqrt{3}} = 20\,V$$

$$I_{L(pri)} = \frac{V_A}{\sqrt{3} \cdot V_{L(pri)}} = \frac{50 \times 1000}{\sqrt{3} \times 100} = 288.67\,A$$

TABLE 4.1

Three-Phase Transformer Line-Voltage and Current

Primary-Secondary Configuration	Line-Voltage Primary or Secondary	Line-Current Primary or Secondary
Delta-Delta	$V_L \rightarrow nV_L$	$I_L \rightarrow \frac{I_L}{n}$
Delta-Star	$V_L \rightarrow \sqrt{3}nV_L$	$I_L \rightarrow \frac{I_L}{\sqrt{3}.n}$
Star-Delta	$V_L \rightarrow \frac{nV_L}{\sqrt{3}}$	$I_L \rightarrow \sqrt{3}.\frac{I_L}{n}$
Star-Star	$V_L \rightarrow nV_L$	$I_L \rightarrow \frac{I_L}{n}$

$$I_{sec} = \frac{V_{L(pri)}}{\sqrt{3} \cdot n} = \frac{288.67}{\sqrt{3} \times 0.2} = 833.33\,A$$

Then the secondary side of the transformer supplies a line-voltage, V_L, of about 35 V giving a phase voltage, V_P, of 20 V at 0.834 A.

4.10.4 Three-Phase Transformer Construction

We have said previously that the three-phase transformer is effectively three interconnected single-phase transformers on a single laminated core and considerable savings in cost, size, and weight can be achieved by combining the three windings onto a single magnetic circuit, as shown in Figure 4.24.

A three-phase transformer generally has the three magnetic circuits that are interlaced to give a uniform distribution of the dielectric flux between the high- and low-voltage windings. The exception to this rule is a three-phase shell-type transformer. In the shell-type of construction, even though the three cores are together, they are non-interlaced.

The three-limb core-type three-phase transformer is the most common method of three-phase transformer construction allowing the phases to be magnetically linked. The flux

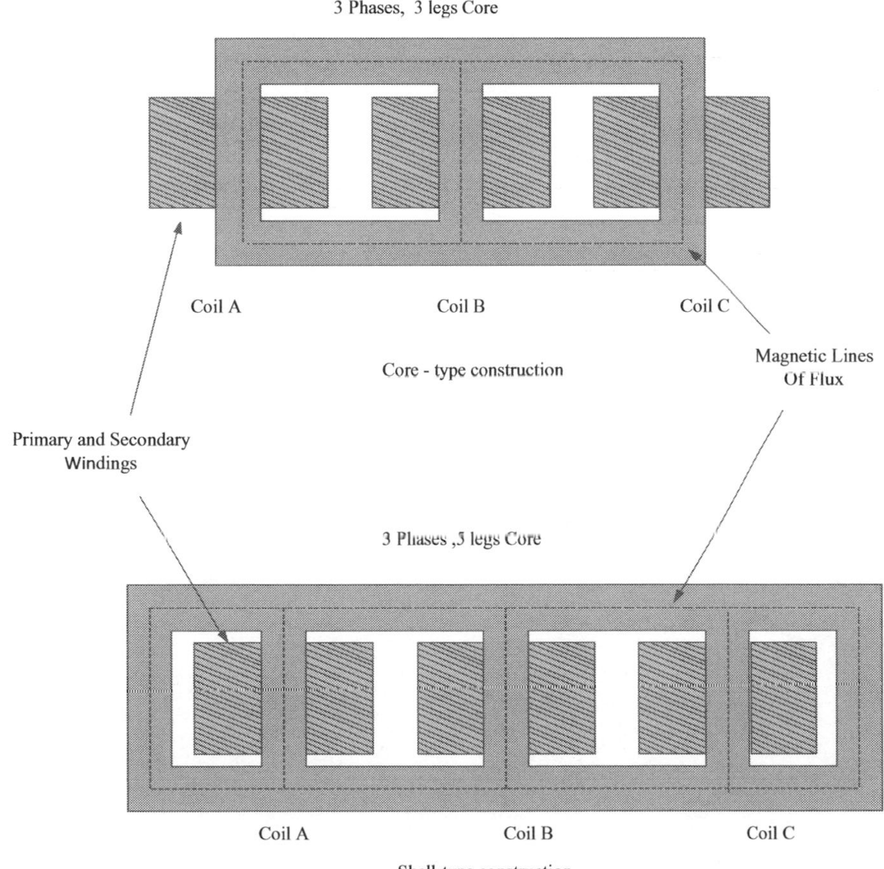

FIGURE 4.24
Three phase transformer construction.

of each limb uses the other two limbs for its return path with the three magnetic flux in the core generated by the line-voltages differing in time-phase by 120°. Thus, the flux in the core remains nearly sinusoidal, producing a sinusoidal secondary supply voltage.

The shell-type five-limb type three-phase transformer construction is heavier and more expensive to build than the core-type. Five-limb cores are generally used for very large power transformers as they can be made with reduced height. A shell-type transformers core material, electrical windings, steel enclosure, and cooling are much the same as for the larger single-phase types.

Problems

4.1 Explain the construction of electrical transformers.

4.2 What are the types of transformers in terms of iron-core shape?

4.3 Derive transformation ratio in electrical transformers.

4.4 A step-up transformer, if the number of high-voltage side is 500 turns, while the number of turns of the low-voltages sides 100 turns, calculate the transformation ratio (α).

4.5 A single-phase transformer of 25 kVA, number of primary turn 500 turns and secondary 40 turns, a voltage source connected to primary windings side of 3000 V, calculate primary windings current and secondary windings current at full load, secondary EMF voltage, and maximum magnetic flux in the magnetic circuit.

4.6 How can electric transformers be divided?

4.7 Explain the work of the transformer at no-load.

4.8 What is the sum of the currents in the transformer at full load?

4.9 Draw the phasor diagram of the transformer at full load.

4.10 What is the significance of stray losses, and should it be within some limits of total losses?

4.11 What is the best core material one should use to achieve minimum losses?

4.12 Consider the circuit shown in Figure 4.25. The input to the circuit is the voltage of the voltage source $V_s(t)$. The output is the voltage across the 3 H inductor, $V_o(t)$. Determine the output voltage, $V_o(t)$. Write MATLAB program to verify the answers.

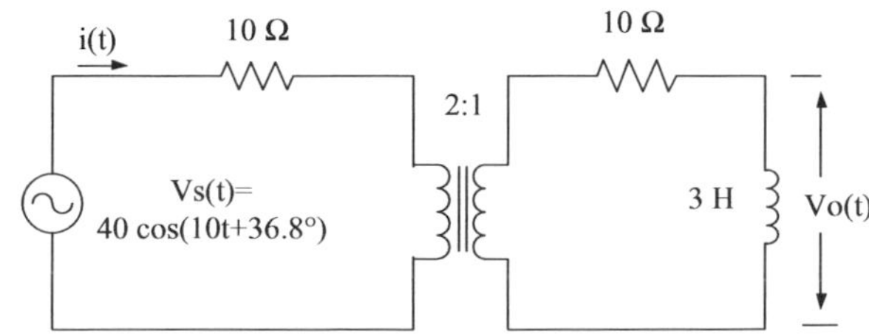

FIGURE 4.25
Circuit diagram for Problem 4.12.

5

Transformer Design

You must learn from your past mistakes, but not lean on your past successes.

Dennis Waitley

This chapter deals with the conventional design of core and shell type for single- and three-phase transformers. It provides all the calculations that are needed to design a transformer. It also gives a design using MATLAB program.

5.1 The Output Equations

It gives the relationship between the electrical rating and physical dimensions of the machines. Let:

V_1 = primary voltage say low voltage side Low Voltage (LV)
V_2 = secondary voltage say high voltage side High Voltage (HV)
I_1 = primary current
I_2 = secondary current
N_1 = primary no. of turns
N_2 = secondary no. of turns
a_1 = sectional area of LV conductors (m^2)

$$= \frac{I_1}{\delta}$$

a_1 = sectional area of HV conductors (m^2)

$$= \frac{I_2}{\delta}$$

δ = permissible current density (A/m^2)
Q = rating in Kilo Volt Amper (KVA).

The transformer winding, the first half of low voltage side on one limb and rest half of LV on another limb to reduce leakage flux. So, the arrangement is low voltage side insulation, then half low voltage side turns, then high voltage side insulation, and then half high voltage side turn.

5.1.1 Single-Phase Core Type Transformer

In single-phase core type transformer, the core is made up of the magnetic core and built with laminations to form a rectangular frame and the windings are arranged concentrically with each other around the legs or limbs. The low voltage windings are wound near the core and high voltage windings are wound over low voltage winding away from the core in order to reduce the number of insulating materials required. Coupling primary and secondary windings are close together to reduce the leakage reactance. In this type of transformer, the rating is given by: (Figure 5.1)

$$Q = V_1.I_1 \times 10^{-3} \quad \text{KVA}$$

$$= \left(4.44 f.\varphi_m.N_1\right).I_1 \times 10^{-3} \quad \text{KVA} \qquad \left(\because V_1 = 4.44 f.\varphi_m.N_1\right)$$

$$\because \varphi_m = A_i.B_m$$

$$\therefore Q = \left(4.44 f.A_i.B_m.N_1\right).I_1 \times 10^{-3} \quad \text{KVA} \tag{5.1}$$

where:

f = frequency
φ_m = maximum flux in the core
A_i = sectional area of core
B_m = maximum flux density in the core.

Window space factor:

$$K_w = \frac{\text{Actual CU section area of winding in window}}{\text{Window area } A_w}$$

$$= \frac{a_1.N_1 + a_2.N_2}{A_w}$$

$$= \frac{(I_1/\delta).N_1 + (I_2/\delta).N_2}{A_w} \qquad (\because a_1 = I/_1\delta \,\&\, a_2 = I/_2\delta)$$

$$= \frac{I_1.N_1 + I_2.N_2}{\delta.A_w}$$

$$= \frac{2I_1N_1}{\delta.A_W}.$$

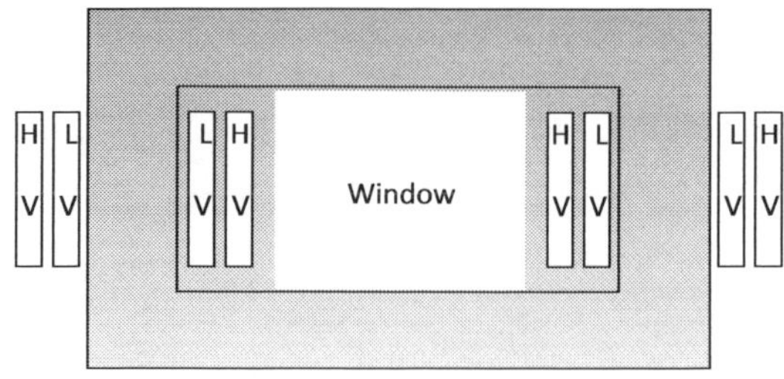

FIGURE 5.1
Phase core type transformer with concentric windings.

For ideal transformer, $I_1N_1 = I_2N_2$. So:

$$\left[N_1.I_1 = \frac{\delta.K_w.A_w}{2}\right] \tag{5.2}$$

Put the equation value of N_1I_1 from Equation (5.2) to Equation (5.1):

$$Q = 4.44\,f.A_i.B_m \frac{\delta.K_w.A_w}{2} \times 10^{-3} \quad KVA$$
$$Q = 2.22f.A_i.B_m.\delta.K_w.A_w \times 10^{-3} \quad KVA \tag{5.3}$$

5.1.2 Single-Phase Shell Type Transformer

In single-phase shell type transformers, the windings are wrap around the central limb and the flux path is completed through two side limbs. The central limb carries total mutual flux while the side limbs forming a part of a parallel magnetic circuit carry half the total flux. The cross-sectional area of the central limb is twice that of each side limbs. In this type of transformer, the window space factor is (Figure 5.2):

$$\begin{aligned} K_W &= \frac{a_1.N_1 + a_2.N_2}{A_w} \\ &= \frac{(I_1/\delta).N_1 + (I_2/\delta).N_2}{A_w} \qquad (\because a_1 = I/_1\delta \,\&\, a_2 = I/_2\delta) \\ &= \frac{I_1.N_1 + I_2.N_2}{\delta.A_w} \\ &= \frac{2I_1N_1}{\delta A_W}. \end{aligned}$$

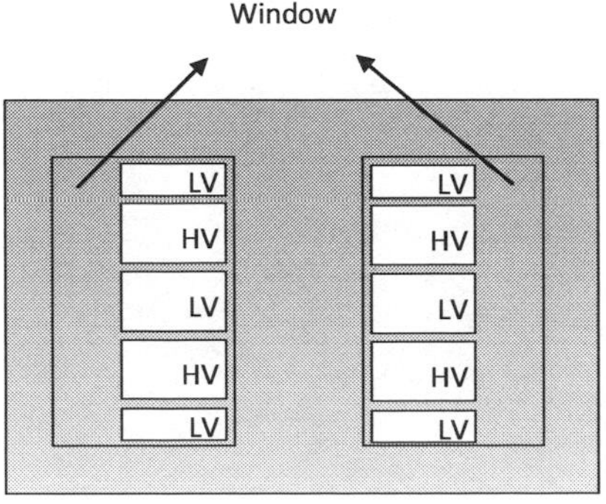

FIGURE 5.2
Phase shell type transformer with sandwich windings.

For ideal transformer $I_1N_1 = I_2N_2$. So:

$$N_1.I_1 = \frac{\delta.K_w.A_w}{2} \tag{5.4}$$

Put the equation value of N_1I_1 from Equation (5.4) to Equation (5.1):

$$\begin{aligned} Q &= 4.44\, f.A_i.B_m \frac{\delta.K_w.A_w}{2} \times 10^{-3} \quad KVA \\ Q &= 2.22\, f.A_i.B_m.\delta.K_w.A_w \times 10^{-3} \quad KVA \end{aligned} \tag{5.5}$$

Note, it is same as for 1-phase core type transformer, i.e., Equation (5.3).

5.1.3 Three-Phase Shell Type Transformer

In three-phase shell type transformer shown in Figure 5.3, the window space factor:

$$\begin{aligned} K_W &= \frac{a_1.N_1 + a_2.N_2}{A_w} \\ &= \frac{(I_1/\delta).N_1 + (I_2/\delta).N_2}{A_w} \qquad (\because a_1 = I/_1\delta \ \& \ a_2 = I/_2\delta) \\ &= \frac{I_1.N_1 + I_2.N_2}{\delta.A_w} \\ &= \frac{2I_1N_1}{\delta A_W}. \end{aligned}$$

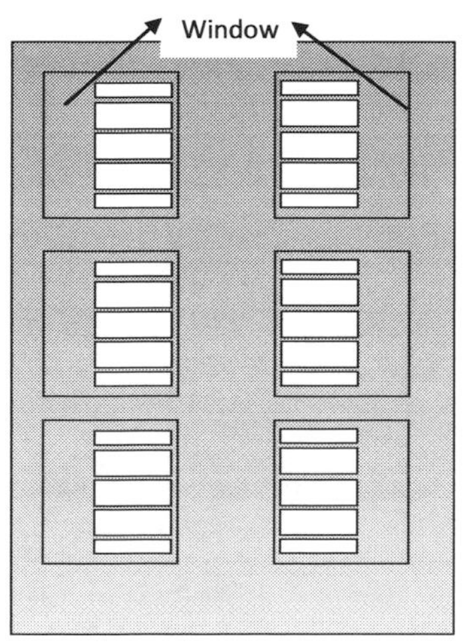

FIGURE 5.3
Three-phase shell type transformer with sandwich windings.

For ideal transformer $I_1N_1 = I_2N_2$. So:

$$N_1.I_1 = \frac{\delta.K_w.A_w}{2} \quad (5.6)$$

Put the equation value of N_1I_1 from Equation (5.6) to Equation (5.3):

$$Q = 3 \times 4.44 f .A_i .B_m \frac{\delta .K_w .A_w}{2} \times 10^{-3} \quad KVA$$
$$Q = 6.66 f .A_i .B_m .\delta .K_w .A_w \times 10^{-3} \quad KVA \quad (5.7)$$

5.2 Choice of Magnetic Loading (B_m)

The selection of magnetic loading depends on the service condition (i.e., distribution or transmission) and the material used for laminations of the core. The flux density decides the magnetic loading, area of the cross-section of core, and core loss.

1. Normal Si-steel B_m = 0.9 T–1.1 T
   35 m thickness, 1.5%–3.5% Si)
2. HRGO B_m = 1.2 T–1.4 T
   (Hot rolled grain oriented Si steel)
3. CRGO B_m = 1.4 T–1.7 T
   (Cold rolled grain oriented Si steel)
   (0.14 mm–0.28 mm thickness)

5.3 Choice of Electric Loading (Δ)

This factor depends upon the cooling method employed:

1. Natural cooling:		1.5 A/mm^2–2.3 A/mm^2
	AN	Air natural cooling
	ON	Oil natural cooling
	OFN	Oil forced circulated with natural air cooling
2. Forced cooling:		2.2 A/mm^2–4.0 A/mm^2
	AB	Air blast cooling
	OB	Oil blast cooling
	OFB	Oil forced circulated with air blast cooling

3. Water cooling: $5.0\ A/mm^2$–$6.0\ A/mm^2$

OW	Oil immersed with circulated water cooling
OFW	Oil forced with circulated water cooling

5.4 Core Construction

The cross-sectional area of core type transformer may be rectangular, square, or stepped. When circular coils are required for distribution and power transformers, the square and stepped cores are used.

For shell type transformer, the cross-section may be rectangular. When rectangular cores are used the coils are also rectangular in shape. The rectangular core is suitable for small and low voltage transformers. Figure 5.4 shows a core construction.

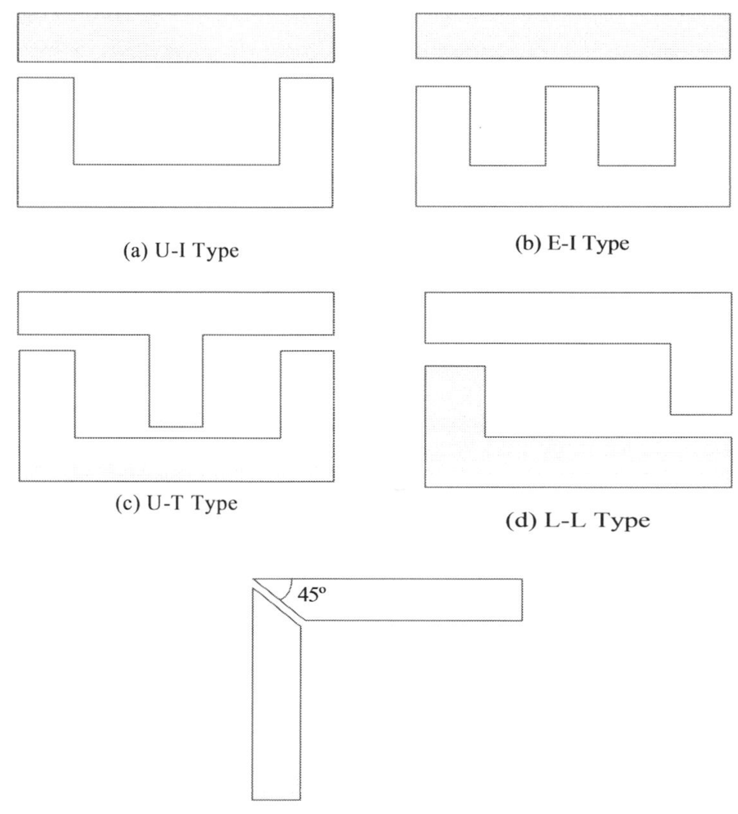

FIGURE 5.4
Core constructions (a) U-I type, (b) E-I type, (c) U-T type, (d) L-L type, and (e) Mitered core construction.

5.5 Electric Motive Force (EMF) per Turn

$$V_1 = 4.44 f.\varphi_m.N_1 \tag{5.8}$$

$$\text{So EMF / Turn} \quad E_t = \frac{V_1}{N_1} = 4.44 f.\varphi_m \tag{5.9}$$

and

$$Q = V_1.I_1 \times 10^{-3} \quad KVA \quad (\textbf{Note:} \text{ Take Q as per phase rating in KVA})$$

$$= (4.44 f.\varphi_m.N_1).I_1 \times 10^{-3} \quad \text{KVA}$$

$$= E_t.N_1.I_1 \times 10^{-3} \quad KVA \tag{5.10}$$

In the design, the ratio of total magnetic loading and electric loading may be kept constant:

$$\text{Magnetic loading} = \varphi_m$$

$$\text{Electric loading} = N_1.I_1.$$

So $\frac{\varphi_m}{N_1 I_1} = const.(say\ "r") \Rightarrow N_1 I_1 = \frac{\varphi_m}{r}$ put in Equation (5.10):

$$Q = E_t \frac{\varphi_m}{r} \times 10^{-3} \quad KVA$$

$$Or\ Q = E_t.\frac{E_t}{4.44 f.r} \times 10^{-3} \quad KVA$$

using Equation (5.9):

$$E_t^2 - (4.44 f.r \times 10^{-3}) \times Q$$

Or $\quad E_t = K_t\sqrt{Q} \quad Volts/Turn,$

where:

$K_t = \sqrt{4.44 f.r \times 10^{-3}}$ is a constant and values are:
$K_t = 0.6–0.7$ for 3-phase core type power transformer
$K_t = 0.45$ for 3-phase core type distribution transformer
$K_t = 1.3$ for 3-phase shell type transformer
$K_t = 0.75–0.85$ for 1-phase core type transformer
$K_t = 1.0–1.2$ for 1-phase shell type transformer.

5.6 Estimation of Core X-Sectional Area A_i

We know:

$$E_t = K_t \sqrt{Q} \tag{5.11}$$

$$E_t = 4.44\, f.\varphi_m$$

$$\text{or} \quad E_t = 4.44\, f.A_i.B_m \tag{5.12}$$

$$\text{so} \quad A_i = \frac{E_t}{4.44\, f.B_m} \tag{5.13}$$

The core may be following types as shown in Figure 5.5,
where d = diameter of the circumscribed circle.
For square core:

$$\text{Gross Area} = \frac{d}{\sqrt{2}} \times \frac{d}{\sqrt{2}} = 0.5d^2$$

Let stacking factor:

$$K_i = 0.9$$

Actual iron area:

$$A_i = 0.9 \times 0.5 \times d^2$$

$$= 0.45 \times d^2$$

(0.45 for square core and take "K" as a general case):

$$= K.d^2$$

So

$$A_i = K.d^2.$$

Or

$$d = \sqrt{\frac{A_i}{K}}$$

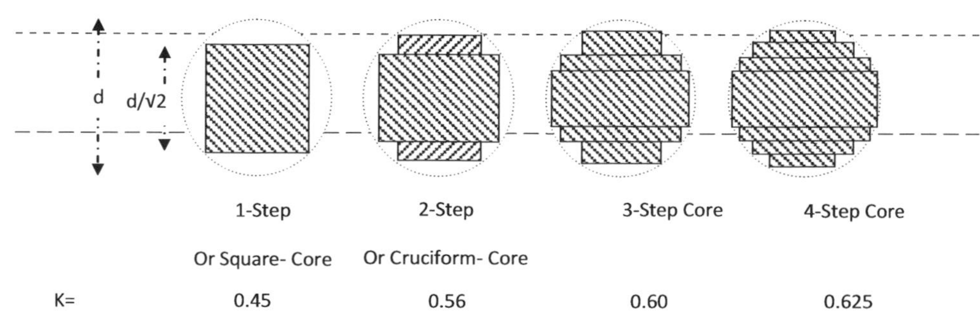

FIGURE 5.5
Types of the core.

5.7 Graphical Method to Calculate Dimensions of the Core

Consider 2 step core (Figure 5.6):

$$\theta = \frac{90^o}{n+1}, \qquad n = No\ \ of\ \ Steps$$

$$\text{i.e } n = 2 \qquad So\ \ a = d\text{Cos}\theta$$

$$\theta = \frac{90^\circ}{2+1} = 30^\circ \quad b = d\text{Sin}\theta$$

Percentage fill

$$= \frac{\text{Gross Area of Stepped core}}{\text{Area of circumcircle}} = \frac{K.d^2/K_i}{\pi.d^2/4}$$

$$= \frac{0.625d^2/0.9}{\frac{\pi}{4}(d^2)} \qquad \text{for 4 Step core}$$

$$= 0.885 \text{ or } 88.5\%.$$

No of Steps	1	2	3	4	5	6	7	9	11
% Fill	63.7%	79.2%	84.9%	88.5%	90.8%	92.3%	93.4%	94.8%	95.8%

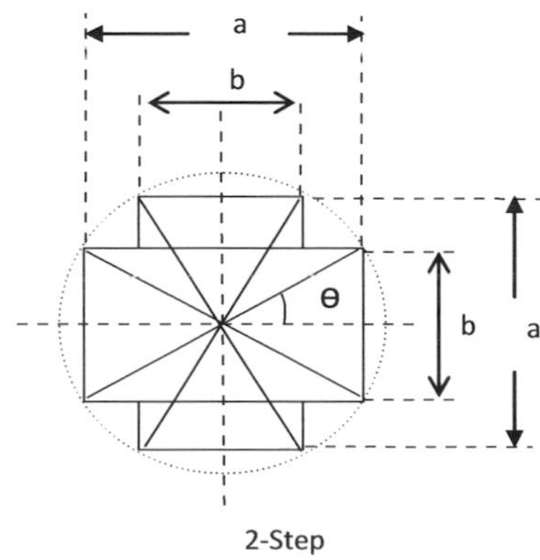

FIGURE 5.6
2-step cruciform-core.

5.8 Estimation of Main Dimensions

Consider a 3-phase core type transformer shown in Figure 5.7.

From the output equation:

$$Q = 3.33 f . A_i . B_m . \delta . K_w . A_w \times 10^{-3} \quad KVA,$$

so, window area:

$$A_w = \frac{Q}{3.33 f . A_i . B_m . \delta . K_w \times 10^{-3}} \ m^2 \tag{5.14}$$

where K_w = window space factor.

$$K_w = \frac{8}{30 + HigherKV} \quad \textit{for upto } 10 \ KVA$$

$$K_w = \frac{10}{30 + HigherKV} \quad \textit{for upto } 200 \ KVA$$

$$K_w = \frac{12}{30 + HigherKV} \quad \textit{for upto } 1000 \ KVA.$$

For higher rating K_w = 0.15 to 0.20.

Assume some suitable range for:

D = (1.7 to 2) d.

Width of the window W_w = D – d.

The height of the window:

$$L = \frac{A_w}{width \ of \ window(W_w)}$$

$$(\because L \times W_w = A_w).$$

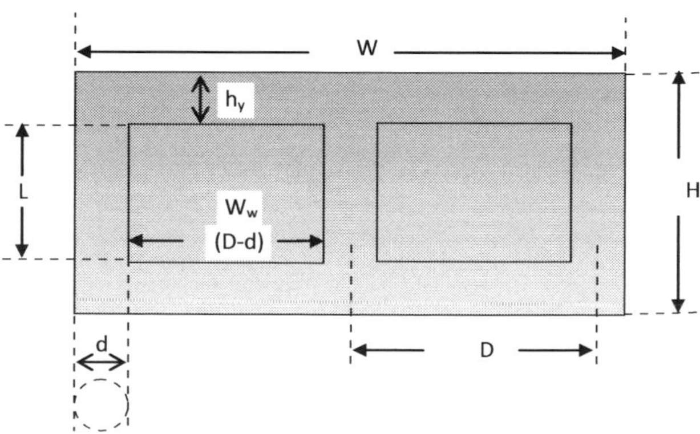

FIGURE 5.7
Three-phase core type transformer.

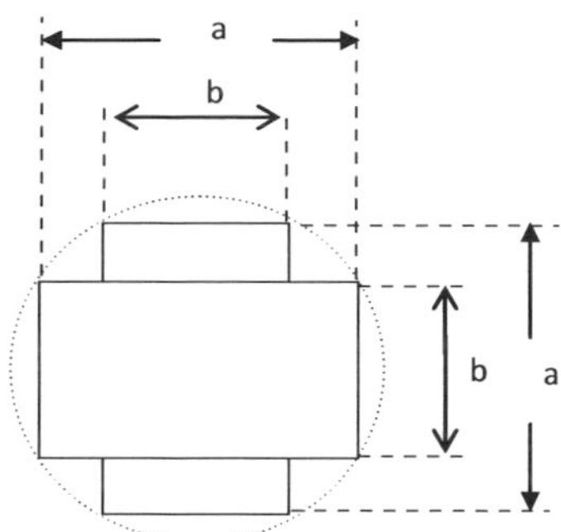

FIGURE 5.8
2-Step or cruciform-core.

Generally, $\frac{L}{W_w} = 2$ *to* 4.

Yoke area A_y is generally taken 10% to 15% higher than core section area (A_i), it is to reduce the iron loss in the yoke section. But if we increase the core section area (A_i), more copper will be needed in the windings and so more cost though we are reducing the iron loss in the core. A further length of the winding will increase resulting in higher resistance to more copper loss (Figure 5.8).

$$A_y = (1.10 \text{ to } 1.15)\ A_i$$

Depth of yoke

$$D_y = a$$

Height of the yoke

$$h_y = A_y/D_y$$

The width of the core:

$$W = 2 \times D + d$$

The height of the core:

$$H = L + 2 \times h_y$$

Flux density in the yoke

$$B_y = \frac{A_i}{A_y} B_m.$$

5.9 Estimation of Core Loss and Core Loss Component of No-Load Current I_c

$$\text{The volume of iron in core} = 3 \times L \times A_i \text{ m}^3$$

$$\text{The weight of iron in core} = \text{density} \times \text{volume}$$

$$= \rho_i \times 3 \times L \times A_i \text{ Kg} \tag{5.15}$$

$$\rho_i = \text{density of iron (kg/m}^3)$$
$$= 7600 \text{ Kg/m}^3 \text{ for normal iron/steel}$$
$$= 6500 \text{ Kg/m}^3 \text{ for M-4 steel.}$$

From the graph we can find out specific iron loss, p_i (Watt/Kg) corresponding to flux density B_m in the core.

So:

the iron loss in core $= p_i \times \rho_i \times 3 \times L \times A_i$ Watt (5.16)

Similarly:

the iron loss in yoke $= p_y \times \rho_i \times 2 \times W \times A_y$ Watt (5.17)

where p_y = specific iron loss corresponding to flux density B_y in yoke.
The total iron loss P_i = iron loss in core + iron loss in the yoke.
The core loss component of no-load current
I_c = core loss per phase/primary voltage:

$$I_c = \frac{P_i}{3V_1}.$$

5.10 Estimation of Magnetizing Current of No-Load Current I_m

Find out magnetizing force H, $at_{core/m}$ corresponding to flux density B_m in the core and at_{yoke} corresponding to flux density in the yoke from B-H curve (Figure 5.9):

$$\left(B_m \Rightarrow at_{core}/m, \quad B_c \Rightarrow at_{yoke}/m\right).$$

So

Magnetic Motive Force (MMF) required for the core $= 3 \times L \times at_{core}$

Magnetic Motive Force (MMF) required for the yoke $= 2 \times W \times at_{yoke}$.

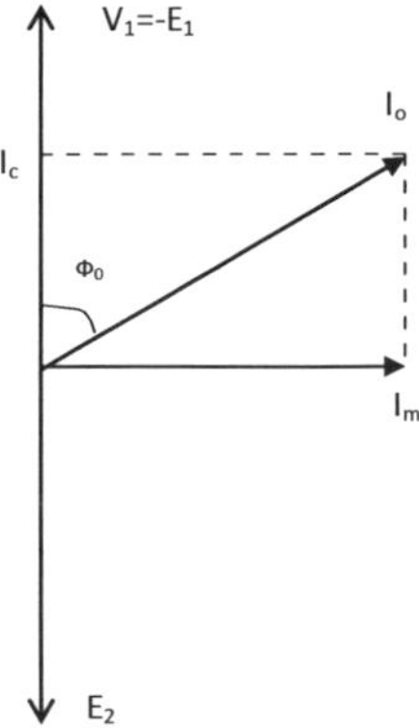

FIGURE 5.9
No-load phasor diagram.

We account for 5% of joints, so

$$\textit{Total MMF required} = 1.05[\textit{MMF for core} + \textit{MMF for yoke}].$$

The peak value of the magnetizing current:

$$I_{m,peak} = \frac{\textit{Total MMF required}}{3N_1}.$$

Root Mean Square (RMS) value of the magnetizing current:

$$I_{m,RMS} = \frac{I_{m,peak}}{\sqrt{2}}$$

$$I_{m,RMS} = \frac{\textit{Total MMF required}}{3\sqrt{2}N_1}.$$

5.11 Estimation of No-Load Current and Phasor Diagram

The no-load current I_o:

$$I_o = \sqrt{I_c^2 + I_m^2},$$

and the no-load power factor:

$$Cos\varphi_o = \frac{I_c}{I_o}.$$

The no-load current should not exceed 5% of the full the load current.

5.12 Estimation of Number of Turns on LV and HV Windings

$$\text{Primary no. of turns} \quad N_1 = \frac{V_1}{E_t}.$$

$$\text{Secondary no. of turns } N_2 = \frac{V_2}{E_t}.$$

5.13 Estimation of Sectional Area of Primary and Secondary Windings

$$\text{Primary current } I_1 = \frac{Q \times 10^{-3}}{3V_1}.$$

$$\text{Secondary current } I_2 = \frac{Q \times 10^{-3}}{3V_2} \quad OR \quad \frac{N_1}{N_2} I_1.$$

$$\text{The sectional area of primary winding } a_1 = \frac{I_1}{\delta}.$$

$$\text{The sectional area of secondary winging } a_2 = \frac{I_2}{\delta},$$

where δ is the current density. Now we can use round conductors or strip conductors.

5.14 Determination of R_1, R_2, and Copper Losses

Let L_{mt} = length of mean turn.

The resistance of the primary winding:

$$R_{1,\,dc,\,75^\circ} = 0.021 \times 10^{-6} \frac{L_{mt}.N_1(m)}{a_1(m^2)} \tag{5.18}$$

$$R_{1,\,ac,\,75^\circ} = (1.15 \ \ to \ \ 1.20) R_{1,\,dc,\,75^\circ}. \tag{5.19}$$

The resistance of the secondary winding:

$$R_{2,\,dc,\,75^\circ} = 0.021 \times 10^{-6} \frac{L_{mt}.N_2(m)}{a_2(m^2)} \tag{5.20}$$

$$R_{2,\,ac,75^\circ} = (1.15 \ \ to \ \ 1.20) R_{2,\,dc,\,75^\circ} \tag{5.21}$$

$$\text{The copper loss in the primary winding} = 3I_1^2.R_1. \tag{5.22}$$

$$\text{The copper loss in the secondary winding} = 3I_2^2.R_2 \tag{5.23}$$

$$\begin{aligned} \text{The total copper loss} &= 3I_1^2.R_1 + 3I_2^2.R_2 \\ &= 3I_1^2.(R_1 + R_2') \\ &= 3I_1^2.R_p \end{aligned} \tag{5.24}$$

where R_{o1} is the total resistance:

$$R_{01} = R_p = R_1 + R_2' \tag{5.25}$$

Under no-load condition, there is a magnetic field connecting leads, which causes additional stray losses in the transformer tanks and other metallic parts. These losses may be taken as 7% to 10% of total copper losses.

5.15 Determination of Efficiency

The efficiency of the transformer can be calculated from:

$$\eta = \frac{Output\ Power}{Input\ Power}$$

$$\eta = \frac{Output\ Power}{Output\ Power + Losses} \tag{5.26}$$

$$\eta = \frac{Output\ Power}{Output\ Power + Iron\ Loss\ + Cu\ loss} \times 100\%$$

5.16 Estimation of Leakage Reactance

Assumptions:

1. Consider permeability of iron as infinity that is MMF is needed only for leakage flux path in the window
2. The leakage flux lines are parallel to the axis of the core

Consider an elementary cylinder of leakage flux lines of thickness dx at a distance x as shown in Figure 5.10.

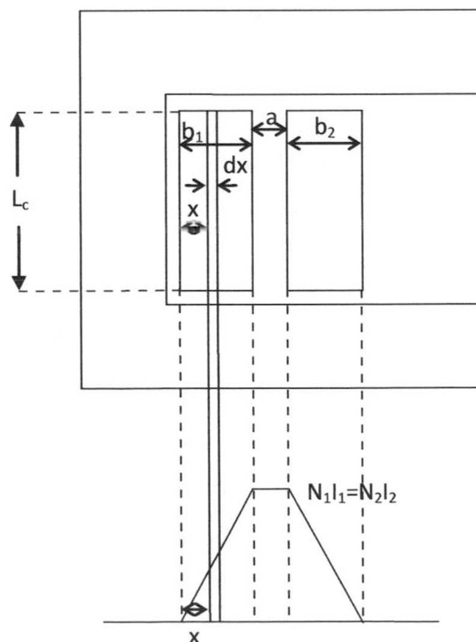

FIGURE 5.10
MMF distribution.

MMF at distance x:

$$M_x = \frac{N_1.I_1}{b_1}.x$$

The Permeance of this elementary cylinder:

$$= \mu_o \frac{A}{L}$$

$$= \mu_o \frac{L_{mt}.dx}{L_c} \text{ (L}_c = \text{length of winding)}$$

$$\left(\because S = \frac{1}{\mu_o}\frac{L}{A} \quad \& \quad Permeance = \frac{1}{S} \right).$$

The leakage flux lines associated with the elementary cylinder:

$$d\varphi_x = M_x \times Permeance$$

$$= \frac{N_1.I_1}{b_1} x \times \mu_o \frac{L_{mt}.dx}{L_c}.$$

The flux linkage due to this leakage flux:
$d\psi$ = number of turns which is associated. $d\varphi_x$

$$= \frac{N_1.I_1}{b_1} \times \frac{N_1.I_1}{b_1} x \times \mu_o \frac{L_{mt}.dx}{L_c}$$

$$= \mu_o.N_1^2.\frac{L_{mt}}{L_c}.I_1.\left(\frac{x}{b_1}\right)^2.dx. \tag{5.27}$$

The flux linkages (or associated) with the primary winding:

$$\psi_1' = \mu_o.N_1^2.\frac{L_{mt}}{L_c}.I_1.\int_0^{b_1}\left(\frac{x}{b_1}\right)^2 dx = \mu_o.N_1^2.\frac{L_{mt}}{L_c}.I_1.\left(\frac{b_1}{3}\right) \tag{5.28}$$

The flux linkages (or associated) with space "a" between primary and secondary windings:

$$\psi_o = \mu_o.N_1^2.\frac{L_{mt}}{L_c}.I_1.a \tag{5.29}$$

We consider half of this flux linkage with primary and rest half with the secondary winding. So total flux linkages with the primary winding:

$$\psi_1 = \psi_1' + \frac{\psi_o}{2}$$

$$\psi_1 = \mu_o.N_1^2.\frac{L_{mt}}{L_c}.I_1.\left(\frac{b_1}{3} + \frac{a}{2}\right) \tag{5.30}$$

Similarly, total flux linkages with the secondary winding:

$$\psi_2 = \psi_2' + \frac{\psi_o}{2} \tag{5.31}$$

$$\psi_2 = \mu_o . N_2^2 . \frac{L_{mt}}{L_c} . I_2 . \left(\frac{b_2}{3} + \frac{a}{2} \right) \tag{5.32}$$

The primary and secondary leakage inductance:

$$L_1 = \frac{\psi_1}{I_1} = \mu_o . N_1^2 . \frac{L_{mt}}{L_c} . \left(\frac{b_1}{3} + \frac{a}{2} \right) \tag{5.33}$$

$$L_2 = \frac{\psi_2}{I_2} = \mu_o . N_2^2 . \frac{L_{mt}}{L_c} . \left(\frac{b_2}{3} + \frac{a}{2} \right) \tag{5.34}$$

The primary and secondary leakage reactance:

$$X_1 = 2\pi . f . L_1 = 2\pi . f . \mu_o . N_1^2 . \frac{L_{mt}}{L_c} . \left(\frac{b_1}{3} + \frac{a}{2} \right) \tag{5.35}$$

$$X_2 = 2\pi . f . L_2 = 2\pi . f \, \mu_o . N_2^2 . \frac{L_{mt}}{L_c} . \left(\frac{b_2}{3} + \frac{a}{2} \right) \tag{5.36}$$

The total leakage reactance referred to the primary side:

$$X_{01} = X_P = X_1 + X_2' = 2\pi . f . \mu_o . N_1^2 . \frac{L_{mt}}{L_c} . \left(\frac{b_1 + b_2}{3} + a \right) \tag{5.37}$$

The total leakage reactance referred to the secondary side:

$$X_{02} = X_S = X_1' + X_2 = 2\pi . f . \mu_o . N_2^2 . \frac{L_{mt}}{L_c} . \left(\frac{b_1 + b_2}{3} + a \right) \tag{5.38}$$

It must be 5% to 8% or maximum 10%.

To control X_P: If increasing the window height (L), L_c will increase and the following will decrease b_1, b_2 & L_{mt} and so we can reduce the value of X_P.

5.17 Calculation of Voltage Regulation of Transformer

$$V.R. = \frac{I_2 . R_{o2} . \text{Cos}\varphi_2 \; \pm \; I_2 . X_{o2} . \text{Sin}\varphi_2}{E_2} \times 100 \tag{5.39}$$

$$= \frac{R_{o2} . \text{Cos}\varphi_2}{E_2 / I_2} \times 100 \pm \frac{X_{o2} . \text{Sin}\varphi_2}{E_2 / I_2} \times 100 \tag{5.40}$$

$$= \left(R_{o2} \cos\varphi_2 \pm X_{o2} \sin\varphi_2 \right)\%.$$

5.18 Transformer Tank Design

The width of the transformer (tank) in Figure 5.11:

$$W_t = 2D + D_e + 2b \tag{5.41}$$

where:

D_e = external diameter of HV winding
b = clearance widthwise between HV and tank.

The depth of the transformer (tank):

$$l_t = D_e + 2a \tag{5.42}$$

where a = clearance depth wise between HV and tank.

The height of the transformer (tank):

$$H_t = H + h \tag{5.43}$$

where $h = h_1 + h_2$ = clearance height wise of top and bottom.

Figure 5.12 shows the transformer tank design outside.

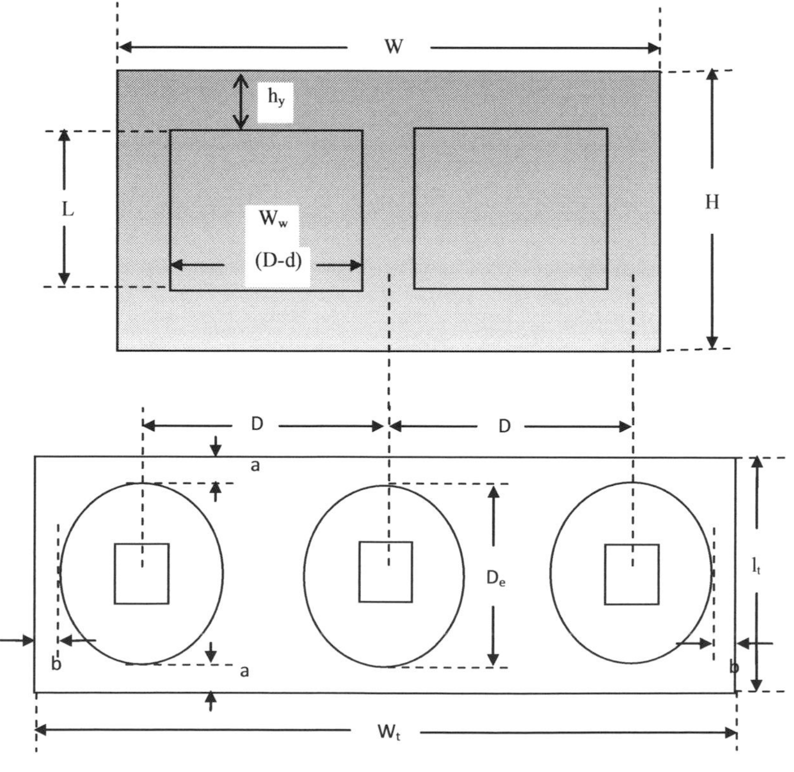

FIGURE 5.11
Transformer tank design.

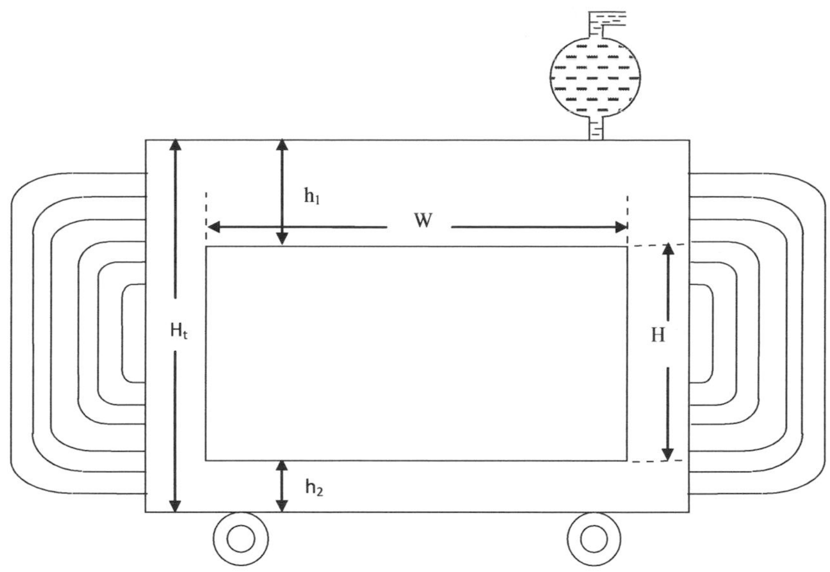

FIGURE 5.12
Transformer tank design outside (http://eed.dit.googlepages.com).

5.19 Calculation of Temperature Rise

The surface area of four vertical sides of the tank (Heat is considered to be dissipated from four vertical sides of the tank):

$$S_t = 2(W_t + l_t)\, H_t \text{ m}^2 \tag{5.44}$$

(Excluding area of top and bottom of the tank)

Let:

$$\theta = \text{temperature rise of oil } (35°\text{C}–50°\text{C})$$

$$12.5 S_t\theta = \text{total full load losses (iron loss + Cu loss)}$$

$$\text{So, the temperature rise in °C } \theta = \frac{\text{Total full load losses}}{12.5\, S_t}.$$

If the temperature rise so calculated exceeds the limiting value, the suitable no. of cooling tubes or radiators must be provided. Specific heat dissipation 6 Watt/m^2.°C by radiation.

5.20 Calculation Cooling Tubes Numbers

Let xS_t = surface area of all cooling tubes, then the losses to be dissipated by the transformer walls and cooling tube:

= total losses

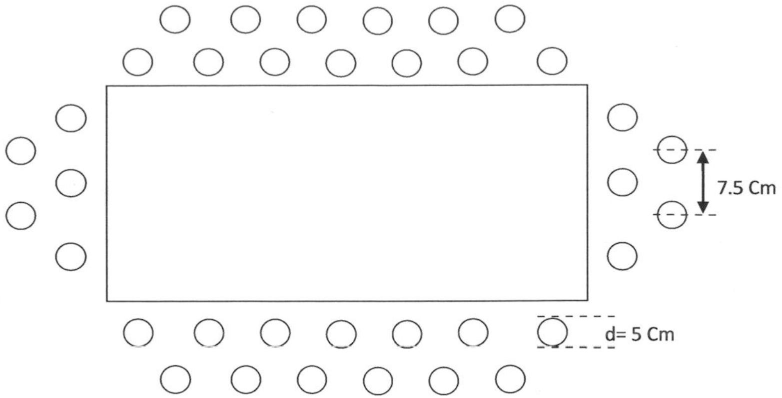

FIGURE 5.13
Tank and arrangement of cooling tubes (http://eed.dit.googlepages.com).

$$6\text{ W-radiation} + 6.5\text{ W} = 12.5\text{ convection}$$

$$6.5 \times 1.35\text{ W} \approx 8.5\ (\approx 35\%\text{ more})\text{ convection only.}$$

So from the above equation, we can find out the total surface of cooling tubes (xS_t) (Figure 5.13).

Normally we use 5 cm diameter tubes and keep them 7.5 cm apart.

$$A_t = \text{surface area of one cooling tube}$$

$$= \pi \times d_{tube} \times l_{tube,\ mean}.$$

Hence:

$$\text{No. of cooling tubes} = \frac{xS_t}{A_t}.$$

5.21 The Weight of Transformer

Let W_i = weight of iron in core and yoke (core volume × density + yoke volume × density).

And W_c = weight of copper in the winding (volume × density)

$$(\text{density of cu} = 8900\text{ Kg/m}^3)$$

weight of oil = volume of oil × 880.

Add 20% of ($W_i + W_c$) for fittings, tank etc.

Total weight is equal to the weight of above all parts.

5.22 MATLAB Programs

5.22.1 Single-Phase Transformer Design Using MATLAB Program

Following a MATLAB program to design of single-phase transformer 125 kVA,11000/120 V:
clc

k = 0.75, S = 125, Q = 2 × S, f = 60, Bm = 1.1, D = 0.34, V1 = 11000, V2 = 120, del = 2.2, Kw = 0.29,

c1 = 0.42, c2 = .1, d1 = 0.7, d2 = 0.14, e1 = 0.9, e2 = .42

%main dimension

Et = k × (S)^0.5 %emf per turn

Ai = Et/(4.44 × f × Bm) %core area

d = (Ai/k)^0.5

Aia = k × ((d)^2) %amended net core section

fa = Aia × Bm %flux

Et = 4.44 × f × fa %emf per turn

T2 = V2/Et %no. secondary turn per phase

T1 = T2 × (V1/V2) %no. primary turn per phase

Aw = S/(2.22 × f × Ai × Bm × del × Kw) %window area

L = Aw/(D–d) %window length

W = 2 × D + 0.9 × d %window width

%magnetic circuit

c11 = 2 × (c1 × c2) × d^2

d11 = 2 × (d1 × d2) × d^2

e11 = 1 × (e1 × e2) × d^2

GC = c11 + d11 + e11

ai = 0.9 × GC

netsec = 0.9 × 2 × c1 × d × e1 × d

%core loss

$$cv = 3 \times ai \times L \text{ \%core volume}$$

$$we = cv \times 7.55 \text{ \%core weight}$$

$$pi0 = 2.7 \text{ \%from curve 2.10}$$

$$loss = we \times pi0 \times 10\text{^}(-3) \text{ \%loss of core}$$

$$Y = 2 \times netsec \times W \times 1000000 \text{ \%yoke size}$$

$$weight = Y \times 7.55 \times 10\text{^}(-3)$$

$$den = 1.35 \times Aia/netsec$$

$$pi1 = 1.9$$

$$loss1 = weight \times pi1 \times 10\text{^}(-3)$$

$$losst = (loss + loss1 \times 1.075) \times 1000 \text{ \%total core loss}$$

%magnetizing current

$$ATC = 3 \times L$$

$$ATY = 2 \times W \times 4 \times 100$$

$$ATT = ATC + ATY$$

$$vac = 31; vay = 10;$$

$$VA = we \times vac + weight \times vay$$

$$Im = VA/(V1)$$

$$I2 = S \times 1000/V2$$

$$del2 = 2.8$$

$$a2 = I2/del2$$

$$l2 = pi \times d \times T2$$

$$ro = 0.021 \times 10\text{^}(-6)$$

$$R2 = ro \times l2/(a2 \times 10\text{^}(-6))$$

$$Ploss2 = (I2\text{^}2) \times R2$$

%primary wdg

$$I1 = S \times 1000/V1$$

$$a1 = I1/del2$$

$$d1 = ((4 \times a2/pi)^\wedge 0.5) \times 0.001$$

$$DIA = 2 \times d1 + d$$

$$l1 = DIA \times T1$$

$$R1 = ro \times l1/(a1 \times 10^\wedge{-6})$$

$$Ploss1 = (I1^\wedge 2) \times R1$$

$$Pc = (Ploss1 + Ploss2) \times 1.07 \text{ \%total copper loss}$$

$$Plosstot = Pc + losst$$

$$eff = 1-(Plosstot/(Plosstot + S \times 1000))$$

$$x = (losst/Pc)^\wedge 0.5$$

%Weight of transformer

$$weiron = we + weight$$

$$wecopper = (l1 \times a1 + l2 \times a2) \times 8900 \times (10^\wedge(-6))$$

$$wic = weiron + wecopper$$

$$copiro = weiron/wecopper$$

Results:

Design	d1 = 0.7000	c11 = 0.0032	losst = 1.1315e + 03	d1 = 0.0218
Values	d2 = 0.1400	d11 = 0.0075	ATC = 969.0345	DIA = 0.2389
	e1 = 0.9000	e11 = 0.0144	ATY = 684.6360	l1 = 313.3376
k = 0.7500	e2 = 0.4200	GC = 0.0251	ATT = 1.6537e + 03	R1 = 1.6213
S = 125	Et = 8.3853	ai = 0.0226	VA = 8.4790e + 03	Ploss1 = 209.3665
Q = 250	Ai = 0.0286	netsec = 0.0260	Im = 0.7708	Pc = 799.5532
f = 60	d = 0.1953	cv = 21.8945	I2 = 1.0417e + 03	Plosstot = 1.931e + 3
Bm = 1.1000	Aia = 0.0286	we = 165.3037	del2 = 2.8000	eff = 0.9848
D = 0.3400	fa = 0.0315	pi0 = 2.7000	a2 = 372.0238	x = 1.1896
V1 = 11000	Et = 8.3853	loss = 0.4463	l2 = 8.7817	weiron = 500.7625
V2 = 120	T2 = 14.3108	Y = 4.4432e + 4	ro = 2.1000e – 08	wecopper = 40.3941
del = 2.2000	T1 = 1.3118e + 3	weight = 335.4587	R2 = 4.9571e – 04	wic = 541.1566
Kw = 0.2900	Aw = 46.7308	den = 1.4881	Ploss2 = 537.8795	copiro = 12.3969
c1 = 0.4200	L = 323.0115	pi1 = 1.9000	I1 = 11.3636	
c2 = 0.1000	W = 0.8558	loss1 = 0.6374	a1 = 4.0584	

5.22.2 Three-Phase Transformer Design Using MATLAB Program

Following a MATLAB program to design three-phase transformer 300 kVA,6600/440 V:
clc

$$k = 0.6, S = 300, Q = 2 \times S/3, f = 60, Bm = 1.35, D = 0.34, V1 = 6600,$$

$$V2 = 440,$$

$$del = 2.5, Kw = 0.29,$$

$$c1 = 0.42, c2 = .1, d1 = 0.7, d2 = 0.14, e1 = 0.9, e2 = .42$$

%main dimension

$$Et = k \times (Q)^{\wedge}0.5 \text{ \%emf per turn}$$

$$Ai = Et/(4.44 \times f \times Bm) \text{ \%core area}$$

$$d = (Ai/k)^{\wedge}0.5$$

$$Aia = k \times ((d)^{\wedge}2) \text{ \%amended net core section}$$

$$fa = Aia \times Bm \text{ \%flux}$$

$$Et = 4.44 \times f \times fa \text{ \%emf per turn}$$

$$T2 = (V2/(3)^{\wedge}0.5)/Et \text{ \%no. secondary turn per phase}$$

$$T1 = T2 \times (V1/V2) \text{ \%no. primary turn per phase}$$

$$Aw = S/(3.33 \times f \times Ai \times Bm \times del \times Kw) \text{ \%window area}$$

$$L = Aw/(D-d) \text{ \%window length}$$

$$W = 2 \times D + 0.9 \times d$$

%magnetic circuit

$$c11 = 2 \times (c1 \times c2) \times d^{\wedge}2$$

$$d11 = 2 \times (d1 \times d2) \times d^{\wedge}2$$

$$e11 = 1 \times (e1 \times e2) \times d^{\wedge}2$$

$$GC = c11 + d11 + e11$$

$$ai = 0.9 \times GC$$

$$netsec = 0.9 \times 2 \times c1 \times d \times e1 \times d$$

%core loss

$$cv = 3 \times ai \times L \text{ \%core volume}$$

$$we = cv \times 7.55 \text{ \%core weight}$$

$$pi0 = 2.7 \text{ \%from curve 2.10}$$

$$loss = we \times pi0 \times 10^\wedge(-3) \text{ \%loss of core}$$

$$Y = 2 \times netsec \times W \times 1000000 \text{ \%yoke size}$$

$$weight = Y \times 7.55 \times 10^\wedge(-3)$$

$$den = 1.35 \times Aia/netsec$$

$$pi1 = 1.9$$

$$loss1 = weight \times pi1 \times 10^\wedge(-3)$$

$$losst = (loss + loss1 \times 1.075) \times 1000 \text{ \%total core loss}$$

%magnetizing current

$$ATC = 3 \times L$$

$$ATY = 2 \times W \times 4 \times 100$$

$$ATT = ATC + ATY$$

$$vac = 31; vay = 10;$$

$$VA = we \times vac + weight \times vay$$

$$Im = VA/(3 \times V1)$$

$$s2 = S/3$$

$$I2 = s2 \times 1000/((V2/(3)^\wedge(.5)))$$

$$del2 = 2.8$$

$$a2 = I2/del2$$

$$l2 = pi \times d \times T2$$

$$ro = 0.021 \times 10^\wedge(-6)$$

$$R2 = ro \times l2/(a2 \times 10\text{^}(-6))$$

$$Ploss2 = 3 \times (I2\text{^}2) \times R2$$

%primary wdg

$$I1 = S \times 1000/(3 \times V1)$$

$$a1 = I1/del2$$

$$d1 = ((4 \times a2/pi)\text{^}0.5) \times 0.001$$

$$DIA = 2 \times d1 + d$$

$$l1 = DIA \times T1$$

$$R1 = ro \times l1/(a1 \times 10\text{^}-6)$$

$$Ploss1 = 3 \times (I1\text{^}2) \times R1$$

$$Pc = (Ploss1 + Ploss2) \times 1.07 \text{ \%total copper loss}$$

$$Plosstot = Pc + losst$$

$$eff = 1-(Plosstot/(Plosstot + S \times 1000))$$

$$x = (losst/Pc)\text{^}0.5$$

%Weight of transformer

$$weiron = we + weight$$

$$wecopper = 3 \times (l1 \times a1 + l2 \times a2) \times 8900 \times (10\text{^}(-6))$$

$$wic = weiron + wecopper$$

$$copiro = weiron/wecopper$$

$$besr = wecopper/S$$

$$\%Lmt = 0.5(Lmt1 + Lmt2)$$

$$AT = I2 \times T2$$

Results:

Design values	d1 = 0.7000	c11 = 0.0033	Lost = 1.362e + 03
	d2 = 0.1400	d11 = 0.0077	DIA = 0.2251
	e1 = 0.9000	e11 = 0.0149	l1 = 101.0678
k = 0.6000	e2 = 0.4200	GC = 0.0259	R1 = 0.3922
S = 300	Et = 8.4853	ai = 0.0233	Ploss1 = 270.1268
Q = 200	Ai = 0.0236	netsec = 0.0268	Pc = 1.6748e + 03
f = 60	d = 0.1983	cv = 32.0571	Plosstot = 3.0367e + 03
Bm = 1.3500	Aia = 0.0236	we = 242.0308	eff = 0.9900
D = 0.3400	fa = 0.0319	pi0 = 2.7000	x = 0.9018
V1 = 6600	Et = 8.4853	loss = 0.6535	weiron = 588.8583
V2 = 440	T2 = 29.9382	Y = 4.5937e + 04	wecopper = 84.6124
del = 2.5000	T1 = 449.0731	Weight = 46.8275	wic = 673.4707
Kw = 0.2900	Aw = 65.0213	den = 1.1905	copiro = 6.9595
c1 = 0.4200	L = 458.8672	pi1 = 1.9000	besr = 0.2820
c2 = 0.1000	W = 0.8585	loss1 = 0.6590	AT = 1.1785e + 04

5.22.3 Three-Phase Transformer Design Using MATLAB Program

Following a MATLAB program to design three-phase transformer 125 kVA,2000/440 V:
clc

k = 1, S = 125, Q = 2 × S/3, f = 50, Bm = 1.1, D = 0.34, V1 = 2000, V2 = 440, del = 2.2, Kw = 0.29,

c1 = 0.42, c2 = .1, d1 = 0.7, d2 = 0.14, e1 = 0.9, e2 = .42

%main dimension

Et = k × (S)^0.5 %emf per turn

Ai = Et/(4.44 × f × Bm) %core area

d = (Ai/k)^0.5

Aia = k × ((d)^2) %amended net core section

fa = Aia × Bm %flux

Et = 4.44 × f × fa %emf per turn

T2 = V2/Et %no. secondary turn per phase

T1 = T2 × (V1/V2) %no. primary turn per phase

Aw = S/(2.22 × f × Ai × Bm × del × Kw) %window area

```
L = Aw/(D–d) %window length
W = 2 × D + 0.9 × d %window width
%magnetic circuit
c11 = 2 × (c1 × c2) × d^2
d11 = 2 × (d1 × d2) × d^2
e11 = 1 × (e1 × e2) × d^2
GC = c11 + d11 + e11
ai = 0.9 × GC
netsec = 0.9 × 2 × c1 × d × e1 × d
%core loss
cv = 3 × ai × L %core volume
we = cv × 7.55 %core weight
pi0 = 2.7 %from curve 2.10
loss = we × pi0 × 10^(–3) %loss of core
Y = 2 × netsec × W × 1000000 %yoke size
weight = Y × 7.55 × 10^(–3)
den = 1.35 × Aia/netsec
pi1 = 1.9
loss1 = weight × pi1 × 10^(–3)
losst = (loss + loss1 × 1.075) × 1000 %total core loss
%magnetizing current
ATC = 3 × L
ATY = 2 × W × 4 × 100
ATT = ATC + ATY
vac = 31;vay = 10;
```

$$VA = we \times vac + weight \times vay$$

$$Im = VA/(V1)$$

$$I2 = S \times 1000/V2$$

$$del2 = 2.8$$

$$a2 = I2/del2$$

$$l2 = pi \times d \times T2$$

$$ro = 0.021 \times 10^\wedge(-6)$$

$$R2 = ro \times l2/(a2 \times 10^\wedge(-6))$$

$$Ploss2 = (I2^\wedge 2) \times R2$$

%primary wdg

$$I1 = S \times 1000/V1$$

$$a1 = I1/del2$$

$$d1 = ((4 \times a2/pi)^\wedge 0.5) \times 0.001$$

$$DIA = 2 \times d1 + d$$

$$l1 = DIA \times T1$$

$$R1 = ro \times l1/(a1 \times 10^\wedge{-6})$$

$$Ploss1 = (I1^\wedge 2) \times R1$$

$$Pc = (Ploss1 + Ploss2) \times 1.07 \text{ \%total copper loss}$$

$$Plosstot = Pc + losst$$

$$eff = 1-(Plosstot/(Plosstot + S \times 1000))$$

$$x = (losst/Pc)^\wedge 0.5$$

%Weight of transformer

$$weiron = we + weight$$

$$wecopper = (l1 \times a1 + l2 \times a2) \times 8900 \times (10^\wedge(-6))$$

$$wic = weiron + wecopper$$

$$copiro = weiron/wecopper$$

Results:

Design values	d1 = 0.7000	c11 = 0.0038	losst = 1.2994e+03	d1 = 0.0114
	d2 = 0.1400	d11 = 0.0090	ATC = 834.2855	DIA = 0.2367
	e1 = 0.9000	e11 = 0.0173	ATY = 698.0590	l1 = 42.3427
k = 1	e2 = 0.4200	GC = 0.0301	ATT = 1.5323e+03	R1 = 0.0398
S =125	Et = 11.1803	ai = 0.0271	VA = 9.3986e+03	Ploss1 = 155.6093
Q = 83.3333	Ai = 0.0458	netsec = 0.0312	Im = 4.6993	Pc = 639.3490
f = 50	d = 0.2140	cv = 22.6200	I2 = 284.0909	Plosstot = 1.939e+03
Bm = 1.1000	Aia = 0.0458	we = 170.7809	del2 = 2.8000	eff = 0.9847
D = 0.3400	fa = 0.0504	pi0 = 2.7000	a2 = 101.4610	x = 1.4256
V1 = 2000	Et = 11.1803	loss = 0.4611	l2 = 26.4547	weiron = 581.2238
V2 = 440	T2 = 39.3548	Y = 5.4363e+04	ro = 2.1000e–08	wecopper = 32.3005
del = 2.2000	T1 = 178.8854	weight = 410.4429	R2 = 0.0055	wic = 613.5243
Kw = 0.2900	Aw = 35.0481	den = 1.9841	Ploss2 = 441.9131	copiro = 17.9943
c1 = 0.4200	L = 278.0952	pi1 = 1.9000	I1 = 62.5000	
c2 = 0.1000	W = 0.8726	loss1 = 0.7798	a1 = 22.3214	

Problems

5.1 A 2000 kVA, 3300/208 V, 60 Hz, single-phase core type transformer, maximum flux density of 1.75 Wb/m^2, current density of 2.5 A/mm^2, EMF voltage per turn = 25 V, and window factor = 0.3, calculate:

1. Core area A_i
2. Window area Aw.

5.2 A three-phase, Δ/Y transformer, core type, rated at 400 kVA, 3300/208 V, a suitable core with 3-step having a circumscribing circle of 0.25 m diameter, and a leg spacing of 0.45 m is available. EMF = 8.6 V/turn, $\delta = 2.55$ A/mm^2, Kw = 0.3, and stacking factor $S_f = 0.8$, determine the main dimensions.

5.3 A 250 kVA, 1100/120 V, 60 Hz, single-phase, core type transformer, maximum flux density of 1.25 Wb/m^2, current density of 2.5 A/mm^2, EMF voltage per turn = 8.2 V, window factor = 0.28, ratio of effective cross-section area of core to square of diameter of circumscribing circle is 0.75, and ratio of height to width of window is 2.5, calculate:

1. The dimensions of the core
2. The number of turns
3. The cross-section area of the conductor.

5.4 A 150 kVA, 2200/120 V, 60 Hz, single-phase, shell type transformer, the ratio of magnetic and electric loadings equal to 480×10^{-8}, Bm = 1.25 Wb/m^2, $\delta = 2.5$ A/mm^2, Kw = 0.3, and stacking factor = 0.9, determine:

1. The dimensions of the core
2. The number of primary and secondary windings
3. The cross section area of conductor in the primary and secondary windings.

5.5 Design a single-phase transformer using MATLAB program with the following data:

123 kVA, V1 = 11000 V, V2 = 110 V, 60 Hz, k = 0.75, Kw = 0.29, D = 0.34, $\delta = 2.2$, $c_1 = 0.42$, $c_2 = .1$, $d_1 = 0.7$, $d_2 = 0.14$, $e_1 = 0.9$, and $e_2 = .42$.

5.6 A 150 kVA, 2200/208 V, Y/Δ, 60 Hz, three-phase, shell type transformer, the ratio of magnetic and electric loadings equal to 480×10^{-8}, $B_m = 1.1$ Wb/m², $\delta = 2.5$ A/mm², Kw = 0.3, and stacking factor = 0.9, determine:

1. The dimensions of the core
2. The number of primary and secondary windings
3. The cross-section area of the conductor in the primary and secondary windings.

5.7 Design a three-phase transformer using a MATLAB program with the following data:

600 kVA, 1100/208 V, 60 Hz, three-phase, core type transformer, maximum flux density of 1.25 Wb/m², current density of 2.5 A/mm², EMF voltage per turn = 8.2 V, window factor = 0.28, ratio of effective cross-section area of core to square of diameter of circumscribing circle is 0.75, and ratio of height to width of window is 2.5.

5.8 Design a three-phase transformer using a MATLAB program with the following data:

1200 kVA, 1000/200 V, 60 Hz, three-phase, core type transformer, maximum flux density of 1.2 Wb/m², current density of 2.5 A/mm², EMF voltage per turn = 5.2 V, window factor = 0.28, ratio of effective cross-section area of core to square of diameter of circumscribing circle is 0.75, and ratio of height to width of window is 2.5.

6

Direct Current Machines

You have to think anyway, so why not think big?

Donald Trump

In this chapter, the well-established theory of DC machines is set forth, and the dynamic characteristics of the separate excited and self-excited machines are illustrated. DC machines are generators that convert mechanical energy to DC electric energy and motors that convert DC electric energy into mechanical energy. This chapter will first explain the principles of DC machine operation by using simple examples, and then consider some of the complications that occur in real DC machines.

6.1 DC Machines

Despite the widespread use of AC machines, DC motors are still widespread in the industry for their simplicity of operation and ease of speed regulation. A DC machine, which converts mechanical energy into electricity, is called *the DC generator,* and a DC machine that converts electrical energy into mechanical is called *the DC motor,* and these machines are typically designed according to the nature of its operation.

Although DC generators can be operated as motors and vice versa, the design specification determines whether the machine is a generator or a motor. Figure 6.1 represents a DC machine.

6.2 DC Machine Parts

DC machine consists of the following basic parts (Figure 6.2).

6.2.1 Stator

The stator consists of the outer frame, which is made up of wrought iron or cast iron to generate the magnetic field. The stator is made from the steel plate set and acts as a path to complete the magnetic circuit, and the poles are installed on it as shown in Figure 6.3.

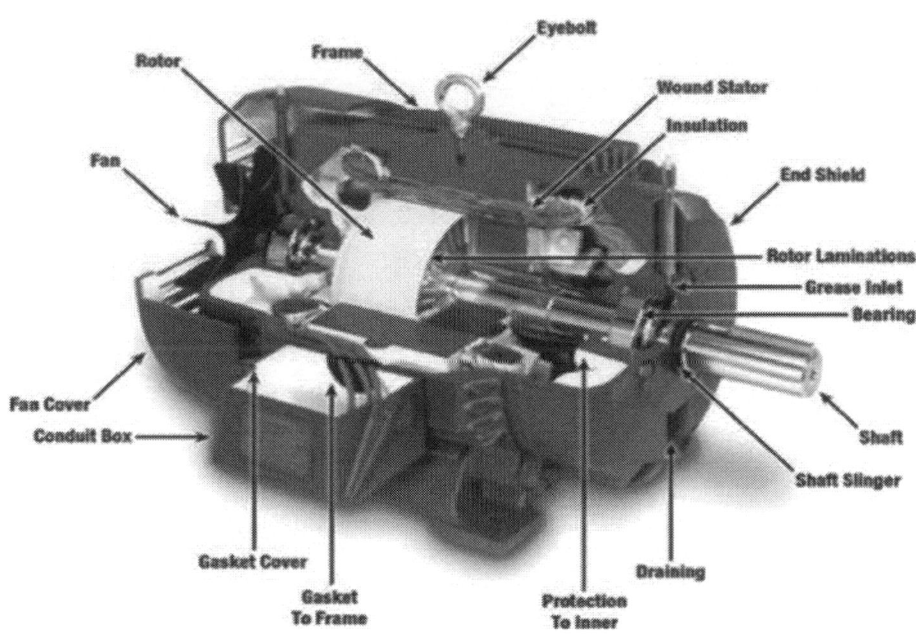

FIGURE 6.1
Direct current machine. (https://newmachineparts.blogspot.com).

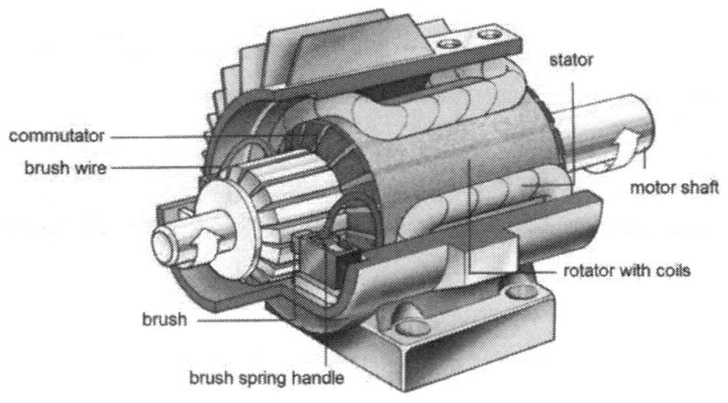

FIGURE 6.2
Parts of DC machine. (https://dumielauxepices.net/wallpaper-1106121).

The main poles are installed in the outer frame and made of steel sheets. The pole interface is called the pole chose. It is useful for attaching the magnetic pole coils and reducing the air gap between the rotor and the iron core of the pole magnetic.

Auxiliary poles are placed between the main poles as shown in Figure 6.4. It is useful to reduce the sparks between the carbon and commutator, and reduce the reaction effect in the armature and directly connect with the armature, and in the outer part of the stator part fix the carbon brushes holder, and the advantage of carbon brushes to transfer the current to and from the armature.

FIGURE 6.3
The stator of the DC machine.

FIGURE 6.4
Magnetic pole of the stator DC machine. (penang-electronic.blogspot.com/2008/06/motors-and-machines-animations-and-java.html).

6.2.2 Rotor

The rotor is called *armature*, it consists of the core which contains the slots where the armature wires are placed, and also has a commutator that alternates AC current into DC current. All the ends of the coils reach the commutator—a set of isolated copper pieces isolated between them and between the axis of rotation and the armature body completely isolated, separating the stator and the rotor the air gap. Figure 6.5 shows parts for DC machine.

6.2.3 Commutator

Commutator shall be cylindrical on the axis of armature rotation by a suitable holder, consisting of a set of copper pieces insulated from each other by mica or steel fiber, and isolated "well" from the axis of rotation as in Figure 6.6.

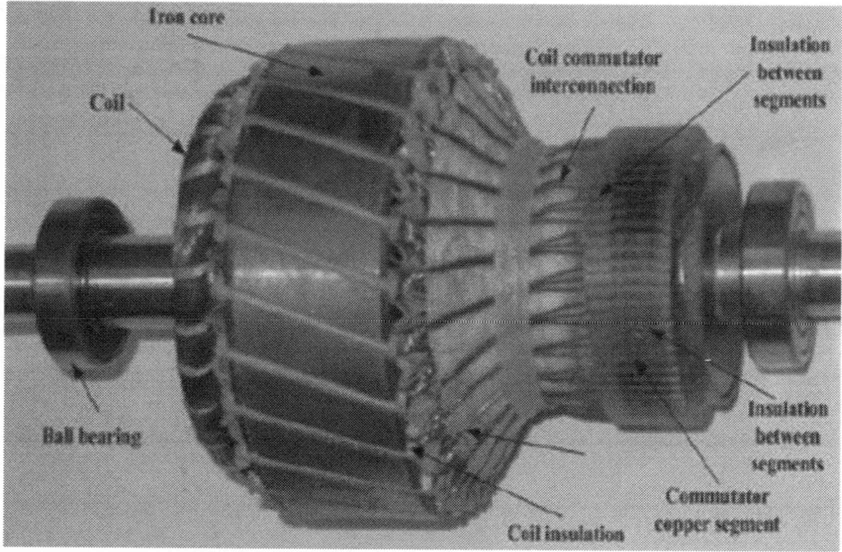

FIGURE 6.5
Rotor for DC machine. (https://slideplayer.com/slide/8758479/).

FIGURE 6.6
The commutator with the carbon brushes installed on it. (https://www.youtube.com/watch?v=cCB0WgpV8Iw).

6.2.4 Armature Coils

The armature is made up of copper or sometimes aluminum wire with different diameters according to the type of application. The wiring is placed in the slots of the armature, and each turn has the beginning and end, reaching the commutator terminals with the pieces of copper, and can be connected in two ways: lap winding and wave winding.

6.2.4.1 Lap Winding

Lap winding coil connects two adjacent pieces of the commutators. The number of parallel circuits will be equal to the number of poles. This type of winding is typically used in low voltage and high current machines. Refer to Figure 6.7 for a single turn lap winding.

2a—represents the number of parallel circuits

2p—represents the number of poles

2a = 2p.

6.2.4.2 Wave Winding

The beginning and end of the coil are connected to two pieces separated by commutator segment. Figure 6.8 shows a wave winding specified for a single turn. The number of parallel circuits is equal to two irrespective of the number of poles. This type of winding is typically used in high voltage and low current machines.

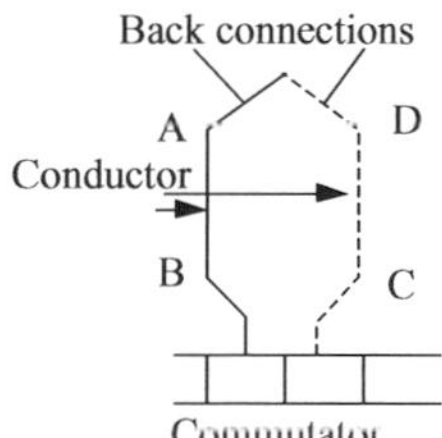

FIGURE 6.7
Lap winding.

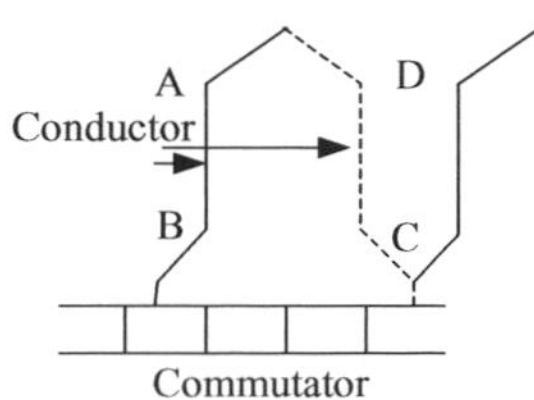

FIGURE 6.8
Wave winding.

6.3 DC Generator

DC generator converts mechanical energy into electrical energy. The operation principle of generators works on Faraday's theory that the Electric Motive Force (EMF) generated in the coil is of the sine waveform. According to Faraday's theory, the rotation of a coil within a magnetic field (magnetic poles) creates an electric motive force in the coil as a result of cutting the magnetic field lines.

6.3.1 Calculate the Motive Force Generated by the Generator (E.M.F)

The electric motive force generated is calculated as follows:

$$E = \frac{N}{60} \times \frac{z}{2a} \times \Phi \times 2p \quad (6.1)$$

where:

E = represents the electric motive force voltage (volt)
N = the speed of the generator (rpm)
Z = the number of conductors in the armature ducts
Φ = the magnetic flux in Weber
2p = the number of poles
2a = number of parallel circuits.

In the case of a lap winding, the formula shall be as follows:

$$2a = 2p$$

$$E = \frac{NZ\Phi}{60}. \quad (6.2)$$

And in the case of a wave winding, the formula shall be as follows:

$$2a = 2$$

$$E = \frac{NZ\Phi \times 2P}{60 \times 2} \quad (6.3)$$

Example 6.1

A DC generator has eight poles, wound as lap then rewound as a wave, calculate the electrical motive force generated in each case if you know that the number of conductors in the armature slots is 240. The magnetic flux of each pole 0.04 Wb. It runs at 1200 rpm.

Solution

In the case of lap winding:

$$2P = 8, Z = 240, \phi = 0.04 \text{ Wb}, N = 1200 \text{ rpm}$$

$$E = \frac{NZ\Phi}{60} = \frac{1200 \times 240 \times 0.04}{60}$$

$$E = 192 \text{ V.}$$

In the case of wave winding:

$$2a = 2$$

$$E = \frac{NZ\Phi 2p}{60 \times 2} = \frac{1200 \times 240 \times 0.04 \times 8}{120}$$

$$E = 768 \text{ V.}$$

6.3.2 Method of Excitation of DC Machines

An external simulated field coil is required to generate magnetic motive force to obtain electric motive force when the field coils continue to generate flux either through an external source or internal source from the machine. The DC generators are divided in terms of methods excitation into two types: separately and self-excited.

6.3.2.1 *Separately Excited Generator*

The magnetic coils are fed by an external DC source (battery or any other source) as in Figure 6.9. This type of characteristic machine generator and the electric motive force generated depends on the amount of current feed, which needs to organize voltage such as (Leonard machines).

6.3.2.2 *Self-Excited Generator*

A self-excited generator requires the armature current feeds of the magnetic pole coils to generate a magnetic motive force. The coils are connected via carbon brushes to the armature circuit. There are three ways to connect these coils with the armature: series, shunt, and compound.

6.3.2.2.1 Series Generator

The field windings resistance R_f is connected in series with the armature windings, as shown in Figure 6.10, and have a large diameter and a few turns to carry the current and the current of the load is same armature current. Under no-load condition, the voltage at the terminals of the generator is zero due to an open circuit on the magnetic field, the voltage

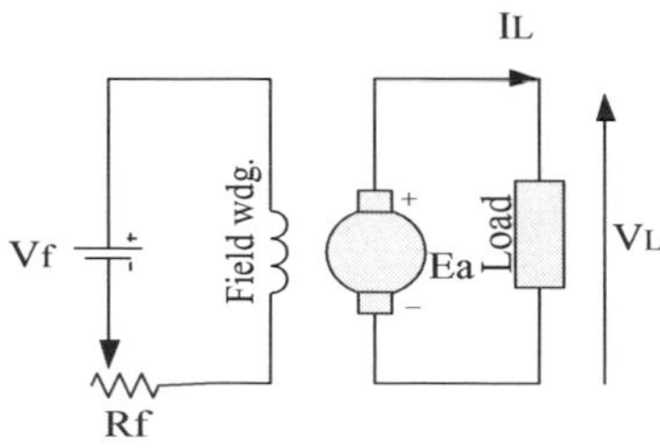

FIGURE 6.9
Electrical circuit of the separately excited generator.

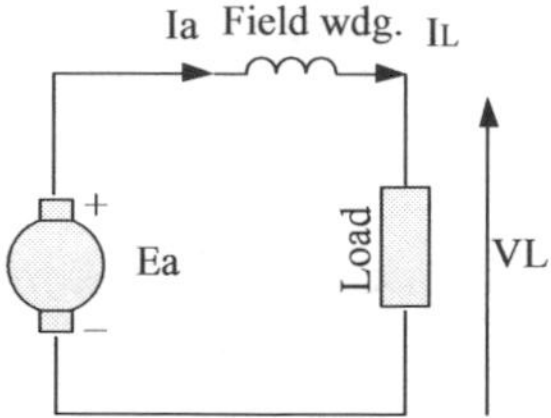

FIGURE 6.10
Series DC generator.

increases depending on the load increase and reaches the maximum value at full load. This type of generator is used as a voltage compensator for the DC power transmission networks (Figures 6.11 through 6.13). The induced electric motive force is calculated as follows:

$$E_a = V_L + I_a (R_a + R_f) \tag{6.4}$$

$$I_a = I_L = I_f \tag{6.5}$$

where:

- I_a = armature current (current produced by the alternator) (A)
- I_L = current load (A)
- I_f = current field (A)
- V_L = load voltage (V)
- E_a = the electric motive force (V)
- R_a = armature windings resistance (Ω)
- R_f = field windings resistance (Ω)
- R_L = load resistance (Ω).

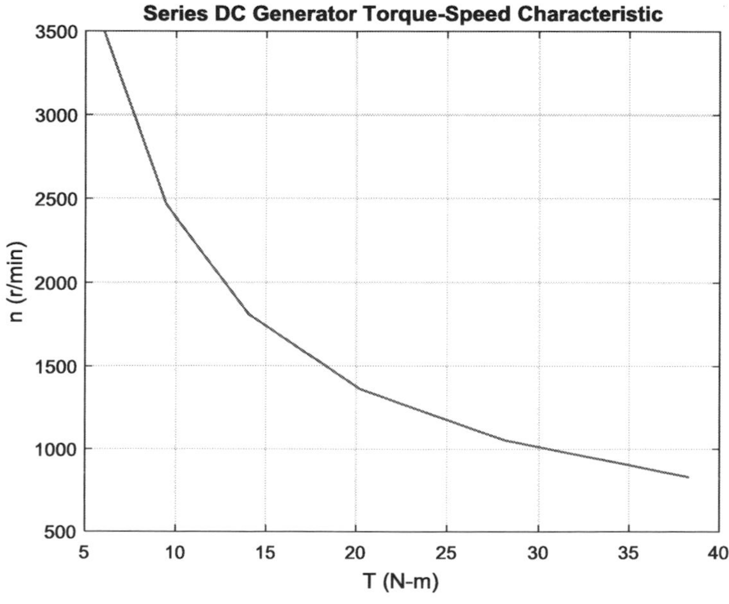

FIGURE 6.11
Input torque-speed characteristics of series DC generator.

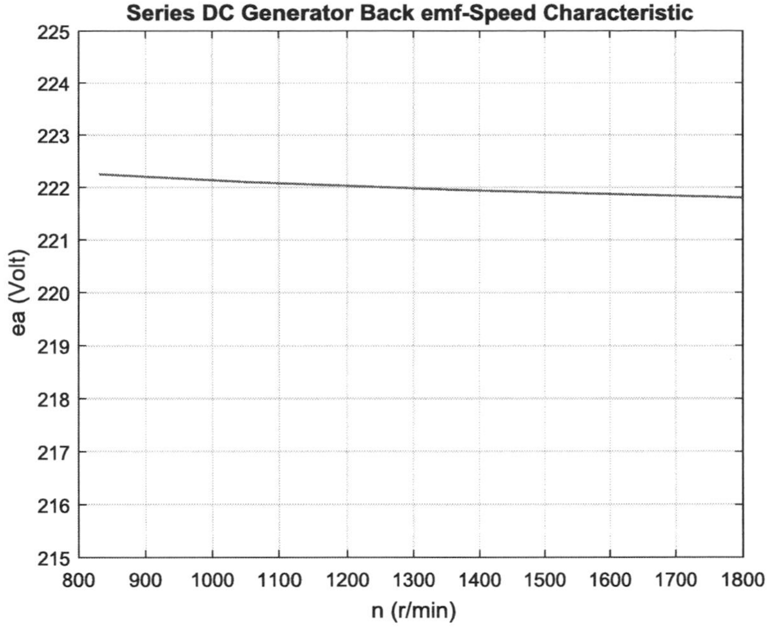

FIGURE 6.12
EMF-speed characteristics of series DC generator.

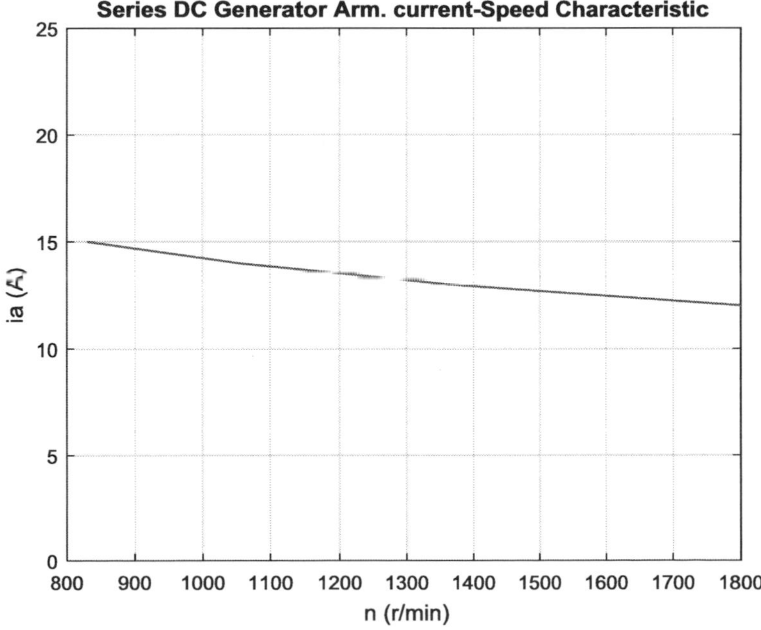

FIGURE 6.13
Armature current-speed characteristics of series DC generator.

Example 6.2

A DC generator feeding a load of current 20 A and voltage 220 V. Calculate the electrical motive force generated in the armature, if it is known that the resistance of the armature 0.02 Ω and the resistance of the field 0.01 Ω.

Solution

$$I_L = I_a = I_f = 20 \text{ A}$$

$$E_a = V_L + I_a (R_a + R_f)$$

$$E_a = 220 + 20 (0.02 + 0.01)$$

$$= 220.6 \text{ V.}$$

Example 6.3

A DC generator feeding a purely resistive load, if it is known that the current produced by the generator is 20 A, the resistance of the field 0.2 Ω, the resistance of the armature 0.1 Ω, and the electric motive force generated 230 V. Calculate load resistance.

Solution

$$I_a = I_L = 20 \text{ A}$$

$$E_a = V_L + I_a . (R_a + R_f)$$

$$230 = V_L + 20 (0.2 + 0.1)$$

$$230 = V_L + 20 \times 0.3$$

$$230 = V_L + 6$$

$$V_L = 230 - 6 = 224 \text{ V}$$

$$V_L = I_L R_L$$

$$224 = 20 \times R_L$$

$$R_L = \frac{224}{20} = 11.2\,\Omega.$$

Example 6.4

A DC generator feeds a load consisting of 22 lamps, each lamp capacity 100 W, 220 V, calculate the amount of electrical motive force generated if the resistance of the armature is known as 0.2 Ω while ignoring the resistance of the field winding.

Solution

$$\text{Total load power} = 22 \times 100 = 2200 \text{ W}$$

$$I_L = P_L/V_L = 2200/220 = 10 \text{ A}$$

$$I_a = I_L = I_f = 10 \text{ A}$$

$$E_a = V_L + I_a (R_a + R_f)$$

$$E_a = 220 + 10 \times 0.2$$

$$E_a = 220 + 2 = 222 \text{ V.}$$

Series DC Generator MATLAB Program

```
ea=[0 10 15 18 31 45 60 75 90 104 120 132 145 160 173 184 200 205 215 223
231];
n0 = 1000;
vt = 220;
ra = 0.15;
ia = 10:1:15;
ea = vt + ia * ra;
ea0 = interp1(if_values, ea_values, ia,'spline');
n1 = 1050;
Eao1 = interp1(if_values, ea_values,58,'spline');
Ea1 = vt + 15 * ra;
Eao2 = interp1(if_values, ea_values, ia,'spline');
n = ((ea./Ea1).* (Eao1./ Eao2)) * n1/100;
T = ea.* ia./ (n * pi / 30);
figure(1);
plot(T, n,'b-','LineWidth',2.0);
hold on;
xlabel('T (N-m)');
ylabel('n (r/min)');
title ('Series DC Generator Torque-Speed Characteristic');
grid on;
hold off;
figure(2);
plot(n, ea,'b-','LineWidth',2.0);
xlabel('n (r/min)');
ylabel('ea (Volt)');
title ('Series DC Generator Back emf-Speed Characteristic');
axis([800 1800 215 225]);
grid on;
%hold off;
figure(3);
plot(n, ia,'b-','LineWidth',2.0);
xlabel('n (r/min)');
ylabel('ia (A)');
title ('Series DC Generator Arm. current-Speed Characteristic');
axis([800 1800 0 25]);
grid on;
%hold off;
```

6.3.2.2.2 The Shunt Generator

The field windings resistance (R_f) is connected in parallel with the armature windings, as in Figure 6.14, and have a large number of turns and a small section area because of the current passing through it is relatively small, and the voltage on both ends of the generator in the case of no-load maximum. Because the circuit of the magnetic field is

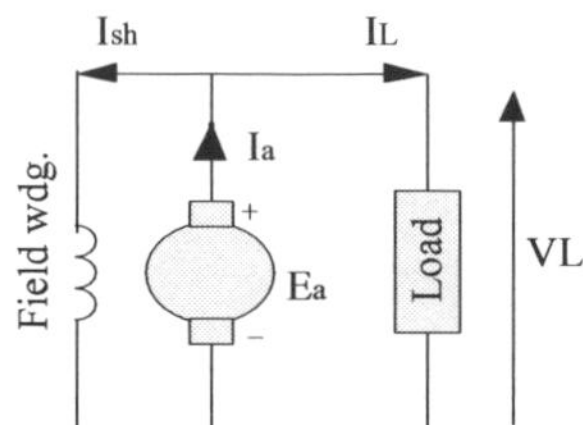

FIGURE 6.14
Shunt generator.

closed, so the voltage does not change on both ends of the generator in the case of a load or no-load. This type of generator is used in cases where the voltage is required for the vehicle and feeding AC generators (Figures 6.15 through 6.17). The electric motive force generated:

$$E_a = V_L + I_a R_a \tag{6.6}$$

$$V_f = V_L = I_f R_f \tag{6.7}$$

$$I_a = I_L + I_f. \tag{6.8}$$

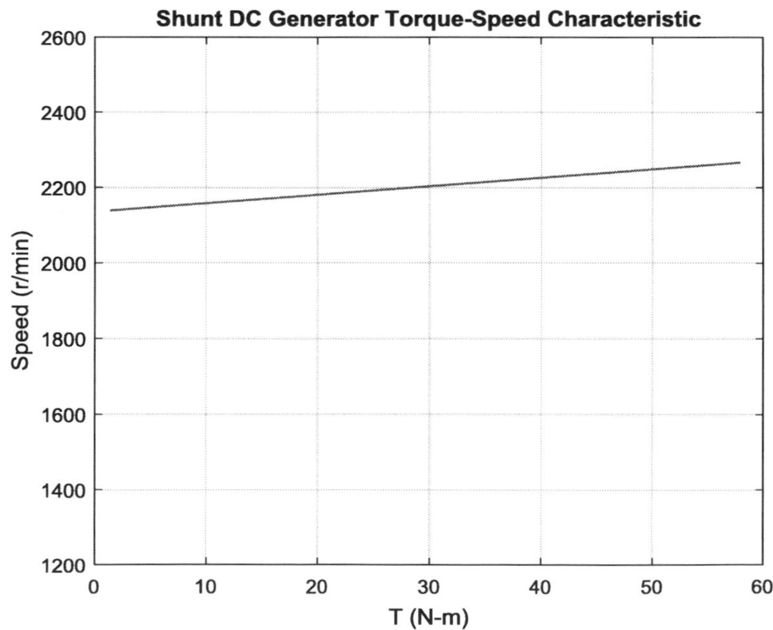

FIGURE 6.15
Input torque-speed characteristics of shunt DC generator.

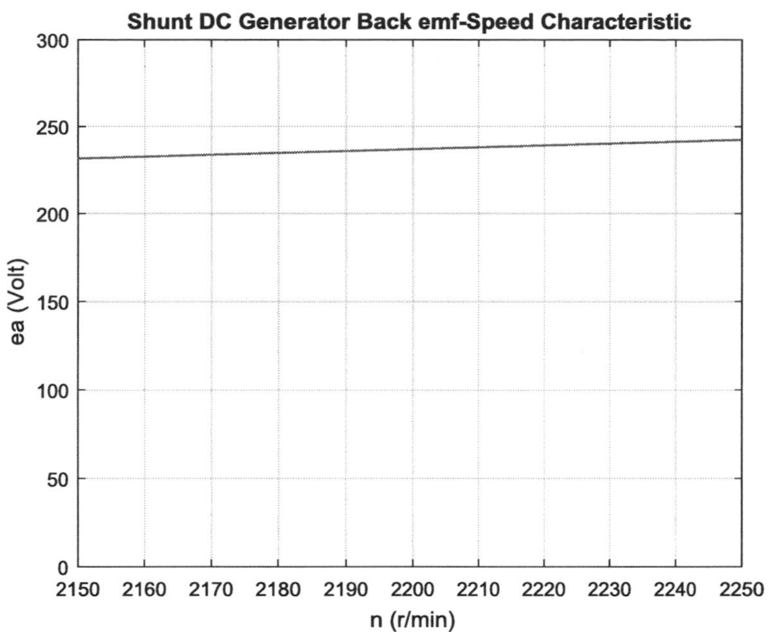

FIGURE 6.16
EMF-speed characteristics of shunt DC generator.

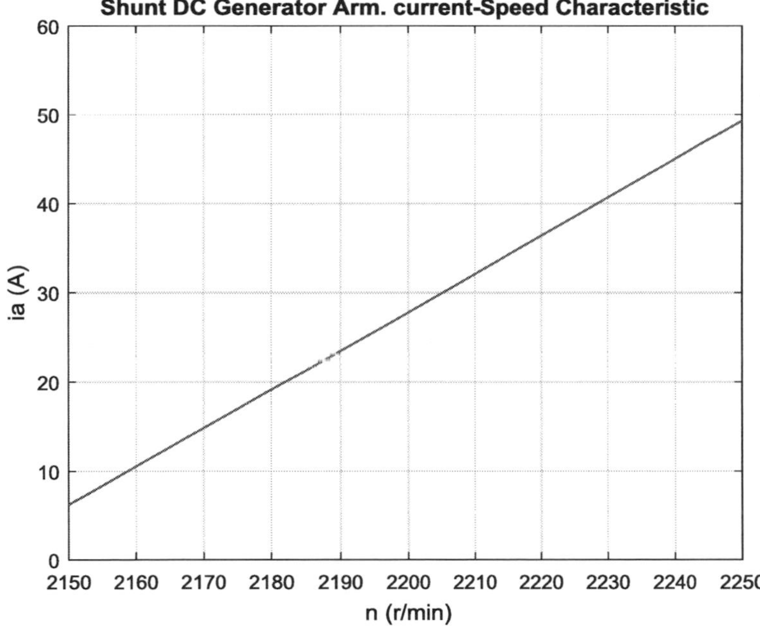

FIGURE 6.17
Armature current-speed characteristics of shunt DC generator.

Example 6.5

A shunt generator feeds a load with a current of 300 A at a voltage of 240 V. If the resistance of the armature windings is 0.02 Ω and the resistance of the field windings is 60 Ω, calculate the generated electrical motive force.

Solution

$$I_f = V/R_f = 300/60 = 4\ A$$

$$I_a = i_l + I_f = 300 + 4 = 304\ A$$

$$E_a = V_L + I_a R_a$$

$$E_a = 240 + 304 \times 0.02$$

$$E_a = 240 + 6.08$$

$$E_a = 246.08\ V.$$

Example 6.6

A shunt DC generator feeds a load of 4068 W, voltage 226 V, armature resistance 0.2 Ω, and the amount of electric motive force generated 230 V, calculate the resistance of field windings.

Solution

$$I_L = P_l/V_L = 4068/226 = 18\ A$$

$$E_a = V_L + I_a R_a$$

$$230 = 226 + I_a \times 0.2$$

$$4 = I_a \times 0.2$$

$$I_a = I_L + I_f$$

$$20 = 18 + I_f$$

$$I_f = 20 - 18 = 2\ A$$

$$V_L = V_f = I_f R_f$$

$$R_f = V/I_f = 226/2 = 113\ \Omega.$$

DC Shunt Generator MATLAB Program

```
if_values=[0 0.1 0.2 0.3 0.4 0.5 0.6 0.7 0.8 0.9 1.0 1.1 1.2 1.3 1.4 1.5
1.6 1.7 1.8 1.9 2.0];
ea_values=[0 10 15 18 31.30 45.46 60.26 75.06 89.74 104.4 118.86 132.86
146.46 159.78 172.18 183.98 195.04 205.18...
```

```
214.52 223.06 231.2];
n0 = 1500;
vt = 230;
rf = 100;
radd = 75;
ra = 0.250;
il = 0:1:55;
nf = 2700;
far0 = 1500;
ia = il + vt / (rf + radd);
ea = vt + ia * ra;
ish= vt / (rf + radd);
ea0 = interp1(if_values, ea_values, ish);
n = (ea./ ea0) * n0;
T = ea.* ia./ (n * 2 * pi / 60);
figure(1);
plot(T, n,'b-','LineWidth',2.0);
xlabel('T (N-m)');
ylabel('Speed (r/min)');
title ('Shunt DC Generator Torque-Speed Characteristic');
axis([0 60 1200 2600]);
grid on;
%hold off;
figure(2);
plot(n, ea,'b-','LineWidth',2.0);
xlabel('n (r/min)');
ylabel('ea (Volt)');
title ('Shunt DC Generator Back emf-Speed Characteristic');
axis([2150 2250 0 300]);
grid on;
%hold off;
figure(3);
plot(n, ia,'b-','LineWidth',2.0);
xlabel('n (r/min)');
ylabel('ia (A)');
title ('Shunt DC Generator Arm. current-Speed Characteristic');
axis([2150 2250 0 60]);
grid on;
%hold off;
```

6.3.2.2.3 Compound Generator

The compound generator contains both series and shunt coils connected to armature windings. If the shunt coils connected directly with the ends of the armature, this method of connection is called *the short compound* as shown in Figure 6.18, or shunt coils are connected across the terminals of the outer circuit (the armature with the series coils), it is called *the long compound* as shown in Figure 6.19.

The self-excited generators polarity depends on the field coils connection with the armature, where the field current helps the remaining magnetism in the machine.

Any inverse connection to the field coils will cause the remaining magnetism to be eliminated, and the construction of potential is not generated on the terminals of the machine. The compound generator shall be composed of two types of short and long compound (Figure 6.20 through 6.22).

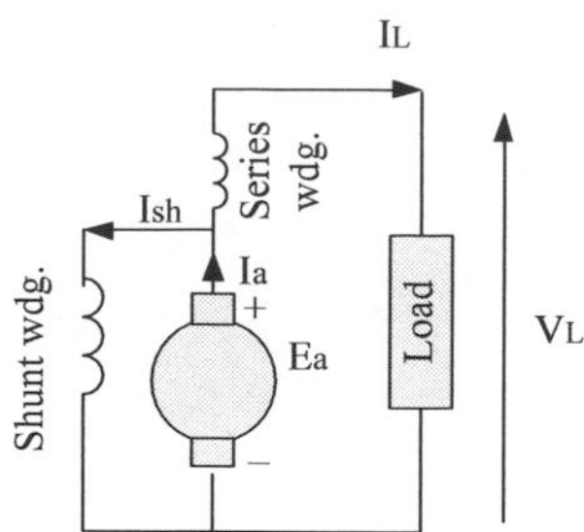

FIGURE 6.18
Short compound DC generator.

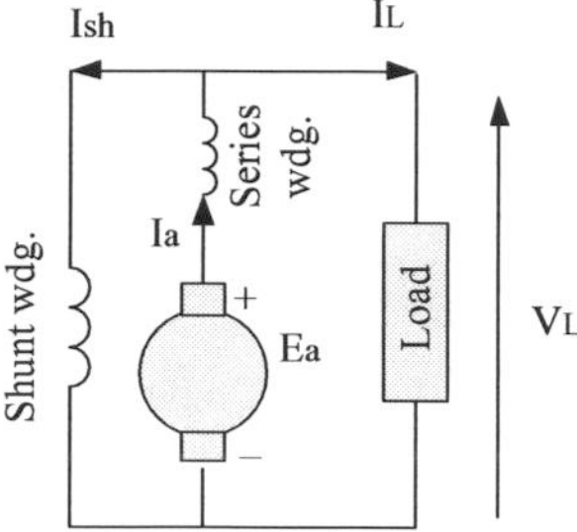

FIGURE 6.19
Long compound DC generator.

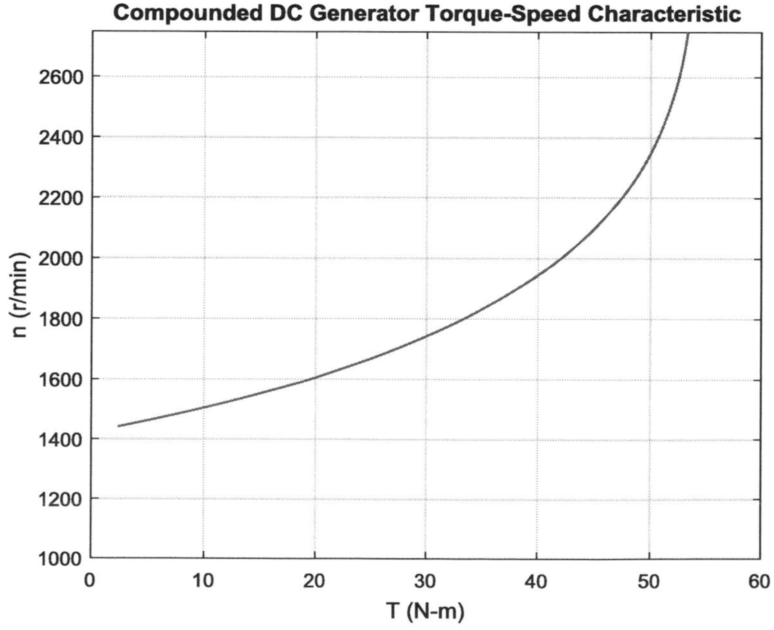

FIGURE 6.20
Input torque-speed characteristics of compound DC generator.

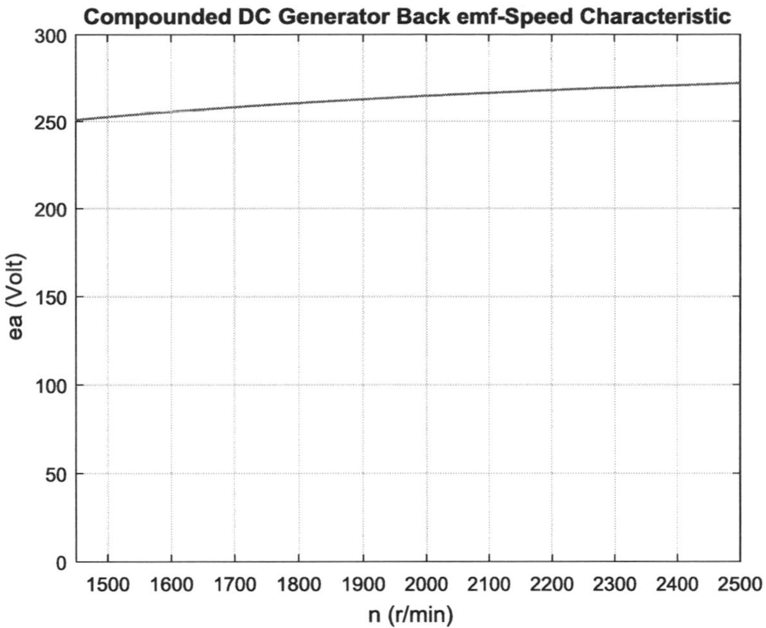

FIGURE 6.21
EMF-speed characteristics of compound DC generator.

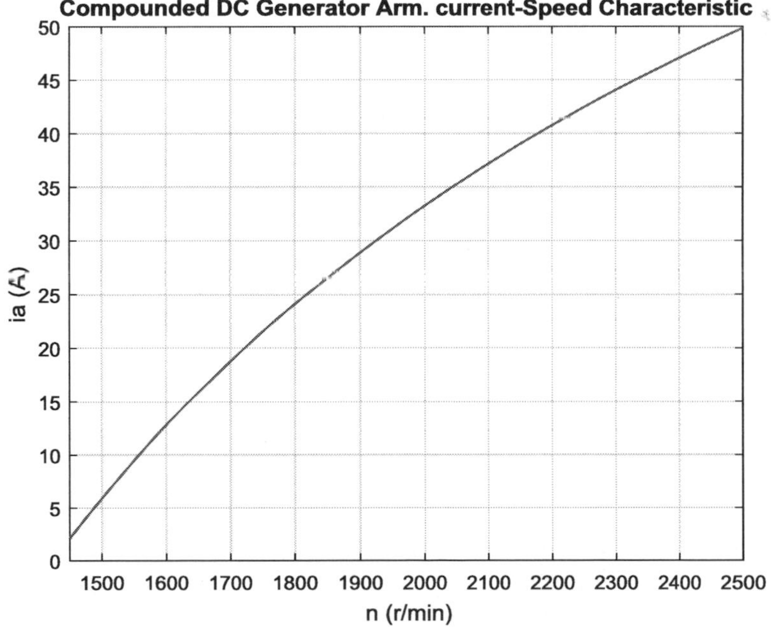

FIGURE 6.22
Armature current-speed characteristics of compound DC generator.

i. The short compound DC generator

From Figure 6.18, the general equations are:

$$E_a = V_L + I_a R_a + I_L R_{se} \tag{6.9}$$

$$I_a = I_L = I_{sh} \tag{6.10}$$

$$I_{se} = I_L \tag{6.11}$$

$$I_{sh} = \frac{Vl + Il\,Rse}{Rsh} \tag{6.12}$$

where:

Rse = series resistant (Ω)

R_{sh} = resistant of parallel coils (Ω)

I_{sh} = shunt current (A)

I_{se} = series windings current (A).

ii. The long compound DC generator

From Figure 6.19, the general equations are:

$$E_a = V_L + I_a (R_a + R_{se}) \tag{6.13}$$

$$I_a = i_L + I_{sh} \tag{6.14}$$

$$I_{se} = i_a \tag{6.15}$$

$$I_{sh} = V_L/R_{sh} \tag{6.16}$$

The voltage difference on both ends of the generator can be extracted as follows:

Voltage difference on both ends of the generator = $V_L + I_L R_{se}$.

Example 6.7

A long compound DC generator feeding a load of current 100 A, at a voltage of 230 V. Calculate the generated electrical motive force if the resistance of the armature is 0.04 Ω, the resistance of 0.01 Ω, and the resistance of the shunt coils 115 Ω.

Solution

$$E_a = V_L + I_a (R_a + R_{se})$$

$$E_a = 230 + I_a (0.04 + 0.01)$$

$$I_a = I_L + I_{sh}$$

$$I_{sh} = \frac{VL}{Rsh}$$

$$I_{sh} = \frac{230}{115} = 2A$$

$$I_a = 100 + 2 = 102$$

$$E_a = 230 + 102 \times 0.05$$

$$E_a = 230 + 501 = 235.1 \; volt.$$

Example 6.8

A short compound DC generator feeding power of 22 kW, 220 V, armature resistance of 0.2 Ω, the series resistance of 0.1 Ω, and shunt resistance of 120 Ω. Calculate the generated electric motive force.

$$E_a = V_L + I_a R_a + I_L R_{se}$$

$$P_L = 22 \times 1000 = 22000 \text{ W}$$

$$I_L = P_L/V_L = 22000/220 = 100 \text{ A}$$

$$I_{sh} = (V_L + I_L R_{se})/R_{sh} = (200 + 100 \times 0.2)/120 = 2 \text{ A}$$

$$I_a = IL + Ish = 100 + 2 = 102 \text{ A}$$

$$E_a = V_L + I_a R_a + I_L R_{se}$$

$$E_a = 220 + 102 \times 0.2 + 100 \times 0.1$$

$$E_a = 220 + 20.4 + 10 = 250.4 \; volt.$$

Compound DC Generator MATLAB Program

```
If=[0 0.1 0.2 0.3 0.4 0.5 0.6 0.7 0.8 0.9 1.0 1.1 1.2 1.3 1.4 1.5 1.6 1.7
1.8 1.9 2.0];
ea=[0 10 15 18 31.30 45.46 60.26 75.06 89.74 104.4 118.86 132.86 146.46
159.78 172.18 183.98 195.04 205.18...
214.52 223.06 231.2];
n0 = 1000;
vt = 250;
rf = 100;
radd = 75;
ra = 0.44;
il = 0:55;
nf = 2700;
nse = 27;
ia = il + vt / (rf + radd);
ea = vt + ia * ra;
ifi = vt / (rf + radd) - (nse / nf) * ia;
ea0 = interp1(if_values, ea_values, ifi);
n = (ea./ ea0) * n0;
T = ea.* ia./ (n * pi / 30);
figure(1);
plot(T, n,'b-','LineWidth',2.0);
xlabel('T (N-m)');
```

```
ylabel('n (r/min)');
title ('Compounded DC Generator Torque-Speed Characteristic');
axis([0 60 1000 2750]);
grid on;
figure(2);
plot(n, ea,'b-','LineWidth',2.0);
xlabel('n (r/min)');
ylabel('ea (Volt)');
title ('Compounded DC Generator Back emf-Speed Characteristic');
axis([1450 2500 0 300]);
grid on;
%hold off;
figure(3);
plot(n, ia,'b-','LineWidth',2.0);
xlabel('n (r/min)');
ylabel('ia (A)');
title ('Compounded DC Generator Arm. current-Speed Characteristic');
axis([1450 2500 0 50]);
grid on;
%hold off;
```

6.3.3 Losses in DC Generator

When the generator converts mechanical energy into electrical energy, the part of this energy is lost and is usually dissipated as heat in resistance. The generated heat in the machine may cause damage to insulation materials and a short circuit between the windings. Heat calculations must be done to reduce the loss in the machine so that the coefficient of quality (efficiency) high means reducing the cost of operation of the machine.

During the conversion of mechanical energy into electrical energy by the generator, a portion of the energy in the magnetic circuit and part of the electrical circuit is lost, as well as a part of the mechanical process (friction) when the rotor is rotated.

A. The Loss in the Magnetic Circuit (Iron Losses)

The magnetic **losses** including: eddy current loss and hysterics loss. It is considered to be a fixed loss.

Hysteresis Losses in DC Machine

Hysteresis losses occur in the armature winding due to the reversal of magnetization of the core. When the core of the armature exposed to a magnetic field, it undergoes one complete rotation of magnetic reversal. The portion of the armature which is under S-pole, after completing half electrical revolution, the same piece will be under the N-pole, and the magnetic lines are reversed in order to overturn the magnetism within the core. The constant process of magnetic reversal in the armature consume some amount of energy which is called hysteresis loss. The percentage of loss depends upon the quality and volume of the iron.

The Frequency of Magnetic Reversal

$$f = \frac{P.n}{120} \tag{6.17}$$

where:

P = number of poles

n = speed in rpm.

The Steinmetz formula is used for the hysteresis loss calculation.

$$P_h = \eta . B_{max}^{1.6} . f . V \tag{6.18}$$

where:

η = Steinmetz hysteresis co-efficient

B_{max} = maximum flux density in armature winding

f = frequency

V = volume of the armature in m^3.

Eddy Current Loss in DC Machine

According to Faraday's law of electromagnetic induction, when an iron core rotates in the magnetic field, an EMF is also induced in the core. Similarly, when the armature rotates in a magnetic field, a small amount of EMF induced in the core which allows the flow of charge in the body due to the conductivity of the core. This current is useless for the machine. This loss of current is called eddy current. This loss is almost constant for the DC machines. It could be minimized by selecting the laminated core. The eddy current can calculate by using the formula:

$$P_e = K_h . B_{max}^2 . f^2 . t^2 \tag{6.19}$$

where:

t = lamination thickness.

B. Losses in the Electrical Circuit (Copper Losses):

It is caused by the passage of the current in the parts of the circuit in the windings of the rotor and the windings of the field, which is a variable loss and according to:

$I_a^2 R_a$ = armature copper losses

$I_{se}^2 R_{se}$ = series windings copper losses

$I_f^2 R_f$ = field windings copper losses.

C. Mechanical Loss:

The losses associated with the mechanical friction of the machine are called mechanical losses. These losses occur due to friction in the moving parts of the machine-like bearing, brushes, etc., and windage losses occur due to the air inside the rotating coil of the machine. These losses are usually very small about 15% of full load loss. Figure 6.23 shows the power stage diagram of the generator.

$$P_g = E_a . I_a$$

$$P_g = P_{in} - (P_{mech} + P_i)$$

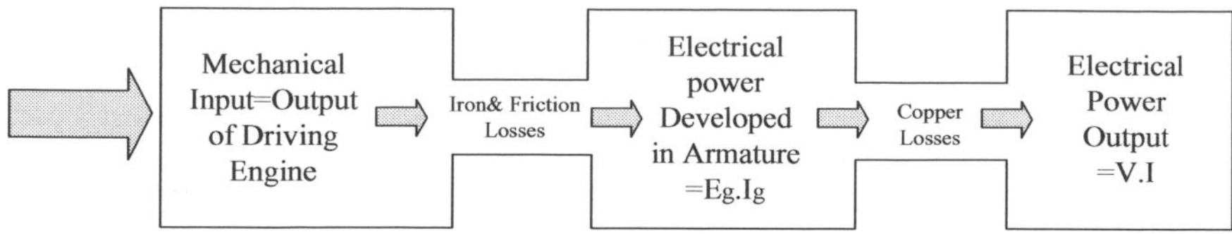

FIGURE 6.23
The power stage diagram of the generator.

Pg = armature capacity generated in the air gap

Pin = input capacity (mechanical capacity) (horsepower)

Each 1 Hp = 746 W

O/P = output power $(V_L.I_L)$.

6.3.4 Efficiency Calculation

With reference to the power path within the DC generator, the overall efficiency can be calculated as follows:

$$\eta = \frac{o/p}{i/p} = (\text{V}_\text{L}\ \text{I}_\text{L}) / \text{HP} \times 746 \tag{6.20}$$

$$\eta = \text{P}_\text{out} \ / \ (\text{p}_\text{out} + \text{Losses}) \tag{6.21}$$

$$\eta = (\text{P}_\text{in} - \text{losses}) \ / \ \text{P}_\text{in} = 1 - (\text{Losses} / \text{P}_\text{in}) \tag{6.22}$$

Example 6.9

A long compound DC generator, rotating at a speed of 1000 rpm, fed a load capacity of 22 kW at 220 V, if the resistance of the armature coils 0.02 Ω, the resistance of the series coils 0.01 Ω, and the resistance of the shunt coils 110 Ω, calculate generator efficiency if the iron and mechanical losses are 2500 W.

Solution

$$P_i + P_{mec} = 2500\ W$$

$$I_L = P_{out} / VL = \frac{22 \times 1000}{220} = 100\text{A}$$

$$I_{sh} = V_{sh}/R_{sh} = V_l/R_{sh} = 220/110 = 2\text{ A}$$

$$I_a = I_l + i_{sh} = 100 + 2 = 102\text{ A}$$

$$P_{cu} = I_a^2\, R_a + I^2{}_a\, R_{se} + I^2{}_{sh}\, R_{sh}$$

$$P_{cu} = (102)^2 \times 0.02 + (102)^2 \times 0.01 + (2)^2 \times 110$$

$$P_{cu} = 752.12\ \text{W}$$

$$\text{the power losses} = P_{cu} + (P_i + P_{mech}.)$$

$$\text{so, power losses} = 752.1 + 2500 = 3252.1\ \text{W}$$

$$\eta = \frac{P_{out}}{P_{out} + losses} = \frac{22000}{22000 + 3252.1} = \frac{22000}{25252.1} = \eta = 0.87 = 87\%.$$

Example 6.10

A shunt DC generator feeds a load with a current of 20 A, 200 V. Calculate generator efficiency if the armature resistance is 0.2 Ω, the shunt resistance is 100 Ω, and the mechanical and iron losses are 201 W.

Solution

$$I_f = \frac{vf}{Rf} = \frac{Vl}{Rf} = 200/100 = 2A$$

$$I_a = I_L + I_f = 20 + 2 = 22\ \text{A}$$

$$P_{cu} = I_a^2R_a + I_f^2R_f = (22)^2 \times 0.02 + (2)^2 \times 100 = 409.68$$

$$P_{cu} = 9.68 + 400 = 409.68\ \text{W}$$

$$\text{the power losses} = P_{cu} + (P_i + P_{mech}) = 409.68 + 201$$

$$= 610.68\ \text{W}$$

$$P_{out} = I_L \cdot V_L = 20 \times 200 = 4000\ \text{W}$$

$$\eta = \frac{P_{out}}{P_{out} + losses} = \frac{4000}{4000 + 610.68} = 86.7 = 86.7\%.$$

Example 6.11

A shunt DC generator has a shunt field resistance of 60 Ω. When the generator delivers 60 KW to a resistive load at a terminal voltage of V = 120 V, while the generated Generator voltage (Eg) is 135 V. Determine (a) the armature circuit resistance Ra and (b) determine the generated voltage Eg when the output is changed to 20 KW and the terminal voltage is V = 135 V. Armature consists of generated voltage Eg in series with armature resistance Ra. (c) Write MATLAB code program to repeat a and b above.

Solution

a) $R_{sh} = 60$ ohm
   the terminal voltage V = 120 V
   the field current in the winding $I_f = V/R_{sh} = 120/60 = 2A$
   when the connected load is 60 kW,
   $V \times I_L = 60000$
   $I_L = 60000/120 = 500A$,

for a shunt generator the
armature current = load current + field current

$$Ia = 500 + 2 = 502\ A,$$

the generated EMF Eg = V + Ia. Ra

$$=> 135 = 120 + 502 \times Ra$$

$$Ra = 0.0298\ \Omega.$$

b) When the output is 20 kW and terminal voltage V = 135 V,

$$V \times I_l = 20000$$

$$I_l = 20000/135 = 148\ A$$

$$I_f = 135/60 = 2.25\ A$$

$$Ia = 148 + 2.25 = 150.25\ A$$

$$Eg = V + Ia.\ Ra$$

$$=135 + 150.25 \times 0.0298 = 139.47\ V.$$

c) MATLAB code

```
R_sh=60;
Eg1=135;
V_l1=120;
i_f=V_l1/R_sh;
load1=60000;
i_l1=load1/V_l1;
i_a=i_f+i_l1;
R_a=(Eg1-V_l1)/i_a;
disp(R_a)
Load2=20000;
Vl2=135;
I_f2=Vl2/R_sh;
Il2=Load2/Vl2;
Ia2=Il2+I_f2;
Eg2=Vl2+Ia2*R_a;
disp(Eg2)
Ra=0.0299
Eg2=139.4940
```

6.4 DC Motors

The DC motor consists of the same parts as the DC generator.

The theory of operation is as follows:

When a wire is placed in a magnetic field, a mechanical force affects the wire, depending on the intensity of the magnetic flux, the length of the wire and the current strength, as in Figure 6.24.

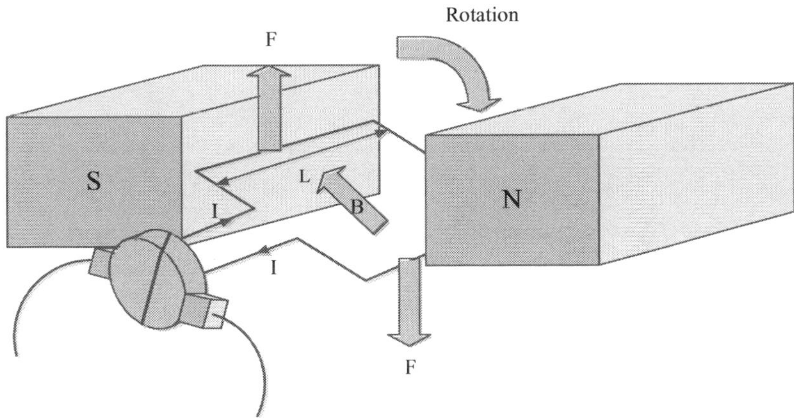

FIGURE 6.24
The diagram shows the effect of magnetic field forces.

$$F = B.\ L.\ I \tag{6.23}$$

where:
F = mechanical force in Newton
B = flux density in Weber/m^2
I = current in A.

When a conductor wire is placed in the form of a coil that carries an electric current, to rotate around a specific axis within a magnetic field with two poles, the direction of the magnetic lines around the wire is opposite to the direction of the lines resulting from the magnetic poles from one end of the coil, a mechanical force moves the coil in a direction that we can assign according to the left-hand rule as in Figure 6.25. The difference between the motor and the generator is illustrated in Figure 6.26.

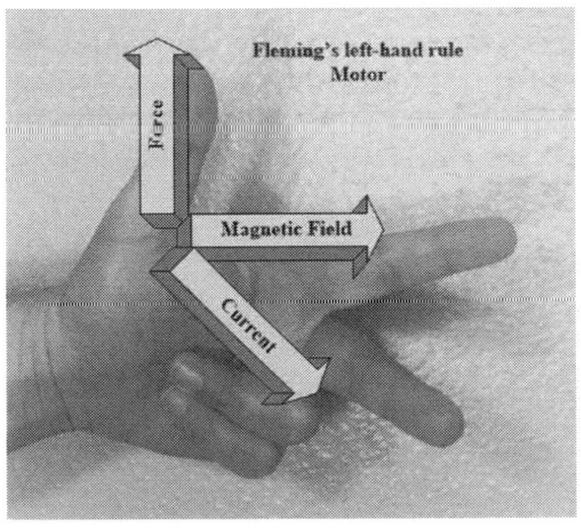

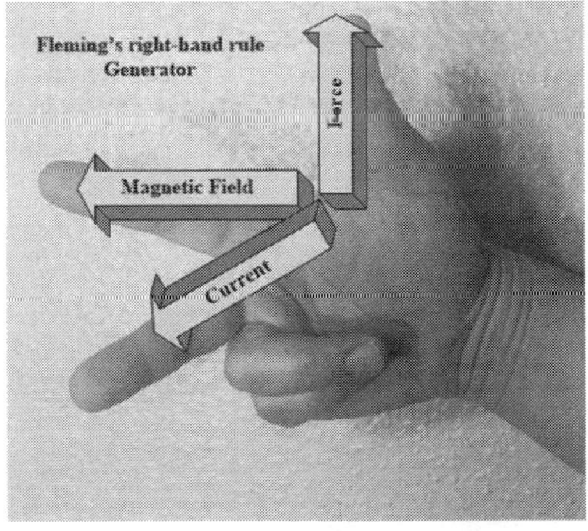

FIGURE 6.25
Fleming's hand rule for generator and motor.

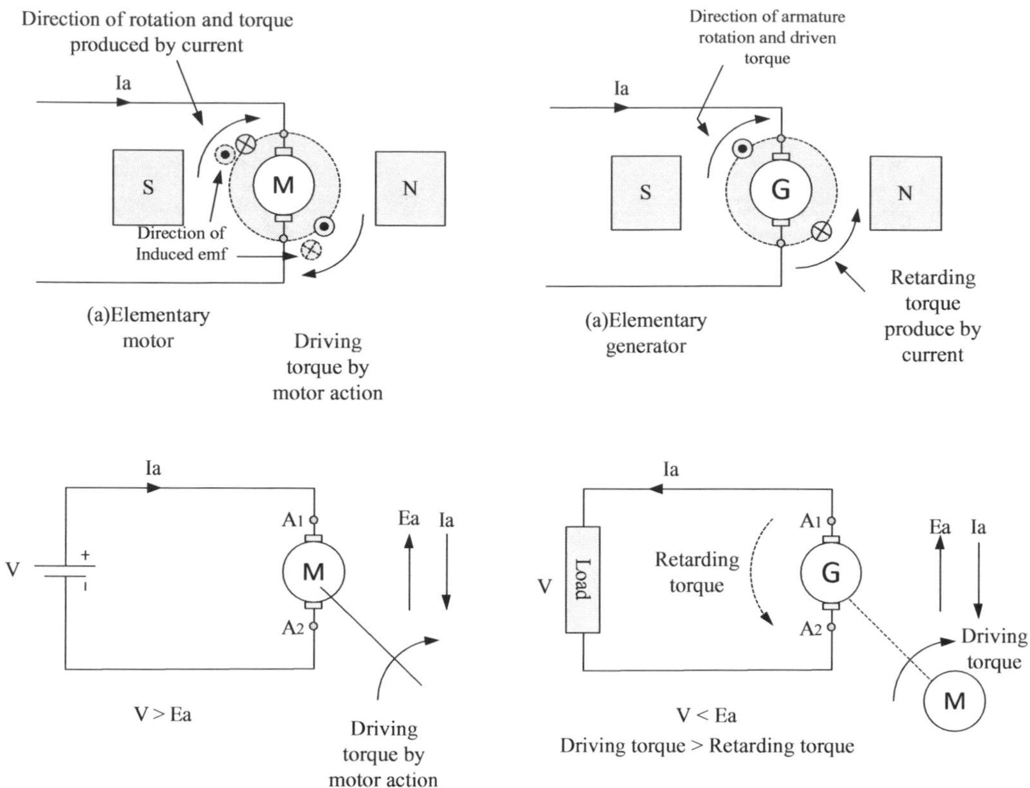

FIGURE 6.26
Comparisons between the generator and the motor, noting the direction of the current in each.

6.4.1 Types of DC Motors

6.4.1.1 Series Motor

The magnetic poles coils are connected to the armature coils, with a few turns and a thick section. The motor is used in cases where high torque is required, such as moving electric trains and cranes, starting the motor drive. The direction of rotation can be reversed as opposed to the ends of the magnetic poles windings (Figure 6.27).

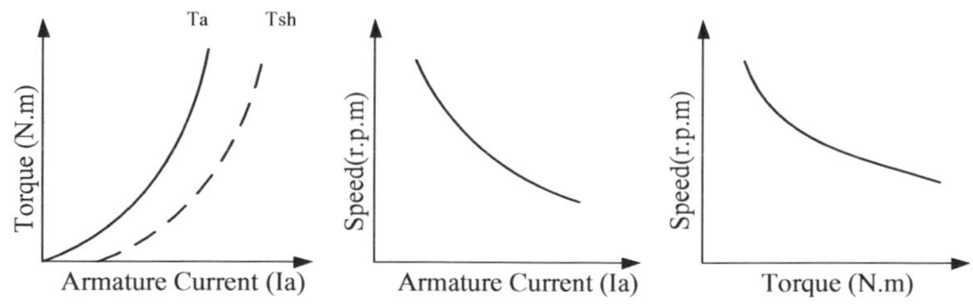

FIGURE 6.27
The relationship between armature current, speed, and torque.

It is noticeable from the characteristic curve that speed increases very significantly at no-load (T = 0). Therefore, it is preferable to not use the series motor at no-load because this causes mechanical problems related to speed up (Figures 6.28 through 6.30).

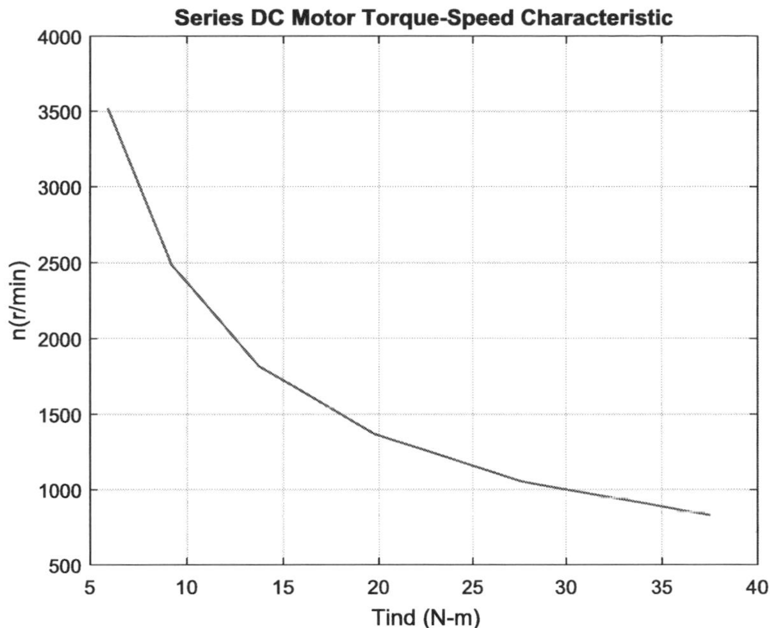

FIGURE 6.28
Torque-speed characteristics series DC motor.

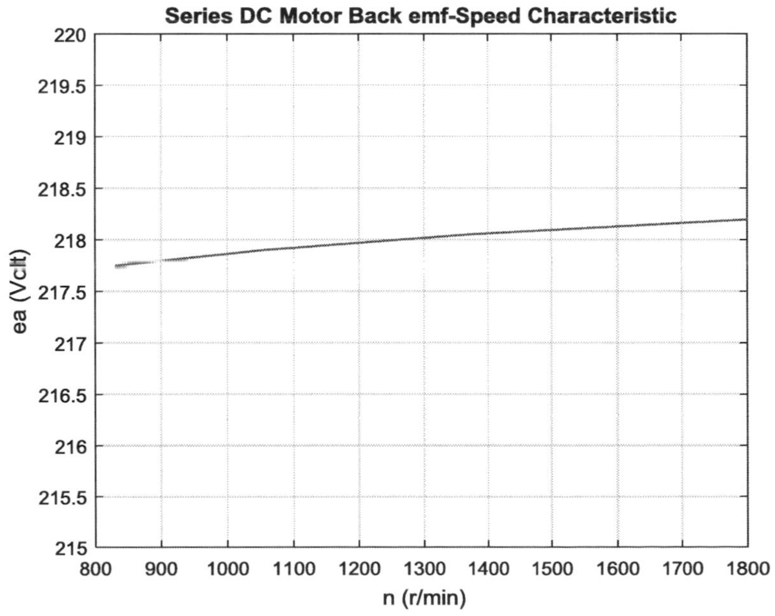

FIGURE 6.29
EMF-speed characteristics of series DC motor.

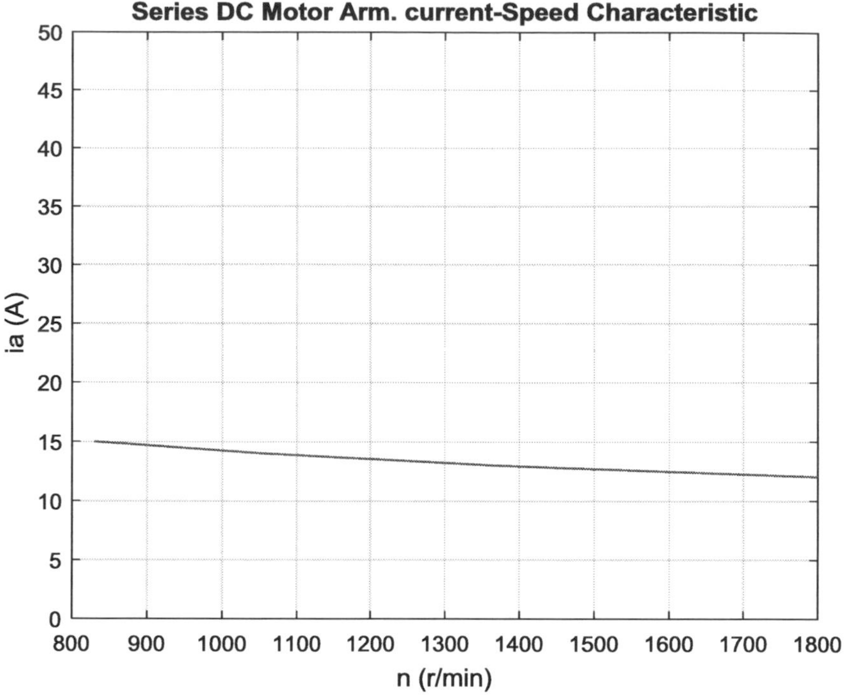

FIGURE 6.30
Armature current-speed characteristics of series DC motor.

The back electrical voltage is given as follows:

$$E_b = V_{in} - I_a (R_a + R_{se}) \tag{6.24}$$

$$I_a = I_{in} = I_{se} \tag{6.25}$$

where:

I_{in} = the current drawn by the motor (A)
I_{se} = the field current (A)
I_a = the armature current (A)
R_{se} = the series resistance (Ω)
R_a = the armature resistance (Ω)
E_b = the back electric motive force (V)
V_{in} = source voltage (V)
T = torque (N. m)
n = speed (rpm).

Example 6.12

A DC series motor, with series resistance 0.2 Ω and armature resistance 0.1 Ω, calculate the back-electric motive force if the value of the drawn current is 50 A and the voltage of the source is 220 V.

Solution

$$E_b = V_{in} - I_a (R_a + R_{se})$$

$$I_a = I_{in} = I_{se} = 50 \text{ A}$$

$$E_b = 220 - 50\,(0.1 + 0.2)$$

$$E_b = 220 - 50 \times 0.3$$

$$E_b = 220 - 15$$

$$E_b = 205 \text{ V.}$$

Series DC Motor MATLAB Program

```
ea=[0 10 15 18 31 45 60 75 90 104 120 132 145 160 173 184 200 205 215 223
231];
n0 = 1000;
vt = 220;
ra = 0.15;
ia = 10:1:15;
ea = vt - ia * ra
ea0 = interp1(if_values, ea_values, ia,'spline');
n1 = 1050;
Eao1 = interp1(if_values, ea_values,58,'spline');
Ea1 = vt - 15 * ra;
Eao2 = interp1(if_values, ea_values, ia,'spline');
n = ((ea./Ea1).* (Eao1./ Eao2)) * n1/100;
tind = ea.* ia./ (n * pi / 30);
figure(1);
plot(tind, n,'b-','LineWidth',2.0);
hold on;
xlabel('Tind (N-m)');
ylabel('n(r/min)');
title ('Series DC Motor Torque-Speed Characteristic');
grid on;
hold off;
figure(2);
plot(n, ea,'b-','LineWidth',2.0);
xlabel('n (r/min)');
ylabel('ea (Volt)');
title ('Series DC Motor Back emf-Speed Characteristic');
axis([800 1800 215 220]);
grid on;
%hold off;
figure(3);
plot(n, ia,'b-','LineWidth',2.0);
xlabel('n (r/min)');
ylabel('ia (A)');
title ('Series DC Motor Arm. current-Speed Characteristic');
axis([800 1800 0 50]);
grid on;
%hold off;
```

6.4.1.2 Shunt Motor

The magnetic pole windings are connected in parallel to the armature by carbon brushes. The selection of magnetic pole windings have many turns and a small section area to obtain relatively high resistance. Therefore, the speed of the motor does not change with the change of load, and on this basis, it is used in situations requiring constant speed and when the load changes such as electric locomotives, elevators, printing machines, and papermaking machines. Figure 6.31 represents the equivalent electric circuit of the shunt motor, and Figure 6.32 represents the relationship between speed, torque, and armature current (Figures 6.33 through 6.35).

$$E_b = V_{in} - I_a R_a \tag{6.26}$$

$$I_a = I_{in} - I_{sh} \tag{6.27}$$

$$I_{sh} = V_{in}/R_{sh} \tag{6.28}$$

Motor speed,

$$n = \frac{V_{in}}{K_b \Phi} \tag{6.29}$$

$$K_b = \frac{z.2p}{60 \times 2a} \tag{6.30}$$

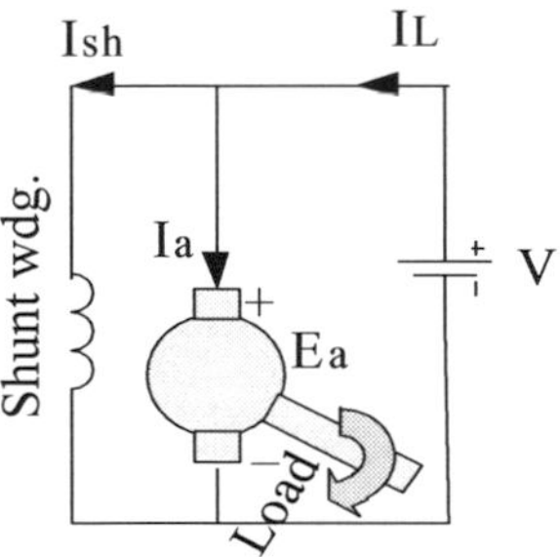

FIGURE 6.31
Equivalent electric circuit of the shunt motor.

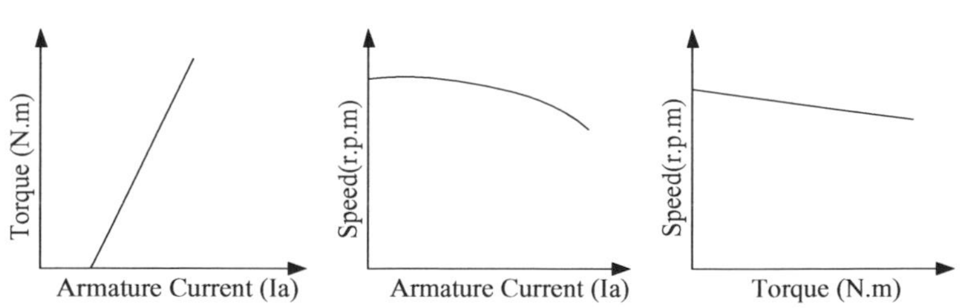

FIGURE 6.32
Relationship between speed, torque, and armature current.

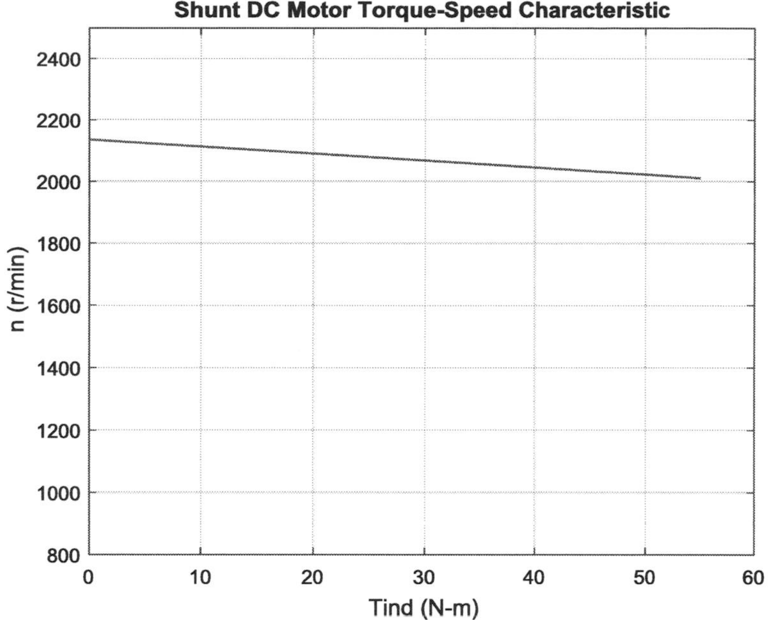

FIGURE 6.33
Torque-speed characteristics of shunt DC motor.

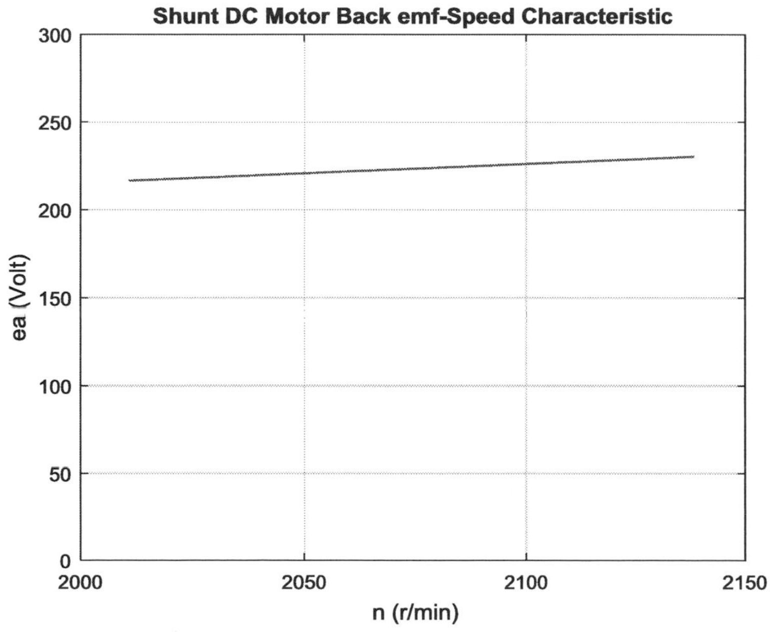

FIGURE 6.34
EMF-speed characteristics of shunt DC motor.

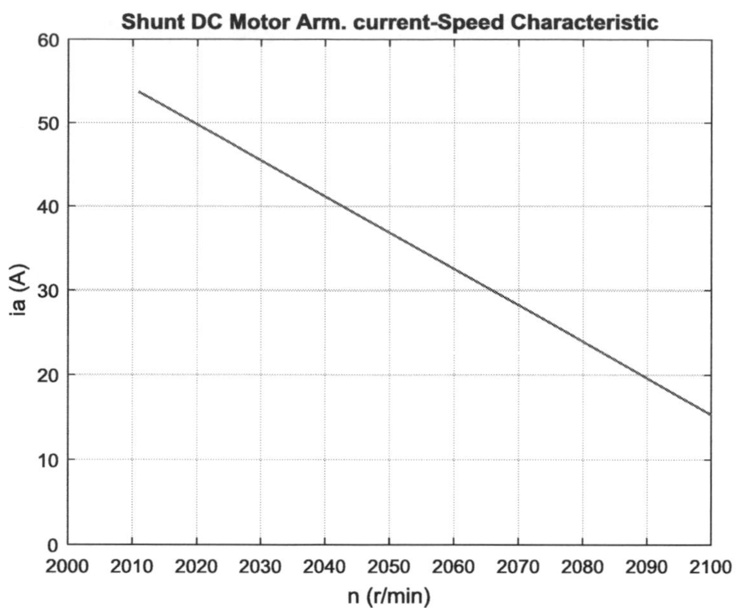

FIGURE 6.35
Armature current-speed characteristics of shunt DC motor.

Example 6.13

A shunt motor operates at 220 V and draws a current of 22 A. If the resistance of shunt 110 Ω and of armature 0.2 Ω, calculate the back-electric motive force.

Solution

$$I_{sh} = V_{in}/R_{sh} = 220/110\ 2\ A$$

$$I_a = I_{in} - I_{sh} = 22 - 2 = 20\ A$$

$$E_b = V_{in} - I_a\ R_a$$

$$E_b = 220 - 20 \times 0.2 = 220 - 4 = 216\ V.$$

Example 6.14

A four-pole DC shunt motor, 220 V, the number of conductors in the armature 1000, the typical winding coil, the motor draws 52 A, the magnetic flux per pole 0.02 Wb, the armature resistance 0.2 Ω, and shunt resistance 110 Ω. Calculate motor speed.

Solution

$$I_{sh} = V_{in}/R_{sh} = 220/110 = 2A$$

$$I_a = I_{in} - I_{sh} = 52 - 2 = 50\ A$$

$$E_b = V_{in} - I_a\ R_a$$

$$E_b = 220 - 50 \times 0.2 = 210\ V$$

$$E_b = \frac{nz\Phi 2p}{120}$$

$$210 = \frac{n \times 1000 \times 0.02 \times 4}{120}$$

$$n = \frac{210 \times 120}{1000 \times 0.02 \times 4} = 315 \text{ rpm}$$

Shunt DC Motor MATLAB Program

```
if_values=[0 0.1 0.2 0.3 0.4 0.5 0.6 0.7 0.8 0.9 1.0 1.1 1.2 1.3 1.4 1.5
1.6 1.7 1.8 1.9 2.0];
ea_values=[0 10 15 18 31.30 45.46 60.26 75.06 89.74 104.4 118.86 132.86
146.46 159.78 172.18 183.98 195.04 205.18...
214.52 223.06 231.2];
n0 = 1500;
vt = 230;
rf = 100;
radd = 75;
ra = 0.250;
il = 0:1:55;
nf = 2700;
far0 = 1500;
ia = il - vt / (rf + radd);
ea = vt - ia * ra;
ish= vt / (rf + radd);
ea0 = interp1(if_values, ea_values, ish);
n = (ea./ ea0) * n0;
Tind = ea.* ia./ (n * 2 * pi / 60);
figure(1);
plot(Tind, n,'b-','LineWidth',2.0);
xlabel('Tind (N-m)');
ylabel('n (r/min)');
title ('Shunt DC Motor Torque-Speed Characteristic');
axis([0 60 800 2500]);
grid on;
%hold off;
figure(2);
plot(n, ea,'b-','LineWidth',2.0);
xlabel('n (r/min)');
ylabel('ea (Volt)');
title ('Shunt DC Motor Back emf-Speed Characteristic');
axis([2000 2150 0 300]);
grid on;
%hold off;
figure(3);
plot(n, ia,'b-','LineWidth',2.0);
xlabel('n (r/min)');
ylabel('ia (A)');
title ('Shunt DC Motor Arm. current-Speed Characteristic');
axis([2000 2100 0 60]);
grid on;
%hold off;
```

6.4.1.3 Compound DC Motor

The compound motor is the basis of a shunt motor with series windings, in which the source current is passed in the short compound motor in Figure 6.36 or the current of the armature in the long compound motor as shown in Figure 6.37. In a certain direction so that the effect of the magnetic field given by these coils is in the magnetic field of shunt windings. Thus, the motor acquires certain properties for speed and torque. The compound motor can be used to obtain high torque and constant speed, which are not significantly affected by the change in load as in moving locomotives, electric buses, and printing machines (Figures 6.38 through 6.40).

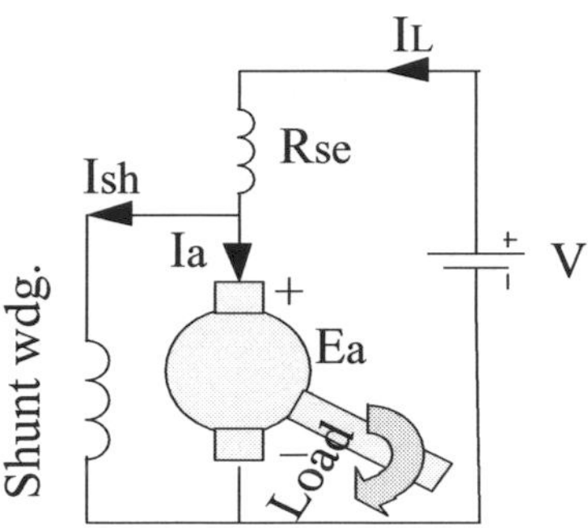

FIGURE 6.36
Short compound motor.

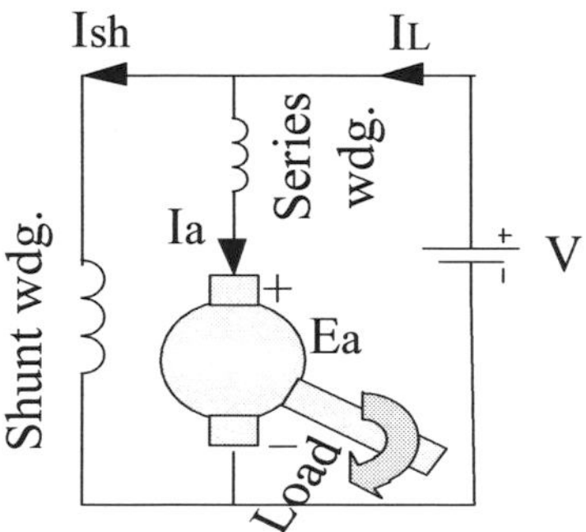

FIGURE 6.37
Long compound motor.

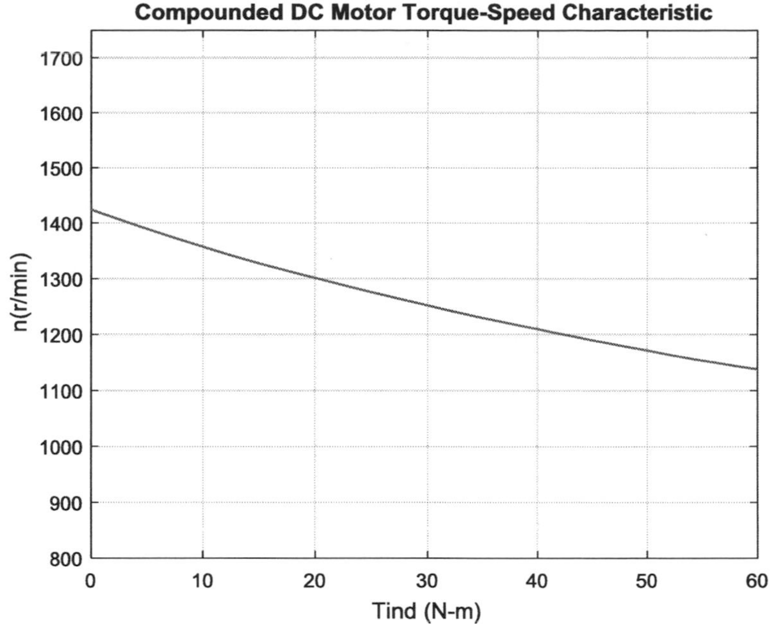

FIGURE 6.38
Torque-speed characteristics of compound DC motor.

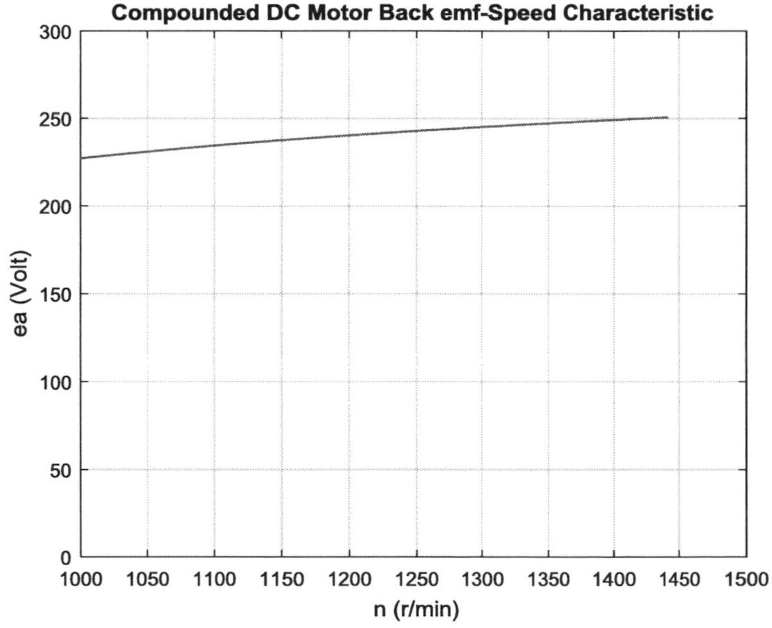

FIGURE 6.39
EMF-speed characteristics of compound DC motor.

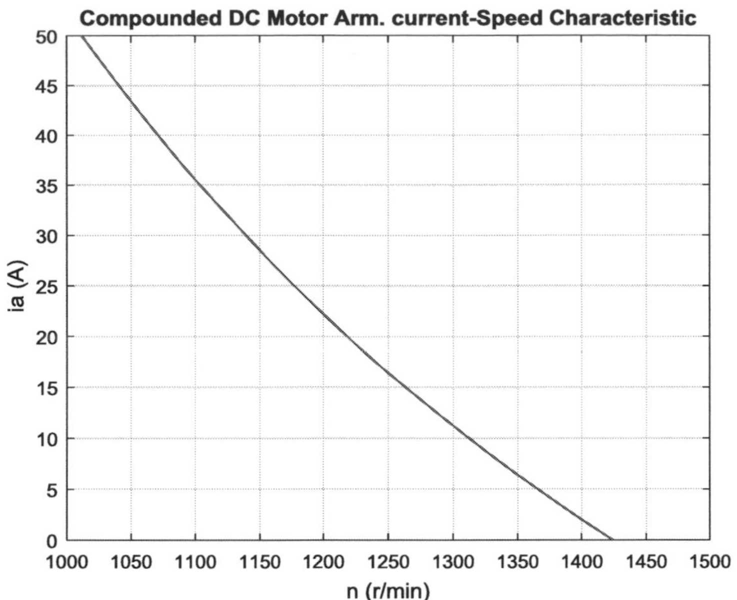

FIGURE 6.40
Armature current-speed characteristics of compound DC motor.

Compound DC Motor MATLAB Program

```
If=[0 0.1 0.2 0.3 0.4 0.5 0.6 0.7 0.8 0.9 1.0 1.1 1.2 1.3 1.4 1.5 1.6 1.7
1.8 1.9 2.0];
ea=[0 10 15 18 31.30 45.46 60.26 75.06 89.74 104.4 118.86 132.86 146.46
159.78 172.18 183.98 195.04 205.18 214.52 223.06 231.2];
n0 = 1000;
vt = 250;
rf = 100;
radd = 75;
ra = 0.44;
il = 0:55;
nf = 2700;
nse = 27;
ia = il - vt / (rf + radd);
ea = vt - ia * ra;
ifi = vt / (rf + radd) + (nse / nf) * ia;
ea0 = interp1(if_values, ea_values, ifi);
n = (ea./ ea0) * n0;
tind = ea.* ia./ (n * pi / 30);
figure(1);
plot(tind, n,'b-','LineWidth',2.0);
xlabel('Tind (N-m)');
ylabel('n(r/min)');
title ('Compounded DC Motor Torque-Speed Characteristic');
axis([0 60 800 1750]);
grid on;
figure(2);
plot(n, ea,'b-','LineWidth',2.0);
```

```
xlabel('n (r/min)');
ylabel('ea (Volt)');
title ('Compounded DC Motor Back emf-Speed Characteristic');
axis([1000 1500 0 300]);
grid on;
%hold off;
figure(3);
plot(n, ia,'b-','LineWidth',2.0);
xlabel('n (r/min)');
ylabel('ia (A)');
title ('Compounded DC Motor Arm. current-Speed Characteristic');
axis([1000 1500 0 50]);
grid on;
%hold off;
```

6.4.2 DC Motor Speed Control

When using DC motors for industrial purposes, it is necessary to control the start of its movement and regulate its speed to suit the purposes used.

The speed changes either by a resistance connected to the armature or by voltage exerted on the ends of the motor or by changing the magnetic flux through the circuit of the field. Both the shunt motor and the compound are like the speed control methods.

6.4.2.1 Speed Control of the Shunt Motor

The speed control methods of the shunt motor are:

1. By using variable resistance:

   Variable resistance is connected with the armature circuit, the speed changes by changing the value of the resistance by a switch that controls variable resistance values. One of the disadvantages of this method is to reduce the efficiency of the motor. As illustrated in Figure 6.41
2. Speed control by using voltage control:

   The speed of the shunt motor can be controlled by controlling the voltage applied to it, as in the case of Ward Leonard. However, this method is very expensive (Figure 6.42)
3. Speed control by field:

   This method is simple and low cost, as low-power field resistance is used to control the current of the field and then the magnetic flux (Figure 6.43).

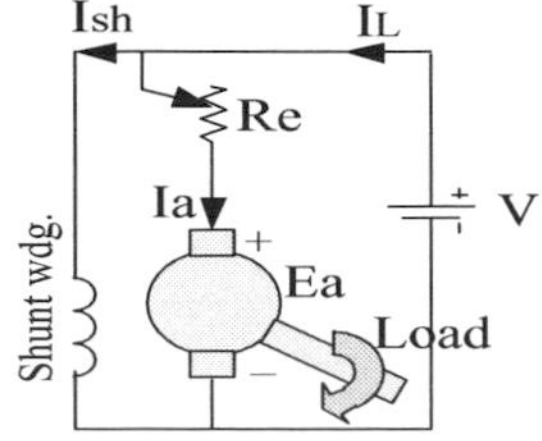

FIGURE 6.41
The speed regulation of a shunt motor using resistance with the armature.

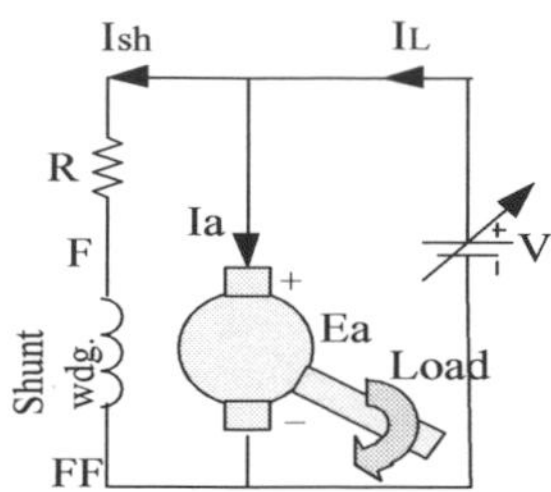

FIGURE 6.42
Speed control by the voltage control method.

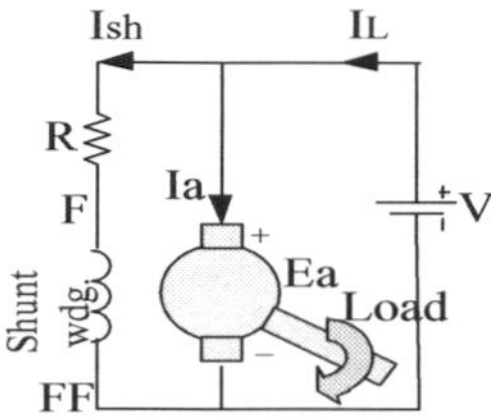

FIGURE 6.43
Speed control by field.

6.4.2.2 Speed Control of the Series Motor

The speed control methods of the series motor are:

1. By connecting series resistance with the motor circuit:

   The motor speed can be changed by adding resistance to the armature circuit (Figure 6.44).
2. Connect resistance in parallel with the field windings:

   Controlling the field current value can only be achieved by connecting resistance in parallel with the field windings. Thus, we can control the current of the field and thus the speed of the motor (Figure 6.45).
3. Speed control by using armature diverter:

   A diverter across the armature can be used for giving speeds lower than the normal speed as shown in Figure 6.46.

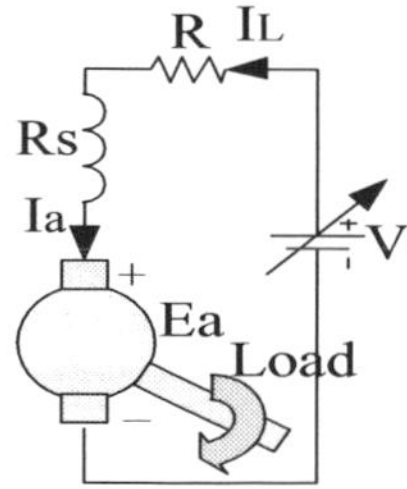

FIGURE 6.44
Connect series resistance with the motor circuits.

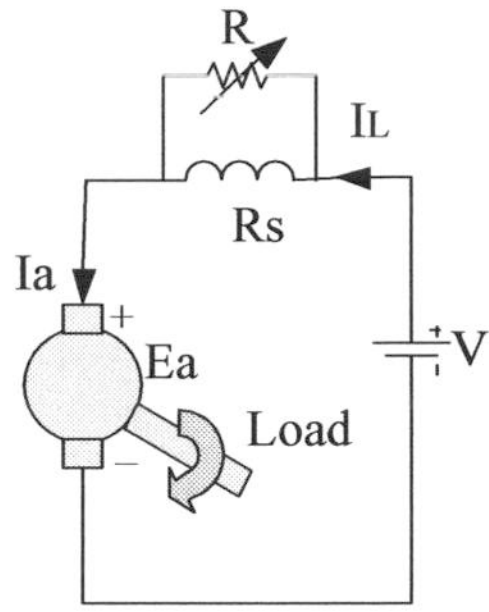

FIGURE 6.45
Resistance parallel with the field windings control the motor speed.

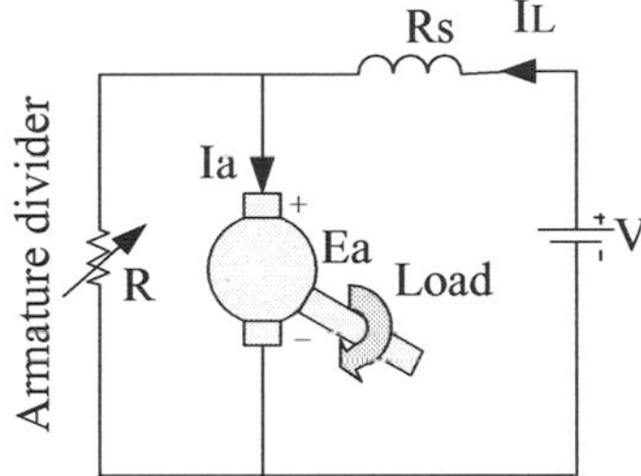

FIGURE 6.46
Armature diverter.

6.4.3 Starting Methods

The purpose of using different ways to start the DC motors is to reduce the current at the start, where this current is very high, and this is illustrated by the current equations in the below:

$$I_a = (V_{in} - E_b)/R_a \tag{6.31}$$

$$I_a = (V_{in} - E_b)/(R_a + R_{se}) \tag{6.32}$$

This means that the current will be very high because of the low resistance to its windings, so the resistance to start the movement until the arrival of the motor to 75% of its actual speed after up growth of the back-electric motive force, which reduces the resistance gradually until the value of zero.

A manual or automatic initiator is used, and the initiator of the motion is a set of conductors that are conductively connected and of which any number can be added or separated from the motor by a variable key.

Problems

6.1 What does the stator part of DC machines contain?

6.2 What are the ways to turn the armature in the DC machine?

6.3 What are the methods of feeding magnetic pole windings in DC generators?

6.4 What is the use of the commutator in the DC machine?

6.5 What is the benefit of the auxiliary poles in the DC machine?

6.6 Draw the DC generator circuit and the motor of the shunt type indicating where the direction of the current.

6.7 Why is it not allowed to operate a DC series motor without a load?

6.8 What types of power losses in DC generators?

6.9 Draw a diagram of the power path in the DC generators.

6.10 What are the ways to speed control of the DC motor?

6.11 How can reverse the rotation of DC motors?

6.12 What is the beneficial use of the resistance to start a movement in the DC motors?

6.13 What is the theory of running DC motors?

6.14 What is the beneficial use of carbon brushes in DC generators?

6.15 What is the effect of the feeding current in the DC motor rotation speed?

6.16 A DC generator have the number of poles four, wave winding, and the number of conductors in the armature 1000 conductor, if the electric motive force generated 200 V, the value of the magnetic flux 0.02 Wb calculated its speed.

6.17 A DC generator feeding a load of 10 Ω, 200 V, the series resistance of 0.2 Ω. Calculate the resistance of the armature.

6.18 A DC generator that feeds a load of 22 kW. The current in the load is 88 A. Calculate the current passing through the shunt windings if you know that the resistance is 125 Ω.

6.19 A short compound generator feeding a load with a current of 98 A, a shunt resistance of 100 Ω, a series resistance of 0.2 Ω, and the current passing through the shunt resistance 2 A. Calculate the armature resistance.

6.20 A short compound DC generator with eight poles, and a lap windings, the number of conductors is 1200, the speed is 600 rpm, the armature current 50 A, magnetic flux 0.02 Wb, resistance of armature windings 0.4 Ω, series resistance 0.1 Ω, shunt resistance 110 Ω, find efficiency if the mechanical and iron losses 1330 W.

6.21 A DC power motor with four poles, the armature has a typical coil, the number of conductors is 1000 conductor, the resistance of the series windings 0.2 Ω, resistance of the armature windings 0.4 Ω, calculate its speed if it is known that the magnetic flux generated in each pole 0.02 Wb, the current drawn 50 amp, and working on the source 230 V.

6.22 A DC motor is designed to reduce its speed, adding resistance in series with the shunt windings, calculate the value of this added resistance, if the armature resistance is known to be 0.5 Ω, and the shunt resistance is 10 Ω, the current is pulled by the motor 60 A, the current that passes the armature resistance is 10 A, and operates at 220 V.

6.23 A long compound DC generator that feeds a load of 2200 W, 220 V, armature resistance of 0.2 Ω, 44 Ω of parallel resistance, and 227.5 V generated electric motive force. Calculate the resistance of the series windings.

7

AC Motors

Your world is a living expression of how you are using and have used your mind.

Earl Nightingale

The machine that converts electrical energy into mechanical energy is the electric motor. Electric motors are characterized by the ability to classify them in sizes and capacities ranging from very small that can be placed in the clocks to the very large used in large cranes and in various industries. It is also characterized by different speeds and the possibility of controlling the speed of rotation, and the possibility of rotation in two directions, and the need for maintenance and no exhaust from them and use electric motors in various devices such as refrigerators, washing machines, vacuum cleaners, mixers, air conditioners, fans, and elevators and many devices. The alternative electric car is a vehicle that contains internal combustion motors that now exist that contribute to environmental pollution by a large amount.

7.1 Single-Phase Motor

The electric motors have been used in the operation of various types of modern devices and equipment that emerged with the emergence of these motors such as audio and video recorders, computer drives, printers, games, and industrial robots of uses.

The principle of the work of the electric motor is the opposite of the principle of the work of the generator when following an electric current in a wire located within a fixed magnetic field generates mechanical force affect the wire and push it to the movement, and thus could convert electricity to mechanical energy (Figure 7.1).

The single-phase electric motors are of many types depending on the nature of their work and design.

1. Induction-motors that decrease in speed with increased load
2. Synchronous motors that do not affect the speed when the load is changed on the axis of rotation
3. General motors (universal-motors).

7.1.1 Induction Motors

These machines are the most widely used and used in daily use for household, office, medical, and other purposes, because they are simple in design, low cost, good efficiency, low power, and rotate at different speeds (Figure 7.2).

FIGURE 7.1
Different types of single-phase motor.

(a)Squirrel cage rotor motor.

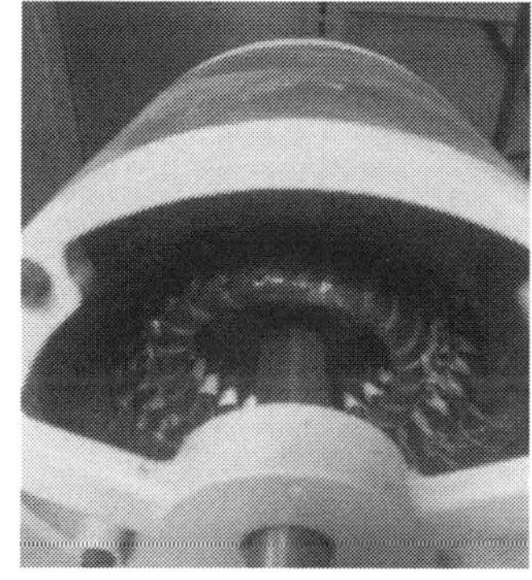
(b)Slip ring motor.

FIGURE 7.2
Single-phase induction motor.

7.1.1.1 Motor Construction

These motors consist of major parts that are present in all types, and additional parts are only present in some of them. The main parts are:

1. **The stator**: It consists of three basic parts:
   a. **Yoke (outer frame)**

      It is made up of steel (cast iron) and contains fins on the outer surface, which work to cool the coils during the air pump from the cooling fan and uses the frame to carry the sheets (the chips) of the core of the iron, as well as to install the side sinks. It lists motor specifications such as current, voltage, frequency, speed, capacity, and so on (Figure 7.3).

   b. **Iron core**

      It is made up of iron-silicon wafers or sheets, isolated from each other by varnishes and compressed to form the iron core.

      Figure 7.4a shows the iron core flakes of the stator and the rotor, and Figure 7.4b shows the iron core after the collection of chips.

FIGURE 7.3
The outer shape of the single-phase induction motor. (https://www.indiamart.com/proddetail/induction-motor-11915087030.html).

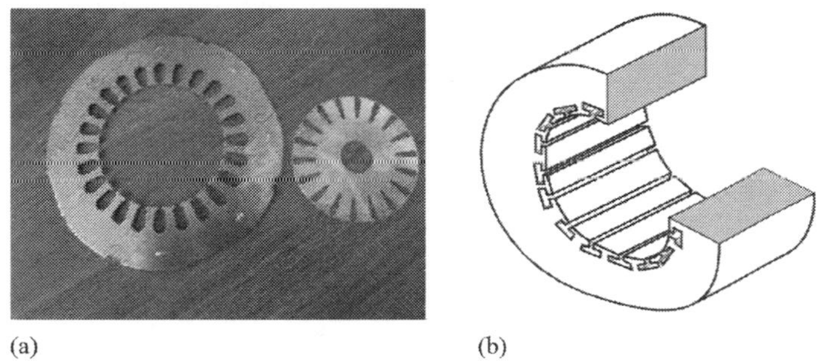

(a) (b)

FIGURE 7.4
Iron core of single-phase induction motor. (a) The iron core chips of the stator and the rotor. (b) The iron core after collecting the chips (https://www.indiamart.com/proddetail/induction-motor-11915087030.html).

c. **Running winding coils**

Made of copper wires isolated by varnish, thick wires, called *operating coils,* and the number of turns proportional to the capacity of the motor and connect the AC source (Figure 7.5).

2. **The rotor**

The rotor is made up of a spindle, which is based on the side rails in parallel to the occupant. It presses on the shaft of iron sheets that are isolated from each other by varnishes. The outer perimeter has long, straight, or sloping ducts where copper or aluminum rods are placed. Aluminum is usually poured directly into these slots.

These coils are not fed by an electrical current. This kind of motor is called *squirrel cage motors*. Figure 7.6a shows the rotor in the induction motors in which the slots are slanted, and Figure 7.6b shows where the slots are fair inside the stator.

Figure 7.7 shows different sizes of the iron cores of the rotor and the stator.

From the observation of the construction of these motors that there is no electrical connection between the stator and rotor, and the only link is the electromagnetic bond, that is, this type of motor works in magnetic induction.

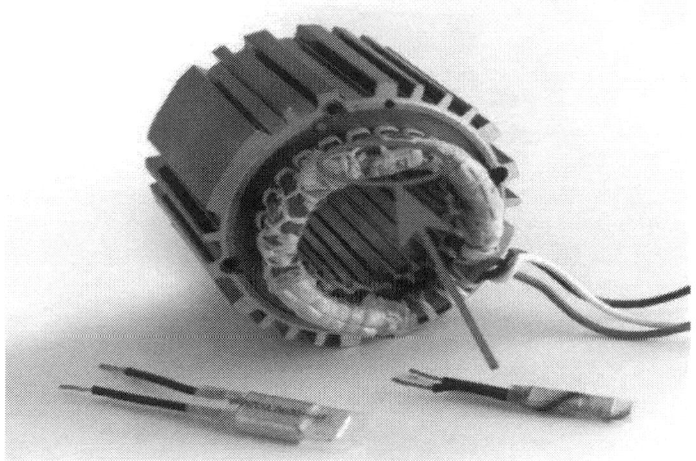

FIGURE 7.5
The stator coils of a single-phase induction motor.

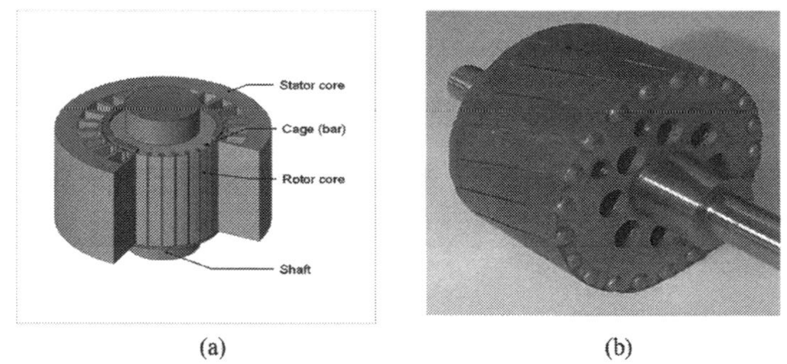

FIGURE 7.6
(a) The rotor in induction motors with skewed slots. (b) The slots are fair inside the stator.

FIGURE 7.7
Different sizes of the iron cores of the rotor and the stator.

FIGURE 7.8
The side cover of the motor.

3. **The two sides cover**

   They are made of solid steel, i.e., of the metal frame itself, are fixed by bolts, and the central part of the centrifuge switch is placed in the front cover.

   The two covers are mounted on the spindle, balancing the rotor, facilitating rotation, and making it in a position to move freely. Figure 7.8 shows the motor side cover.

4. **Ventilation fan**

   It is an important part of aluminum or plastic and operates during the rotation of the motor, the air flux between the fins of the frame. So, the temperature resulting from following the current in the fixed iron core coils will reduce.

7.1.1.2 The Speed

In induction motors, the speed of the magnetic field of the stator is higher than the actual speed of the rotor. The reason is the air resistance of the rotor, and the friction between the axis of rotation and the supports, and this difference is called *slip*.

The speed of the magnetic field in the stator is proportional to the frequency of the source, and inversely with the number of poles of the motor (2P), i.e.:

$$N_S = 60 \times f/p \tag{7.1}$$

where:

N_S = the speed of the magnetic field and is measured in (rpm)
F = source frequency, measured in (Hz)
P = the number of pairs of poles
2P = the number of poles.

Example 7.1

A motor that has four poles connected to a frequency source 50 Hz, calculate its harmonic speed?

Solution

$$N_S = 60 \times f/P$$

$$N_S = 60 \times 50/2 = 1500 \text{ rpm.}$$

Example 7.2

An induction motor has the speed of 3000 rpm, connect to a source of frequency 50 Hz. Find the number of its poles.

Solution

$$N_S = 60\ f/P$$

$$N_S\ P = 60\ f$$

$$P = 60 \times f/N_s$$

$$P = 60 \times 50/3000 = 1 \text{ pole pair}$$

$$2p = 2 \text{ poles.}$$

7.1.1.3 The Theory of Work

When a single-phase alternating voltage is applying on the stator windings, the current following through it will generate a magnetic field that moves to the rotor windings through the air gap, then cuts it and electromotive force (EMF) voltage will generate, then lead to the passage of an electric current that constitutes another magnetic field of mutual interaction between the two magnetic fields of the stator and the rotor. Electromagnetic forces are composed of the sum of these forces, and as a result, the rotor's torque that is given on the rotor is zero.

To operate the motor, rotor rotation is required by creating a primary torque. This is done by displacing the magnetic field in the rotor from its position to create an angle between it and the magnetic field in the constant.

7.1.1.4 Starting Methods of the Motor

There are several ways to start single-phase induction motors, including:

1. **Starting Windings Method**

   These windings are placed in parallel with the stator's windings (in the stator's slots) have a small section area and high resistance, occupy one-third of the total number of slots, reach parallel to the operating coils, and are usually designed for a limited period not exceeding several seconds. The centrifuge switch is connected sequentially to the auxiliary coils to separate it from the main circuit. When the motor speed reaches (75%) of the synchronous speed. The torque in this type is relatively low, efficiency and power factor are low.

The usefulness of these windings is when the motor is connected to the source. A current in the start coils passes at an angle from the current of the main operating coils and then two different fields occur, thus creating a rotary magnetic field that produces a primary torque for the motor. Figure 7.9 shows the connection of the circuit containing the auxiliary coils and the centrifuge switch. This method is not enough to give a high starting torque, so another factor must be added.

2. Capacitor Start Method

In this method, the capacitor is connected continuously with the start coils and the centrifuge switch. This circuit is connected in parallel with the main operation coils, and Figure 7.10 shows the circuit connection.

When the motor is connected to the source, a current follow through both the start and run windings, but the presence of the capacitor provides the flowing current in the start windings from the current in the operating windings at a given angle, thus creating a starting torque due to the two-phase differentials. The starting torque of the motor is high and very efficient and also reduces the consumption of current. This type of motors is used in washing machines and large refrigerators.

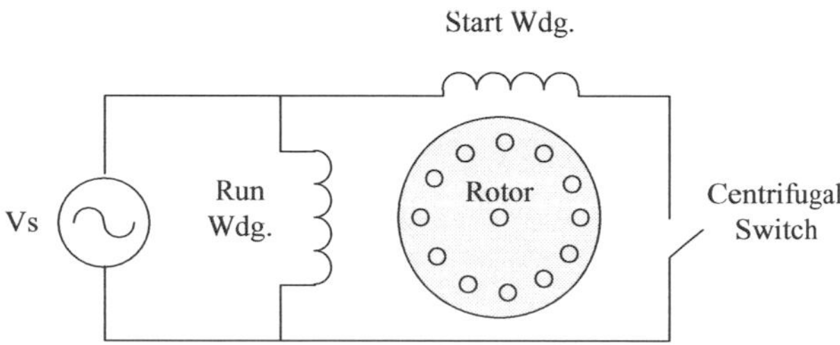

FIGURE 7.9
The connection of the circuit containing the auxiliary coils and the centrifuge switch.

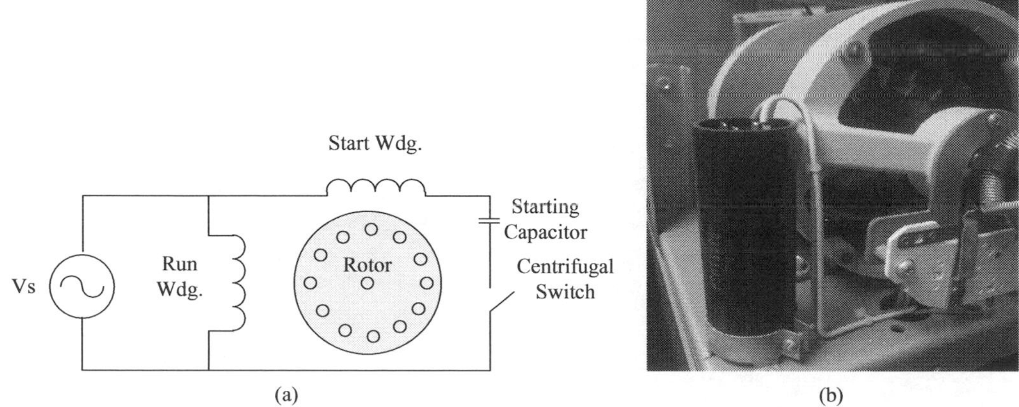

FIGURE 7.10
Capacitor start motor. (a) Circuit diagram. (b) Start capacitor.

The capacitor may remain connected to the starting coils to the source and continuously when the motor is operating. On this basis, the auxiliary coils are designed so that they are not damaged and, in this case, you do not need a centrifuge switch. The result is a low noise during operation and a power factor improvement. This type of motor is used in applications that require low noise during operation as in ceiling and desktop fans and called *capacitor run motor*. Figure 7.11 shows the connection of this type of motor.

There are motors that contain capacitors. The first is run capacitor, a small paper type saturated with oil and connected in parallel with the auxiliary coil circuit, and the second is the start capacitor which is connected in series with the auxiliary coils and the centrifuge switch.

Figure 7.12a shows the method of connecting the capacitors in the motor circuit, and Figure 7.12b shows the form of motor containing capacitors.

3. **Shaded Pole Motor**

In these motors the starting torque is low, and the stator is different in which the motors containing the starting motion coils. The stator is made of thin sheets of iron assembled together with prominent poles placed around the operating coils, and to obtain a magnetic field rotary to start the movement, a slit occurs on one side of the pole and placed around it a ring of copper wire is a bit thick and represent a restricted coil called *shaded pole*. This circuit is compensated for starting coils in single-phase motors. The rotor of these motors is of the type of squirrel cage.

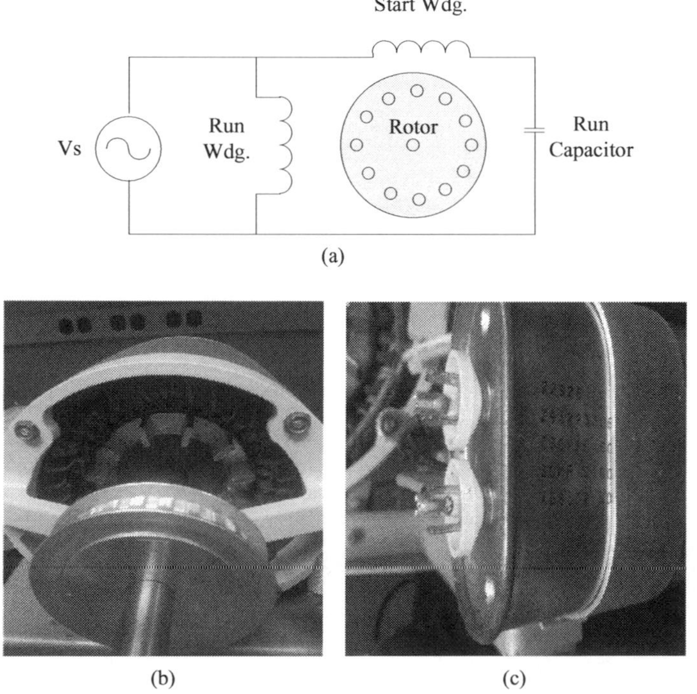

FIGURE 7.11
Capacitor run motor. (a) Circuit diagram. (b) Stator and rotor. (c) Run capacitor.

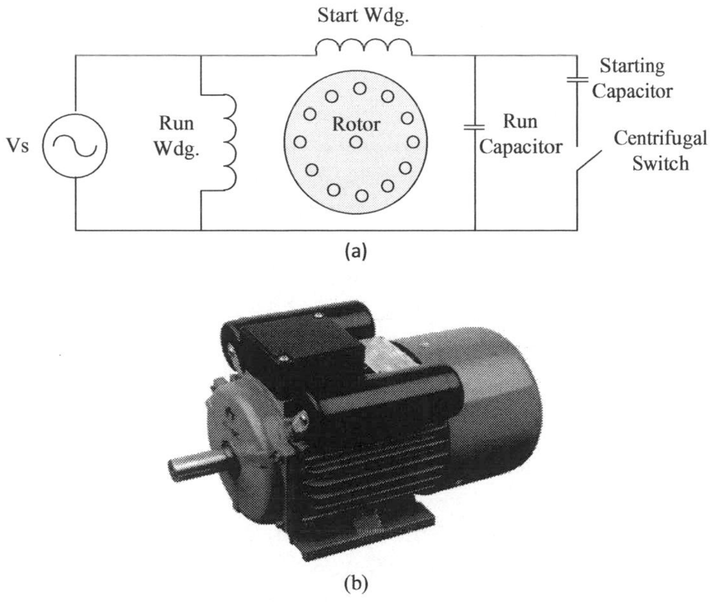

FIGURE 7.12
Capacitor start with capacitor run motor. (a) Capacitor start with capacitor run motor circuit. (b) Motor containing two capacitors.

The electric losses in the shaded ring are relatively large at the rated speed, making so the efficiency low, the motors are often designed at a low power of not more than 40 W. However, due to some improvements in motor design, common uses for this motor are up to 300 W. It also has a low power factor.

Despite the disadvantages of this type of motor, it also finds many uses because of its low cost and simplicity of design, and the noise level is very low due to the lack of slots, as well as the appropriate choice of structural parts and exact calculation of motor dimensions even it is possible to improve the power index and increase the starting torque and may be designed sometimes using two rings are short for each pole, and there are other designs intended to obtain the optimal rotational magnetic field. Figure 7.13 shows the forms of stator and rotor of the shaded pole motor.

FIGURE 7.13
Shaded pole motor. (https://www.indiamart.com/proddetail/shaded-pole-motor).

The working theory of shaded poles motor:

The main coil in the stator acts as an exciting coil in the DC motors, which is responsible for the formation of the magnetic field in the motor. When an alternating current in the coils creates a magnetic field that moves to the copper ring shorted, and as it differs from the main coils in terms of cross-section area for this generates high current as well as a magnetic field opposite to the original field and the latter at an angle, thus obtaining a difference angle between the two fields leads to the emergence of torque by the lag angle. The angle of lag between the two fields in this type of motor is very low, and this is the torque is low, and enough to run light loads.

7.1.2 Synchronous Motors

Synchronous motors are those which synchronize the speed of the magnetic field (N_s) in the stator with rotor speed (N_r), where there is no difference between them, that is, the two speeds are equal, and the slip is zero.

The power of synchronous motors ranges from 1 W to 10 MW or more and operates at any voltage (single or multi-phase) at speeds ranging from 125 to 3600 rpm. Synchronous motors are practical to operate at a speed of less than 500 rpm. When directly connected to the load, as in the case of pumps, mills, and foams, for example, with a power greater than 75 kW, the synchronous motor rated at this low speed is less expensive when compared with the induction motor of the same size. The speed of synchronous motors will not affect by any load. The synchronous motors can not start by itself, it needs an auxiliary external means to rotate them. As soon as their rotation speed reaches the synchronous speed, they remain at a constant speed. Figure 7.14 shows different types of synchronous motors.

The methods used to start the motor are:

1. Put a squirrel-cage coil in the rotor poles called *the starting coil*, so the motor starts as an induction motor until it reaches its synchronous speed. When the DC voltage is placed on the excitation coil, the motor is attracted to the synchronization state

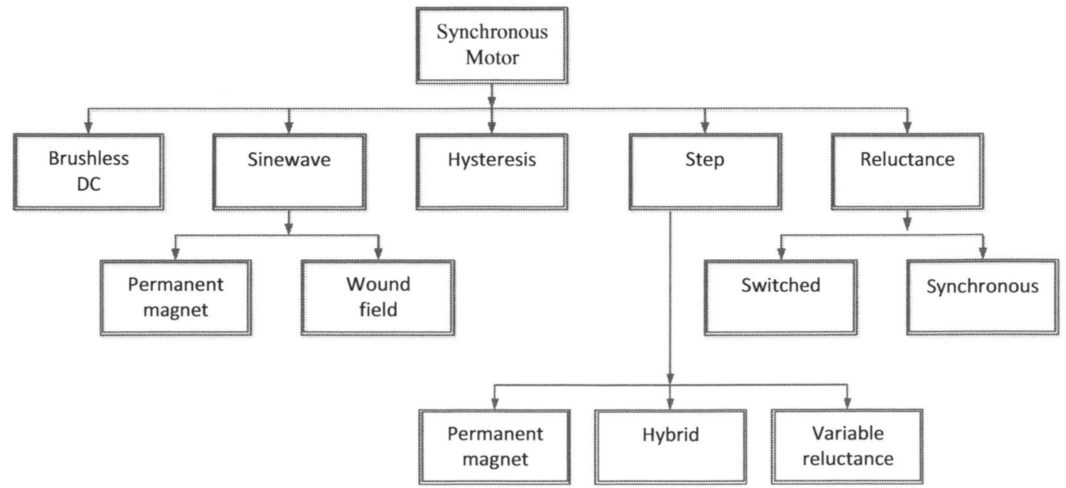

FIGURE 7.14
Types of synchronous motors.

2. Reduce the speed of rotation of the stator's field significantly by reducing the frequency of the voltage applied on the stator windings. Then start the rotor turn slowly to increase with increasing frequency.

7.1.2.1 Synchronous Motor Construction

1. Stator

   It is like a stator in induction motors. The iron core is composed of iron chips and contains internal slots where the main windings connect to an AC source.

2. Rotor

   The magnetic poles that are permanent magnetism, which consists of an alloy of different metals, the hysteresis loop or the power of de-magnetization are high and used in small machines, or coils fed by a direct current follow the slip-rings and carbon brushes.

   The shape of the rotor varies according to the type of poles used, the number and the size of the machine, and its speed. They may be salient poles. These are suitable for slow and medium speed. In high-speed machines, the rotor is a hidden pole, which appears in a cylindrical shape containing slots, the magnetism that feeds from a constant current source through slip-rings and carbon brushes.

7.1.2.2 Synchronous Motor Operation Theory

When the alternating current flows from the source to the stator coils, when positive half-wave follow the polarity of the coils will be positive. Thus, there will be repulsion between the coils and the magnetic poles for the same polarity (positive and positive). This repulsion causes the poles to deviate from their position at a certain angle. In a moment when the half-wave changes to the negative wave, the polarity of the coils will be negative, and thus the polarization will take place between them and the poles. The magnetic poles then return to their original position. Therefore, the starting torque of these motors is zero. For starting torque tracking procedures already mentioned.

The efficiency of these motors is high, employing a high-power factor. Some of these motors are used in wall clocks, timers, and sensitive measuring devices, used in printing machines and textile laboratories. Figure 7.15 shows the stator and the rotor and how the current follows.

7.1.2.3 Synchronous Motor Features

1. Higher power factor.
2. High efficiency and overload capacity.
3. The speed of rotation is fixed within the field of regular load.

7.1.3 Universal Motor

A universal motor (or series AC motor) is a motor that can be operated in direct current, and alternating current about the same speed. It is frequently used in horsepower motors that are used in household uses such as mixers, drill bits, and sewing machines.

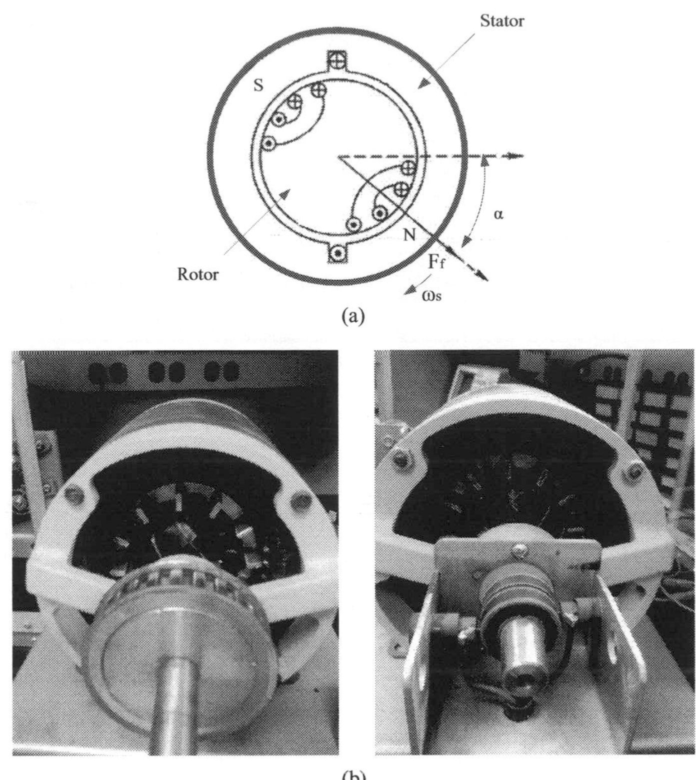

FIGURE 7.15
(a) The stator and the rotor and how the current follow. (b) Synchronous motor.

It has a high starting torque, which is variable speed, and its speed is high when it is not overloaded, and its speed decreases as the load increases. Figure 7.16 shows the shape of the motor and the internal structure of the motor.

7.1.3.1 Universal Motor Construction

1. Stator: It contains the salient magnetic poles placed around the coils of copper and reach the coils of magnetic poles, respectively, with the rotor by carbon brushes.
2. Rotor: Made of a shaft placed around it the chips of iron after the collection to form the iron core, which contains long slots from the outside, where the coils are placed of copper insulated, and reach the ends of the coils to the parts of the commutator according to the type of winding mentioned in Chapter six. Figure 7.17 shows the iron core of the stator and rotor.

7.1.4 Centrifuge Switches

One of the most common switches used in most motors is the one-phase centrifuge switch. Which consists of two main parts:

1. The first part contains two contact points and is fixed in one of the side covers or in the front of the stator.

(a) (https://en.wikipedia.org/wiki/Universal_motor).

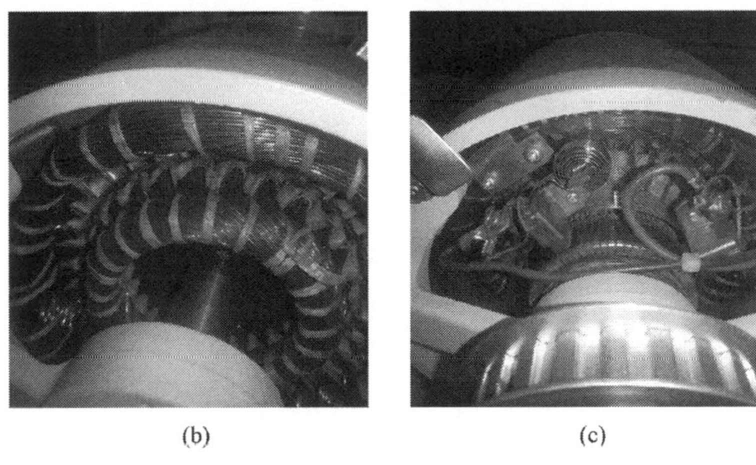

(b) (c)

FIGURE 7.16
The shape of the motor and the internal structure of the universal motor. (a) The stator and rotor (b) The windings (c) The centrifuge switch

FIGURE 7.17
The iron core of the stator and rotor.

FIGURE 7.18
The construction of the centrifuge switch. (beyondelectric.com/Html/productshow.asp).

2. The second part is a movement on the spindle and is influenced by the centrifugal forces resulting from rotor rotation. It contains a set of springs that make the two points of contact touching when the motor is stationary.

The switch works to open and close the two contact points in the help coil circuit.

When the rotation starts, the two points are closed, completing the current circuit in the auxiliary coils, and after the spinner reaches 75% of its actual speed, the two contact points are opened by the centrifugal force influencing the moving part of the switch. When the motor is stopped, the moving part returns to its position, closes the contact points, and completes the auxiliary coil circuit. Figure 7.18 shows the construction of the centrifuge switch.

As well as the main types of motors, which mentioned that there are special types of motors designed to meet the need of some applications, the most important of which.

7.2 Three-Phase Induction Motor

Induction motors are the most widely used motors for appliances, industrial control, and automation. Hence, they are often called the *workhorse* of the motion industry. They are robust, reliable, and durable. When power is supplied to an induction motor at the recommended specifications, it runs at its rated speed. However, many applications need variable speed operations. For example, a washing machine may use different speeds for each wash cycle. Historically, mechanical gear systems were used to obtain variable speed. Recently, electronic power and control systems have matured to allow these components to be used for motor control in place of mechanical gears. These electronics not only control the motor's speed, but can improve the motor's dynamic and steady-state characteristics. In addition, electronics can reduce the system's average power consumption and noise generation of the motor. Induction motor control is complex due to its nonlinear characteristics.

7.2.1 Construction of Induction Machines

The induction machine is a very important AC machine. It is mostly used as a motor. The stator and the rotor are made of laminated steel sheets with stamped in slots. The stator slots contain one symmetrical three-phase winding, which can be connected to the three-phase network in a star or delta connection. The stator of a simple induction machine has six slots per pole pair, in each case one for the forward and one for the backward conductor for each phase winding. Generally, the winding is carried out with many pole pairs ($p > 1$) and distributed in different slots ($q > 1$).

We can distinguish induction machines according to the type of the rotor between squirrel cage motors and slip ring motors. The stator construction is the same in both motors.

7.2.1.1 Squirrel Cage Motor

Almost 90% of induction motors are squirrel cage motors. This is because the squirrel cage motor has a simple and rugged construction. The rotor consists of a cylindrical laminated core with axially placed parallel slots for carrying the conductors. Each slot carries a copper, aluminum, or alloy bar. If the slots are semi-closed, then these bars are inserted from the ends.

These rotor bars are permanently short-circuited at both ends by means of the end rings, as shown in Figure 7.1. This total assembly resembles the look of a squirrel cage, which gives the motor its name. The rotor slots are not exactly parallel to the shaft. Instead, they are given a skew for two main reasons:

1. To make the motor run quietly by reducing the magnetic hum
2. To help reduce the locking tendency of the rotor

Rotor teeth tend to remain locked under the stator teeth due to the direct magnetic attraction between the two. This happens if the number of stator teeth is equal to the number of rotor teeth.

7.2.1.2 Slip Ring Motors

The windings on the rotor are terminated to three insulated slip rings mounted on the shaft with brushes resting on them. This allows an introduction of an external resistor to the rotor winding. The external resistor can be used to boost the starting torque of the motor and change the speed-torque characteristic. When running under normal conditions, the slip rings are short-circuited, using an external metal collar, which is pushed along the shaft to connect the rings. So, in normal conditions, the slip ring motor functions like a squirrel cage motor (Figure 7.19 through 7.22).

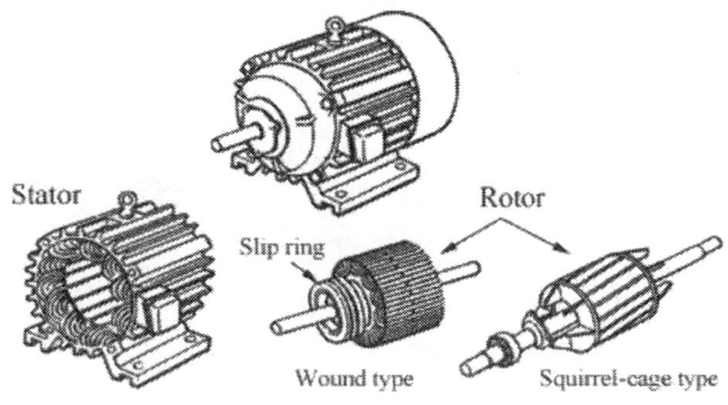

FIGURE 7.19
Rotor structure of induction machines. (https://www.ato.com/three-phase-induction-motor-construction).

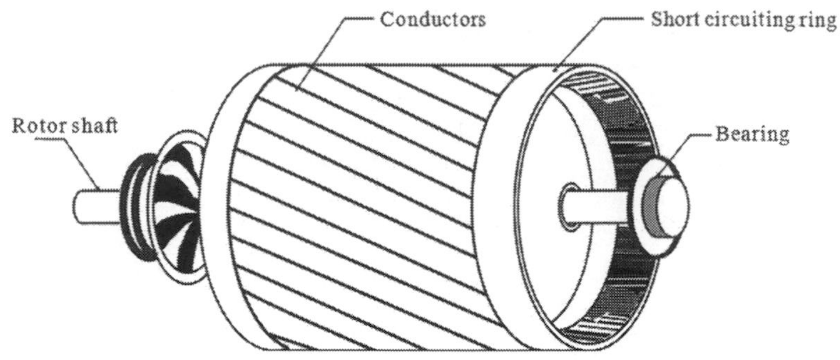

FIGURE 7.20
Typical squirrel cage rotor. (https://www.theengineeringprojects.com).

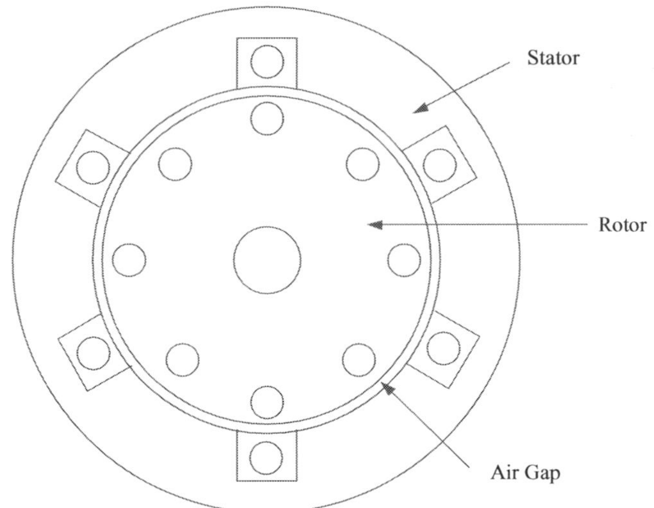

FIGURE 7.21
Schematic construction principle of an induction machine.

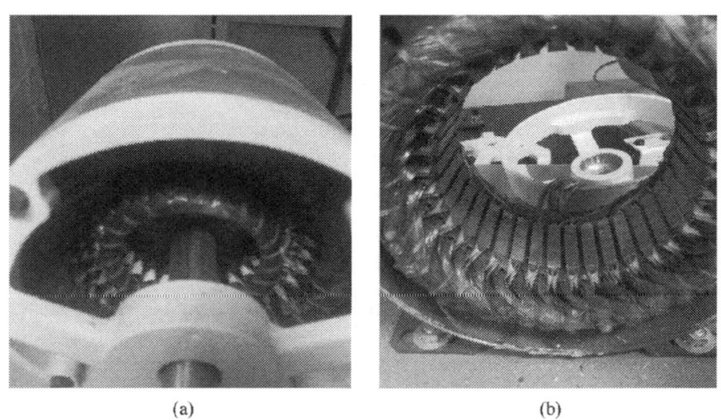

FIGURE 7.22
Three-phase induction motor. (a) Wound rotor. (b) Squirrel cage.

7.2.2 Operation

When the rated AC supply is applied to the stator windings, it generates a magnetic flux rotating at synchronous speed with a constant magnitude. The flux passes through the air gap, sweeps past the rotor surface, and through the stationary rotor conductors. An EMF is induced in the rotor conductors due to the relative speed differences between the rotating flux and stationary conductors. The voltage induced in the bars will be slightly out of phase with the voltage in the next one. Since the rotor bars are shorted at the ends, the EMF induced produces a current in the rotor conductors. The flux linkages will change in it after a short delay. If the rotor is moving at synchronous speed, together with the field, no voltage will be induced in the bars or the windings.

The frequency of the induced EMF is the same as the supply frequency. Its magnitude is proportional to the relative speed between the flux and the conductors. The direction of the rotor current opposes the relative speed between rotating flux produced by the stator and stationary rotor conductors according to Lenz's law.

To reduce the relative speed, the rotor starts rotating in the same direction as that of flux and tries to catch up with the rotating flux. But in practice, the rotor never succeeds in "catching up" to the stator field. So, the rotor runs slower than the speed of the stator field. This difference in speed is called *slip speed*. This slip speed depends upon the mechanical load on the motor shaft (Figure 7.23).

The frequency and speed of the motor, with respect to the input supply, is called *the synchronous frequency* and *synchronous speed*. Synchronous speed is directly proportional to the ratio of supply frequency and a number of poles in the motor. The rotor speed n_r of an induction motor is given by:

$$n_r = \frac{120f}{P}(1-S) \tag{7.2}$$

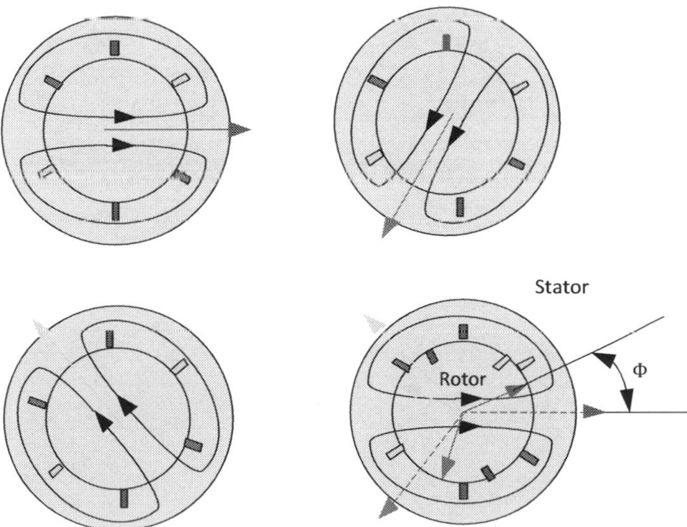

FIGURE 7.23
Concept of operation of three-phase induction motor.

7.2.3 Speed-Torque Characteristics of Induction Motor

Figure 7.24 shows the typical speed-torque characteristics of an induction motor. The *X*-axis shows speed and slip. The *Y*-axis shows the torque and current. The characteristics are drawn with rated voltage and frequency supplied to the stator.

During start-up, the motor typically draws up to seven times the rated current. This high current is a result of stator and rotor flux, the losses in the stator and rotor windings, and losses in the bearings due to friction. This high starting current overcomes these components and produces the momentum to rotate the rotor.

At start-up, the motor delivers 1.5 times the rated torque of the motor. This starting torque is also called *locked rotor torque* (LRT). As the speed increases, the current drawn by the motor reduces slightly (see Figure 7.24). The current drops significantly when the motor speed approaches to 80% of the rated speed. At base speed, the motor draws the rated current and delivers the rated torque.

At base speed, if the load on the motor shaft is increased beyond its rated torque, the speed starts dropping and slip increases. When the motor is running at approximately 80% of the synchronous speed, the load can increase up to 2.5 times the rated torque. This torque is called *breakdown torque*. If the load on the motor is increased further, it will not be able to take any further load and the motor will stall.

In addition, when the load is increased beyond the rated load, the load current increases following the current characteristic path. Due to this higher current flow in the windings, inherent losses in the windings increase as well. This leads to a higher temperature in the motor windings. Motor windings can withstand different temperatures, based on the class of insulation used in the windings and cooling system used in the motor. Some motor manufacturers provide the data on overload capacity and load over a duty cycle. If the motor is overloaded for longer than recommended, then the motor may burn out.

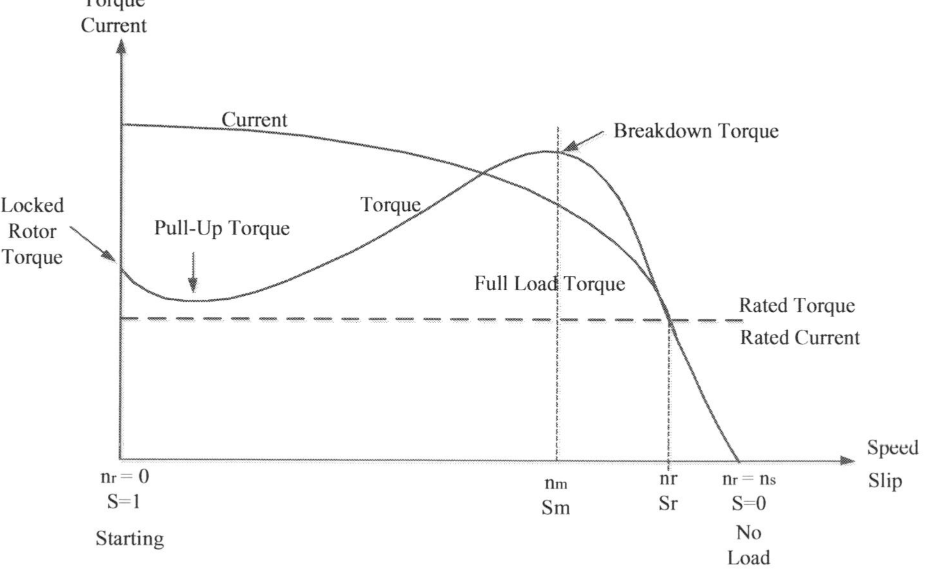

FIGURE 7.24
Speed-torque characteristics of induction motors.

As seen in the speed-torque characteristics, torque is highly nonlinear as the speed varies. In many applications, the speed needs to be varied, which makes the torque vary. We will discuss a simple open loop method of speed control called, variable voltage variable frequency (VVVF or V/*f*) in this application note.

7.2.4 Speed-Torque Characteristics of Induction Motors Using MATLAB Program

%Three phase induction motor

```
r1 = 0.200;
x1 = 0.250;
r2 = 0.10;
x2 = 0.250;
xm = 25.0;
v_phase = 208 / sqrt(3);
nsync = 1800;
wsync = 2*pi*nsync/60;
vth = v_phase * (xm / sqrt(r1^2 + (x1 + xm)^2));
zth = ((j*xm) * (r1 + j*x1)) / (r1 + j*(x1 + xm));
rth = real(zth);
xth = imag(zth);
s = (0:1:50) / 50;
s(1) = 0.001;
nm = (1 - s) * nsync;
for i = 1:51
td(i) = (3 * vth^2 * r2 / s(i)) / (wsync * ((rth + r2/s(i))^2 +
(xth + x2)^2));
end
figure(1);
plot(nm, td,'k-','LineWidth',2.0);
xlabel('n (r/m)');
ylabel('Td (N.m)');
title ('Induction Motor Torque-Speed Characteristics');
grid on;
```

7.2.5 Basic Equations and Equivalent Circuit Diagram

The stator and rotor of the induction machine both are equipped with a symmetrical three-phase winding. Because of the symmetry, it is sufficient to take only one phase into consideration. Every phase of the stator and the rotor winding has an active resistance of R_1 and R_2, as well as a self-inductance of L_1 and L_2. The windings of the stator and the rotor are magnetically coupled through a mutual inductance M. Since the current flowing in the stator winding has the frequency f_1 and the current flowing in the rotor winding has the frequency f_2, then at the rotor speed n,

- Currents induced from the stator into the rotor have $f = f_2$
- Currents induced from the rotor into the stator have $f = f_1$.

According to this, voltage equations for the primary and secondary sides can be derived. The equivalent circuit diagram after the conversion of the rotor parameters on the stator side is presented in Figure 7.25.

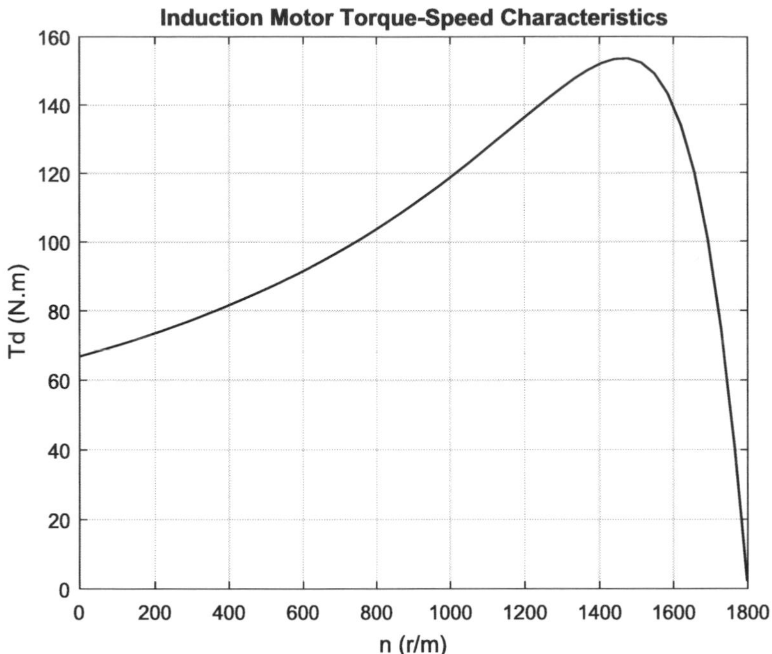

FIGURE 7.25
Speed-torque characteristics of induction motors using the MATLAB program.

To simplify the determination of torque and power equations from the induction machine equivalent circuit, the total real power per phase that crosses the air gap (the air gap power = $P_{air\ gap}$) and is delivered to the rotor is:

$$P_{airgap} = I_2'^2 \left[R'_{w2} + \frac{R'_{w2}}{S}(1-S) \right] = I_2'^2 \frac{R'_{w2}}{S} \tag{7.3}$$

The portion of the air gap power that is dissipated in the form of ohmic loss (copper loss $P_{r.cu}$) in the rotor conductors is:

$$P_{r.cu} = I_2'^2 R'_{w2} \tag{7.4}$$

The total mechanical power (P_{mech}) developed internally to the motor is equal to the air gap power minus the ohmic losses in the rotor which gives:

$$P_{mech} = P_{airgap} - P_{r.cu} = I_2'^2 \frac{R'_{w2}}{S} - I_2'^2 R'_{w2} = P_{airgap}(1-S) \tag{7.5}$$

$$P_{mech} = P_{airgap}(1-S)$$

$$P_{r.cu} = P_{airgap} S \tag{7.6}$$

According to the previous equations, of the total power crossing the air gap, the portion S goes to ohmic losses while the portion (1–S) goes to mechanical power. Thus, the induction machine is an efficient machine when operating at a low value of slip. Conversely, the induction machine is a very inefficient machine when operating at a high value of slip. The overall mechanical power is equal to the power delivered to the shaft of the machine plus losses (windage, friction).

The mechanical power (P_{mech}) is equal to torque (N-m) time's angular velocity (rad/s). Thus, we may write:

$$P_{mech} = T.\omega = I_2'^2 \frac{R'_{w2}}{S}(1-S) \tag{7.7}$$

where T is the torque and ω is the angular velocity of the motor in radians per second given by:

$$\omega = \frac{2\pi.n}{60} = \frac{2\pi.n_s}{60}.(1-S) = \omega_s.(1-S) \tag{7.8}$$

where ω_s is the angular velocity at synchronous speed. Using the previous equation, we may write:

$$(1-S) = \frac{\omega}{\omega_s} \tag{7.9}$$

Inserting this result into the equation relating torque and power gives:

$$P_{mech} = T.\omega = I_2'^2 \frac{R'_{w2}.\omega}{S.\omega_s} \tag{7.10}$$

Solving this equation for the torque yields:

$$T = \frac{I_2'^2 R'_{w2}}{S.\omega_s} = \frac{P_{air\ gap}}{\omega_s} \tag{7.11}$$

Returning to the Thevenin transformed equivalent circuit, we find:

$$I_2'^2 R'_{w2} = \left[\frac{V_{th}}{\left(R_{th} + \frac{R'_{w2}}{S}\right) + j\left(X_{th} + X_2'\right)}\right]^2 R'_{w2}.$$

Note that the previous equation is a phasor while the term in the torque expression contains the magnitude of this phasor. The complex numbers in the numerator and denominator may be written in terms of magnitude and phase to extract the overall magnitude term desired.

$$I_2'^2 R'_{w2} = \left[\frac{V_{th}\angle\theta 1}{\left(R_{th} + \frac{R'_{w2}}{S}\right)^2 + j\left(X_{th} + X'_2\right)^2 \angle\theta 2} \right]^2 R'_{w2}$$

The magnitude of the previous expression is:

$$I_2'^2 R'_{w2} = \left[\frac{V_{th} R'_{w2}}{\left(R_{th} + \frac{R'_{w2}}{S}\right)^2 + j\left(X_{th} + X'_2\right)^2} \right]$$

Inserting this result into the torque per phase equation gives:

$$T = \frac{1}{\omega_s} \cdot \left[\frac{V_{th}^2 \left(R'_{w2} / S\right)}{\left(R_{th} + \frac{R'_{w2}}{S}\right)^2 + \left(X_{th} + X'_2\right)^2} \right] \tag{7.12}$$

This equation can be plotted as a function of slip for a particular induction machine yielding the general shape curve shown in Figure 7.24. At low values of slip, the denominator term of $R_{w2}N/s$ is dominated and the torque can be accurately approximated by:

$$T = \frac{1}{\omega_s} \cdot \left[\frac{V_{th}^2 S}{R'_{w2}} \right] \tag{7.13}$$

where the torque curve is approximately linear in the vicinity of $S = 0$. At large values of slip ($S = 1$ or larger), the overall reactance term in the denominator of the torque equation is much larger than the overall resistance term such that the torque can be approximated. The torque is therefore inversely proportional to the slip for large values of slip. Between $S = 0$ and $S = 1$, a maximum value of torque is obtained. The maximum value of torque with respect to slip can be obtained by differentiating the torque equation with respect to S and setting the derivative equal to zero. The resulting maximum torque (called *the breakdown torque*) is:

$$T_{\max} = \frac{1}{2\omega_s} \left[\frac{V_{th}^2}{R_{th} + \sqrt{R_{th}^2 + \left(X_{th} + X'_2\right)^2}} \right] \tag{7.14}$$

and the slip at this maximum torque is:

$$S_{\max.T} = \left[\frac{R'_{w2}}{\sqrt{R_{th}^2 + \left(X_{th} + X'_2\right)^2}} \right] \tag{7.15}$$

If the stator winding resistance R_{w1} is small, then the Thevenin resistance is also small, so that the maximum torque and slip at maximum torque equations are approximated by:

$$T_{\max} = \frac{1}{2\omega_s}\left[\frac{V_{th}^2}{X_{th} + X_{12}'}\right]$$

$$S_{\max.T} = \frac{R_{w2}^2}{X_{th} + X_{12}'} \tag{7.16}$$

The efficiency of an induction machine is defined in the same way like that for a transformer. The efficiency (η) is the ratio of the output power (P_{out}) to the input power (P_{in}).

$$\eta = \frac{P_{out}}{P_{in}} \times 100\% \tag{7.17}$$

The input power is found using the input voltage and current at the stator. The output power is the mechanical power delivered to the rotor minus the total rotational losses.

$$\begin{aligned} P_{in} &= 3V_\phi I_\phi \cos(\theta_v - \theta_i) \\ P_{out} &= P_{mech} - P_{rot} = (1 - S)P_{airgap} - P_{rot} \end{aligned} \tag{7.18}$$

The *internal efficiency* (η_{int}) of the induction machine is defined as the ratio of the output power to the air gap power which gives:

$$\eta_{\text{int}} = \frac{P_{out}}{P_{airgap}} = (1 - S) \tag{7.19}$$

The internal efficiency gives a measure of how much of the power delivered to the air gap is available for mechanical power.

7.2.6 No-Load Test and Blocked Rotor Test

The equivalent circuit parameters for an induction motor can be determined using specific tests on the motor, just as was done for the transformer.

7.2.6.1 No-Load Test

Balanced voltages are applied to the stator terminals at the rated frequency with the rotor uncoupled from any mechanical load. Current, voltage, and power are measured at the motor input. The losses in the no-load test are those due to core losses, winding losses, windage, and friction. In no-load test, the slip of the induction motor at no-load is very low. Thus, the value of the equivalent resistance:

$$\frac{R_{w2}'}{S}(1 - S).$$

In the rotor branch of the equivalent circuit is very high. The no-load rotor current is then negligible and the rotor branch of the equivalent circuit can be neglected. The approximate equivalent circuit for the no-load test becomes as in Figure 7.26.

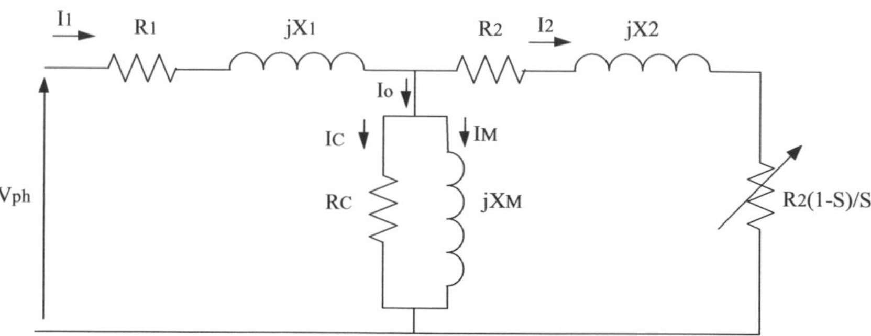

FIGURE 7.26
Equivalent circuit diagram of the induction machine.

Induction machine equivalent circuit for the no-load test note that the series resistance in the no-load test equivalent circuit is not simply the stator winding resistance. The no-load rotational losses (windage, friction, and core losses) will also be seen in the no-load measurement. This is why the additional measurement of the DC resistance of the stator windings is required. Given that the rotor current is negligible under no-load conditions, the rotor copper losses are also negligible. Thus, the input power measured in the no-load test is equal to the stator copper losses plus the rotational losses.

$$P_{NL} = P_{cu1} + P_{rot.} \tag{7.20}$$

where the stator copper losses are given by:

$$P_{cu1} = 3I_{NL}^2 R_{w1} \tag{7.21}$$

From the no-load measurement data (V_{NL}, I_{NL}, P_{NL}) and the no-load equivalent circuit, the value of R_{NL} is determined from the no-load dissipated power.

$$P_{NL} = 3I_{NL}^2 R_{NL} \rightarrow R_{NL} = \frac{P_{NL}}{3I_{NL}^2} \tag{7.22}$$

The ratio of the no-load voltage to current represents the no-load impedance which, from the no-load equivalent circuit, is:

$$\frac{V_{NL}}{I_{NL}} = Z_{NL} = \sqrt{R_{NL}^2 + \left(X_{l1} + X_{ml}\right)^2} \tag{7.23}$$

And the blocked rotor reactance sum $X_{l1} + X_{ml}$ is:

$$X_{l1} + X_{m1} = \sqrt{Z_{NL}^2 - R_{NL}^2} \tag{7.24}$$

Note that the values of X_{l1} and X_{m1} are not uniquely determined by the no-load test data alone (unlike the transformer no-load test). The value of the stator leakage reactance can be determined from the blocked rotor test. The value of the magnetizing reactance can then be determined.

7.2.6.2 Blocked Rotor Test

The rotor is blocked to prevent rotation and balanced voltages are applied to the stator terminals at a frequency of 25% of the rated frequency at a voltage where the rated current is achieved. Current, voltage, and power are measured at the motor input.

In addition to these tests, the DC resistance of the stator winding should be measured to determine the complete equivalent circuit.

The slip for the blocked rotor test is unity since the rotor is stationary. The resulting speed-dependent equivalent resistance:

$$\frac{R'_{w2}}{S}(1-S).$$

Goes to zero and the resistance of the rotor branch of the equivalent circuit becomes very small. Thus, the rotor current is much larger than the current in the excitation branch of the circuit such that the excitation branch can be neglected.

The resulting equivalent circuit for the blocked rotor test is shown in Figure 7.27.

The reflected rotor winding resistance is determined from the dissipated power in the blocked rotor test (Figure 7.28).

$$P_{BR} = 3I_{BR}^2(R_{w1} + R'_{w2}) \rightarrow R'_{w2} = \frac{P_{BR}}{3I_{BR}^2} - R_{w1} \tag{7.25}$$

The ratio of the blocked rotor voltage and current equals the blocked rotor impedance.

$$\frac{V_{BE}}{I_{BR}} = Z_{BR} = \sqrt{(R_{w1} + R'_{w2})^2 + (X_{l1} + X'_{l2})^2} \tag{7.26}$$

The reactance sum is:

$$X_{BR} = X_{l1} + X'_{l2} = \sqrt{Z_{BR}^2 - (R_{w1} + R'_{w2})^2} \tag{7.27}$$

Note that this reactance is that for which the blocked rotor test is performed. All reactances in the induction machine equivalent circuit are those at the stator frequency. Thus, all reactances computed based on the blocked rotor test frequency must be scaled according

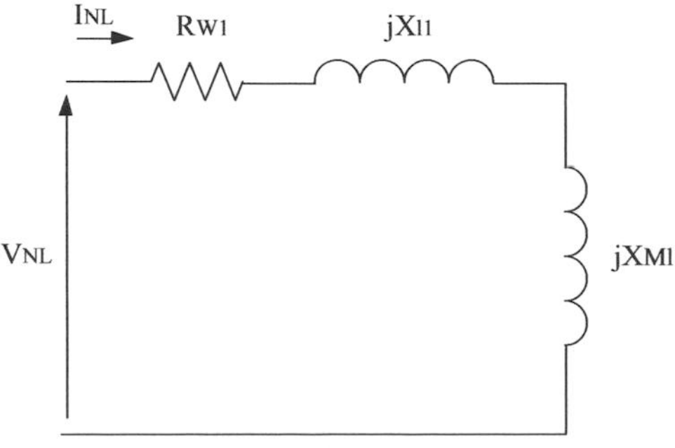

FIGURE 7.27
Equivalent circuit of induction motor at the no-load test.

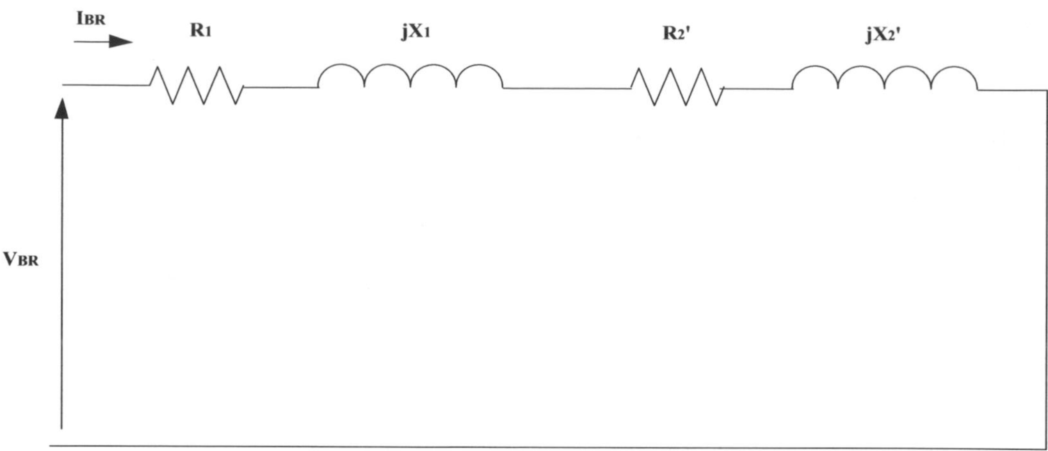

FIGURE 7.28
Equivalent circuit of induction motor at short-circuit test.

to relative frequencies (usually, a factor of 4 since T_{BR} is usually 25% of T_{NL}). The actual distribution of the total leakage reactance between the stator and the rotor is typically unknown, but empirical equations for different classes of motors (squirrel-cage motors) can be used to determine the values of X_{l1} and X_{l2}.

The following is a description of the four different classes of squirrel-cage motors.

Class A Squirrel-Cage Induction Motor: characterized by normal starting torque, high starting current, low operating slip, low rotor impedance, good operating characteristics at the expense of high starting current, common applications include fans, blowers, and pumps.

Class B Squirrel-Cage Induction Motor: characterized by normal starting torque, low starting current, low operating slip, higher rotor impedance than Class A, good general purpose motor with common applications being the same as Class A.

Class C Squirrel-Cage Induction Motor: characterized by high starting torque, low starting current, higher operating slip than Classes A and B, common applications include compressors and conveyors.

Class D Squirrel-Cage Induction Motor: characterized by high starting torque, high starting current, high operating slip, inefficient operation efficiency for continuous loads, common applications are characterized by an intermittent load such as a punch press.

Motor Reactance Distribution

Squirrel-cage Class A $X_{l1} = 0.5X_{BR}, X_{l2} = 0.5X_{BR}$

Squirrel-cage Class B $X_{l1} = 0.4X_{BR}, X_{l2} = 0.6X_{BR}$

Squirrel-cage Class C $X_{l1} = 0.3X_{BR}, X_{l2} = 0.7X_{BR}$

Squirrel-cage Class D $X_{l1} = 0.5X_{BR}, X_{l2} = 0.5X_{BR}$

And for Wound rotor $X_{l1} = 0.5\ X_{BR}, X_{l2} = 0.5X_{BR}$.

Using these empirical formulas, the values of X_{l1} and X_{l2} can be determined from the calculation of X_{BR} from the blocked rotor test data. Given the value of X_{l1}, the magnetization reactance can be determined according to:

$$X_{m1} = \sqrt{Z_{NL}^2 - R_{NL}^2 - X_{l1}} \quad (7.28)$$

Problems

7.1 What is the motor? What types of motors? What is the principle of motor work?

7.2 What are the induction motors? And why it is called by that name?

7.3 An induction motor with 1800 rpm, connect to an AC source frequency 60 Hz, calculate the number of connections.

7.4 An Induction motor with a speed of 600 rpm and the number of poles 10, what is the frequency of the source?

7.5 What are the induction motor parts? What is the nameplate?

7.6 What types of rotor windings, and how to be placed in the iron core?

7.7 Enumerate the starting methods of the single-phase induction motor.

7.8 Describe the properties of the start windings, and how to connect with the running windings.

7.9 What is the advantage use of start capacitor in single-phase motors?

7.10 Is there another capacitor? What specifications?

7.11 What are the advantages and disadvantages of the shaded pole motor? Where do we use them?

7.12 Explain the theory of operation of the shaded pole motor.

7.13 Why are synchronous motors called by this name? Where do we use them?

7.14 What are the means to assist synchronous motor rotation?

7.15 What are synchronous motor parts?

7.16 Explain the theory of synchronous motor operation. With its properties mentioned.

7.17 What is the universal motor? Where do we use them?

7.18 Explain the universal motor construction.

7.19 Where is the centrifuge switch used? Are there ceiling fans?

7.20 Where do you use stepper motors? What are its components?

7.21 What are the uses of control motors, and what are their components?

8

Power Electronics

Formal education will make you a living; self-education will make you a fortune.

Jim Rohn

In this chapter, we cover the most important applications in the power electronics circuits, the most important of which are the rectifiers, uncontrolled and controlled circuits. Also, the DC chopper circuit will explain.

8.1 Rectifiers (AC-DC Converters)

One of the first and most widely used applications of power electronic devices have been in rectification. Rectification refers to the process of converting an AC voltage or current source to DC voltage and current. Rectifiers specially refer to power electronic converters where the electrical power flows from the AC side to the DC side.

8.1.1 Rectifier Types

1. Single-phase rectifier
2. Three-phase rectifier.

Classification of single-phase rectifiers (Figure 8.1)

1. Half-wave rectifiers
    a. Uncontrolled rectifier
    b. Controlled rectifier
2. Full-wave rectifiers
    a. Uncontrolled rectifier
    b. Controlled rectifier.

In the rectifier analysis, following simplifying assumptions will be made.

- The internal impedance of the AC source is zero
- Power electronic devices used in the rectifier are ideal switches.

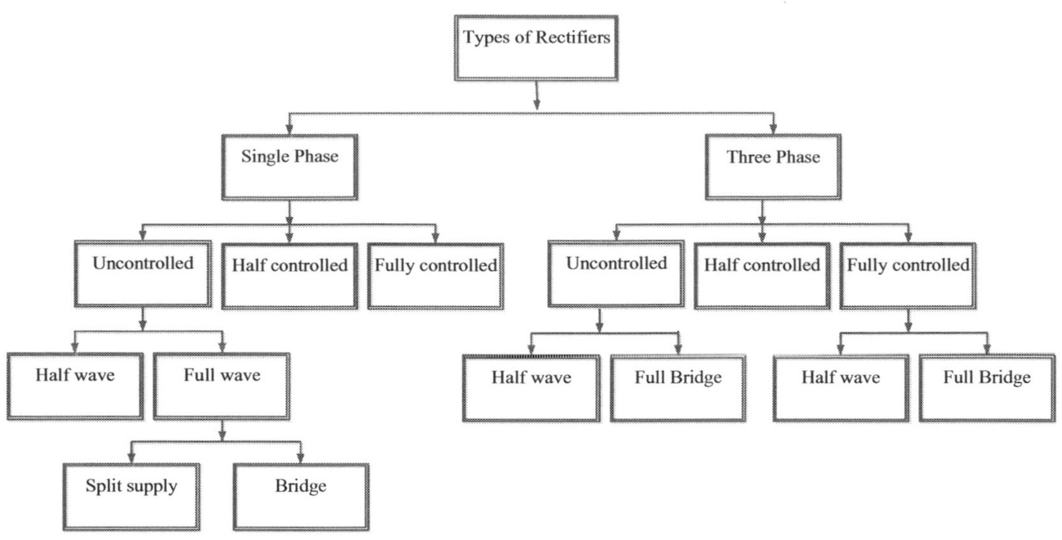

FIGURE 8.1
Classification of rectifiers.

Uncontrolled Rectifiers: Provide a fixed DC the output voltage for a given AC supply where diodes are used only.

Controlled Rectifiers: Provide an adjustable DC output voltage by controlling the phase at which the devices are turned on, where thyristors and diodes are used.

The full-wave rectifiers can divide into:

1. *Half-controlled*
   Allows electrical power flow from AC to DC (i.e., rectification only)
2. *Fully controlled*
   Allow power flow in both directions (i.e., rectification and inversion).

8.1.2 Performance Parameters

The performance of a rectifier can be evaluated in terms of the following parameters:

1. Output DC power

   Output DC power = average output voltage × average output current:

$$P_{dc} = V_{dc} I_{dc} \tag{8.1}$$

2. Total power

   Total power = rms output voltage × rms output current:

$$P_{total} = V_{rms} I_{rms} \tag{8.2}$$

3. The efficiency of a rectifier, *Rectification efficiency* $(\mathcal{J}) = \frac{P_{dc}}{P_{ac}} = \frac{I_{av}.V_{av}}{I_{rms}.V_{rms}}$.

   Let f be the instantaneous value of any voltage or current associated with a rectifier circuit, then the following terms, characterizing the properties of f, can be defined.

   Average (DC) value of f (F_{av}): Assuming f to be periodic over the time period T:

$$F_{av} \text{ or } F_{mean} = \frac{1}{T}\int_0^T f(t)dt \tag{8.3}$$

4. Output AC component

   The output voltage can be considered to have two components: including (1) DC value and (2) the AC components or ripple. The RMS value of the AC component of the output voltage is:

$$V_{ac} = \sqrt{V_{rms}^2 - V_{dc}^2} \tag{8.4}$$

5. Form factor, FF

   RMS (effective) value of f (F_{rms}): For periodic current over the time period T:

$$F_{rms} = \sqrt{\frac{1}{T}\int_0^T f^2(t)dt} \tag{8.5}$$

   The form factor of f (FF): It is a measure of the shape of the output voltage. It is defined as:

$$FF = \frac{F_{rms}}{F_{av}} \tag{8.6}$$

6. Ripple factor (RF)

   It is a measure of the ripple content:

$$RF = \frac{V_{ac}}{V_{dc}} \tag{8.7}$$

   or

$$RF = \sqrt{\left(\frac{V_{rms}}{V_{dc}}\right)^2 - 1} = \sqrt{FF^2 - 1} \tag{8.8}$$

7. Transformer utilization factor (TUF)

   A transformer is most often used both to introduce galvanic isolation between the rectifier input and the AC mains and to adjust the rectifier AC input voltage to a level suitable for the required application. One of the parameters used to define the characteristics of the transformer is the TUF:

$$TUF = \frac{P_{dc}}{V_S\, I_S} \tag{8.9}$$

where:

V_S = RMS voltage of the transformer secondary, and

I_S = RMS current of the transformer secondary.

8. Displacement factor (*DF*):

$$DF = \text{Cos } \Phi \tag{8.10}$$

where Φ is the phase angle between the fundamental of the input current and voltage.

9. Harmonic factor (HF)

$$HF = \sqrt{\frac{I_S^2 - I_1^2}{I_1^2}} \tag{8.11}$$

where I_1 is the fundamental RMS component of the input current.

10. Power factor (PF)

$$PF = \frac{V_S\, I_1 \text{ Cos } \Phi}{V_S\, I_S} = \frac{I_1}{I_S} \text{Cos } \Phi \tag{8.12}$$

$$= \text{Distortion Factor. Displacement Factor}$$

8.1.3 Uncontrolled Rectifiers

The rectifiers converted AC to DC using power diodes and called uncontrolled rectifiers because they give constant out-voltage and fixed value if the value of the input voltage (AC voltage) is constant. The diode is an appropriate component of uncontrolled rectifiers circuits because of one-way conducted properties and classifications of circuits are based on:

1. A number of phases: One phase and three phases
2. Waveform shape: Half-wave or full-wave.

In this chapter, we will look at single-phase circuits, if the diode has ideal properties, i.e., its resistance to the current is zero if the diode is forward biased, and its resistance is very high if it is reverse biased.

8.1.3.1 Half-Wave Uncontrolled Rectifiers

8.1.3.1.1 Single-Phase Half-Wave Rectifier with Resistive Load

This circuit is one of the simplest circuits of the rectifier. During the positive half of the input voltage wave, the rectifier is in the forward bias state and allows the current to pass through to the load resistance. In the case of an ideal diode used, the value of the voltage drops on both ends of the rectifier is zero. Consequently, the voltage on both ends of the load resistance is the same as the positive half the input voltage as shown in Figure 8.2.

There are two advantages of connecting using the transformer. First, it allows raising and reducing the source voltage as needed. Second, it achieves electrical insulation between the source of the alternating current and rectifier, to prevent sudden electric shocks in the secondary coil circuit.

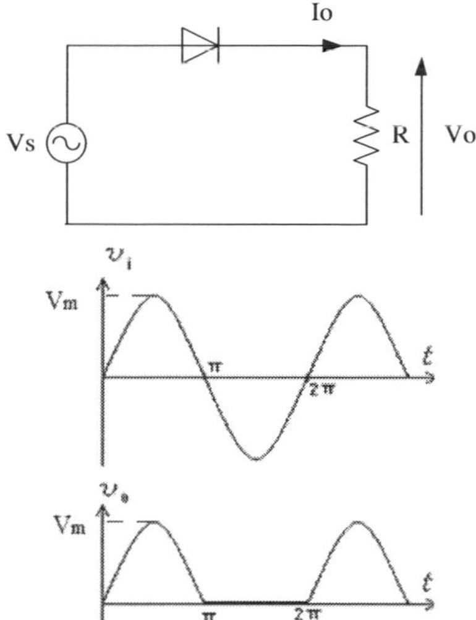

FIGURE 8.2
Half-wave uncontrolled rectifier.

The continuous output voltage can be calculated from the following formula:

$$V_{out} = V_{dc} = \frac{Vm}{\pi} = \frac{\sqrt{2}}{\pi} \times V_{rms} \tag{8.13}$$

The output current is calculated from the following formula:

$$I_{out} = I_{dc} = \frac{V_{dc}}{R} = \frac{V_m}{\pi R} \tag{8.14}$$

where:
V_m = Represents the maximum value of the source voltage, and
V_{rms} = Effective source voltage.

To calculate the average value of the output voltage of the one-half-wave is the value measured by the continuous voltmeter and can be calculated from:

$$V_{av} = V_m/\pi \tag{8.15}$$

where V_{av} is the average value of the output voltage.

Example 8.1

A single-phase single wave has a purely resistive load of 8 Ω and the source voltage of 100 V at 60 Hz. Calculate the output voltage and current.

Solution

$$V_o = V_{dc} = V_m/\pi$$
$$= (\sqrt{2} \times 100)/\pi = 45\ V$$
$$I_o = I_{dc} = V_{dc}/R = 45/8$$
$$= 5.63\ A.$$

Example 8.2

Find the V_{av} of the combined voltage half-wave if the maximum value of the source voltage is 50 V.

Solution

$$V_{av} = V_m/\pi$$
$$= 50/3.14$$
$$= 15.9\ V.$$

Formula derivative:

$$V_{av} \text{ or } V_{dc} \text{ or } V_{lmean} = \frac{1}{T}\int_0^T f(t)dt$$
$$= \frac{1}{2\pi}\int_0^{\pi} V_{s\ max} sin\theta d\theta$$
$$= \frac{V_{s\ max}}{2\pi}\int_0^{\pi} sin\theta d\theta$$
$$= -\frac{V_{s\ max}}{2\pi}[cos\pi - cos0]$$

$$v_{s\ rms} = \frac{V_{s\ max}}{\sqrt{2}}$$

$$\therefore V_{lmean} = \frac{V_{s\ max}}{\pi}$$

$$\therefore I_{lmean} = \frac{V_{lmean}}{R} \text{ or } I_{lmean} = \frac{1}{2\pi}\int_0^{\pi} I_{s\ max} \sin\theta d\theta$$

$$v_{o\ rms} = \sqrt{\frac{1}{T}\int_0^{\pi} V^2(t)dt}$$

$$v_{0\,rms} = \sqrt{\frac{1}{2\pi}\int_0^{\pi} V_{smax}{}^2 sin^2\theta d\theta}$$

$$= V_{smax}\sqrt{\frac{1}{2\pi}\int_0^{\pi}\frac{1}{2}(1-cos2\theta)d\theta}$$

$$= V_{smax}\sqrt{\frac{1}{4\pi}\int_0^{\pi}(1-cos2\theta)d\theta}$$

$$= V_{smax}\sqrt{\frac{1}{4\pi}\int_0^{\pi}1d\theta - \frac{1}{2}\int_0^{\pi}cos2\theta d\theta}$$

$$= V_{smax}\sqrt{\frac{1}{4\pi}\left[\theta - \frac{sin2\theta}{2}\right]_0^{\pi}}$$

$$= V_{smax}\sqrt{\frac{1}{4\pi}\left[\pi - \frac{sin2\pi}{2} - 0 + \frac{sin2\times 0}{2}\right]}$$

$$= V_{smax}\sqrt{\frac{1}{4\pi}[\pi]}$$

$$\therefore i_{0\,rms} = \frac{v_{o\,rms}}{R} \text{ or } i_{0\,rms} = \sqrt{\frac{1}{2\pi}\int_0^{\pi} I_{smax}{}^2 sin^2\theta d\theta}$$

$$P_{dc} = \frac{\left(\frac{Vm}{\pi}\right)^2}{R} - \frac{V_m{}^2}{\pi^2 R}$$

$$P_{total} = \frac{V_{rms}{}^2}{R} = \frac{(V_m / 2)^2}{R} = \frac{1}{4}\frac{V_m{}^2}{R}.$$

1. $\frac{P_{dc}}{P_{total}} = \frac{1}{\pi^2 R} 4R = 0.405\ (40.5\%)$.

2. $FF = \frac{V_{rms}}{V_{dc}} = \frac{\frac{Vm}{2}}{Vm/\pi} = \frac{\pi}{2} = 1.57\ (157\%)$.

3. $RF = \sqrt{FF^2 - 1}$

$= \sqrt{1.57^2 - 1}$

$= 1.21\ (121\%)$.

$$4.\ TUF = \frac{P_{dc}}{V_S.I_S} = \frac{\frac{V_m^{\ 2}}{\pi^2 R}}{\left(\frac{V_m}{\sqrt{2}}\right)*\left(\frac{V_m}{2R}\right)} = \frac{2\sqrt{2}}{\pi^2} = 0.287\#.$$

Example 8.3

For the single-phase half-wave uncontrolled rectifier circuit shown in Figure 8.3, $R = 5\Omega$: $V_s = 150\ sinwt$. Calculate V_{Lmean}, I_{Lmean}, $v_{s_{r.m.s}}$, $v_{o_{r.m.s}}$, $i_{o_{r.m.s}}$ and RF.

Solution

$$V_{lmean} = \frac{V_{s\ max}}{\pi} = \frac{150}{3.14} = 47.7\ V$$

$$I_{lmean} = \frac{V_{lmean}}{R} = \frac{47.7}{5} = 9.5A$$

$$v_{s\ rms} = \frac{V_{s\ max}}{\sqrt{2}} = \frac{150}{\sqrt{2}} = 106.06V$$

$$v_{o\ rms} = \frac{V_{s max}}{2} = \frac{150}{2} = 75\,V$$

$$i_{o\ rms} = \frac{v_{o\ rms}}{R} = \frac{75}{5} = 15\ A$$

$$RF = \sqrt{\frac{V_{o\ rms}^{\ \ 2} - V_{lmean}^{\ \ 2}}{V_{lmean}^{\ \ 2}}} = \sqrt{\frac{(75)^2 - (47.7)^2}{(47.7)^2}} = 1.213.$$

8.1.3.1.2 Single-Phase Half-Wave Rectifier with Inductive Load

Due to the inductive load, the conduction period of the diode will extend beyond 180° until the current becomes zero. Consider the case of the resistance-inductance (R-L) load as shown in Figure 8.4. The voltage source, is a sine wave, the positive half period when $0 < \omega t < \pi$, and the negative half period when $\pi < \omega t < 2\pi$. When the voltage source starts becoming positive, the diode starts conducting and the source keeps the diode in conduction till ωt reaches π radians. At that instant defined by $\omega t = \pi$ radians, the current through the circuit is not zero and there is some energy stored in the inductor.

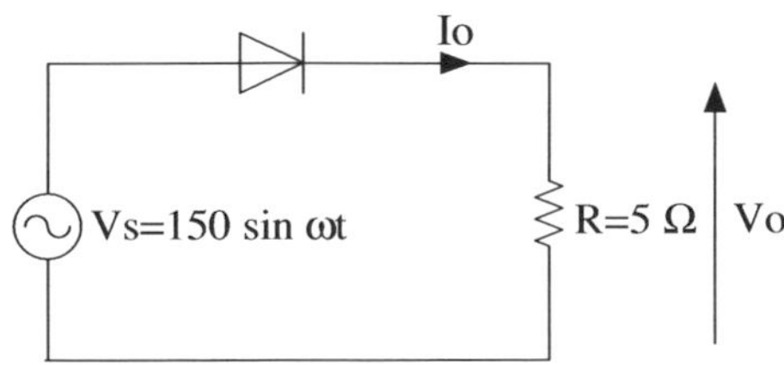

FIGURE 8.3
Single-phase half-wave uncontrolled rectifier circuit diagram.

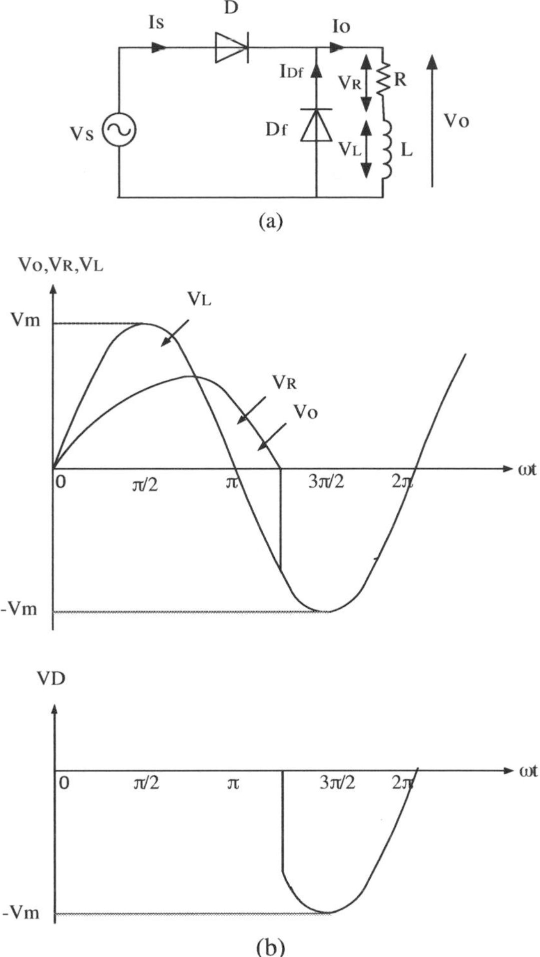

FIGURE 8.4
Single-phase half-wave rectifier with an inductive load. (a) Circuit diagram. (b) Waveforms.

The voltage across an inductor is positive when the current through it is increasing, and it becomes negative when the current through it tends to fall. When the voltage across the inductor is negative, it is in such a direction as to forward-bias the diode. The polarity of the voltage across the inductor is as shown in the waveforms shown in Figure 8.3. When source voltage changes from a positive to a negative value, the voltage across the diode changes its direction and there is current through the load at the instant $\omega t = \pi$ radians and the diode continues to conduct till the energy stored in the inductor becomes zero. After that, the current tends to flow in the reverse direction and the diode blocks conduction. The entire applied voltage now appears across the diode as a reverse bias voltage. An expression for the current through the diode can be obtained by solving the differential equation representing the circuit.

During diode conduction:

$$L\frac{di}{dt} + i\,R = V_m \sin \omega t.$$

A solution of this differential equation is:

$$i = \frac{V_m}{Z}\left[Sin(\omega t - \Phi) + Sin\,\Phi\, e^{\frac{-\omega t}{\tan\Phi}}\right]; \qquad 0 \le \omega t \le \beta$$

$$Z = \sqrt{R^2 + \omega^2 L^2}\,;\ \tan\Phi = \frac{\omega L}{R} \tag{8.16}$$

$At\ \omega t = \beta, i = 0$

$$\therefore 0 = Sin(\beta - \Phi) + \sin\Phi\, e^{\frac{-\beta}{\tan\Phi}}.$$

This is a transcendental equation and can be solved by iterative techniques. The extinction angle can be determined for a given load impedance angle Φ.

The average output voltage is:

$$V_{dc} = \frac{V_m}{2\pi}\int_0^{\beta} \sin\omega t\, d\omega t$$

$$= \frac{V_m}{2\pi}(1 - \cos\beta) \tag{8.17}$$

The average output current is:

$$I_{dc} = \frac{V_m}{2\pi R}(1 - \cos\beta) \tag{8.18}$$

Example 8.4

For the single-phase half-wave uncontrolled rectifier circuit shown in Figure 8.5, $R = 10\,\Omega, L = 9mH, V_s = 150\sin wt$. Calculate and trace V_{Lmean}, I_{Lmean}, $v_{S_{r.m.s}}$ and FF.

Solution

$$\varphi = \tan^{-1}\frac{\omega L}{R} = \tan^{-1}\frac{2\times 3.14\times 50\times 9\times 10^{-3}}{10} = 15.7^0$$

$$V_{lmean} = \frac{1}{2\pi}\int_0^{\pi+\varphi} V_{s\ max}\sin\theta d\theta$$

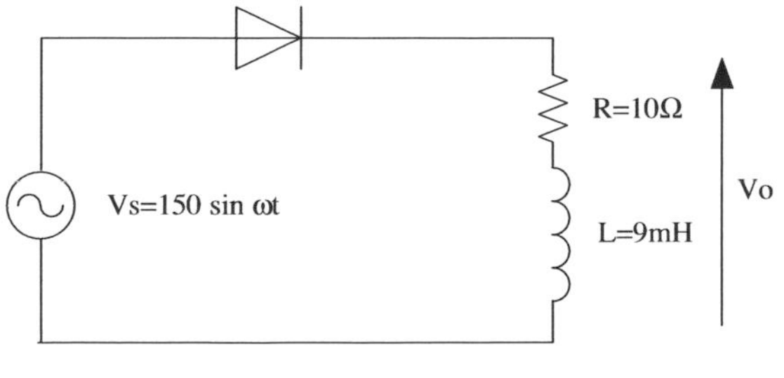

FIGURE 8.5
Circuit of Example 8.4.

$$V_{lmean} = \frac{V_{smax}}{2\pi}[1+\cos\varphi]$$

$$= \frac{150}{2\pi}[1+\cos 15.7\ ] = 46.8\ \text{V}$$

$$\text{so } I_{lmean} = \frac{V_{lmean}}{R} = \frac{46.8}{10} = 4.68\ \text{A}$$

$$V_{s\ rms} = \frac{V_{s\ max}}{\sqrt{2}} = \frac{150}{\sqrt{2}} = 106.06\ \text{V}$$

$$v_{0\,rms} = V_{smax}\sqrt{\frac{1}{4\pi}\left[\pi+\varphi - \frac{sin2(\pi+\varphi)}{2}\right]}$$

$$= 150\sqrt{\frac{1}{4\times 3.14}\left[3.14 + 15.7^{0}\times\frac{\pi}{180} - \frac{sin2(180+15.7)}{2}\right]}$$

$$= 150\sqrt{0.08\left[3.14+0.27-\frac{0.52}{2}\right]} \approx 75.2\,V$$

$$Form\,factor(FF) = \frac{v_{o\,rms}}{V_{lmean}} = \frac{75.2}{46.8} = 1.6$$

8.1.3.2 Full-Wave Rectifiers

Although the rectifier half-wave has some applications, the use of a rectifier full-wave is more prevalent in the sources of power for DC, and the difference between full-wave and the half-wave rectifier is that the rectifier full-wave allows one-way current through the load in each half cycle. While a single half-wave allows the current to pass through the positive half of the wave only, and as a result, we get the positive pulse number from the output of the full-wave rectifier equal to twice the number of positive pulses we get from a single half-wave output over the same time period. Figure 8.6 shows a full-wave rectifier.

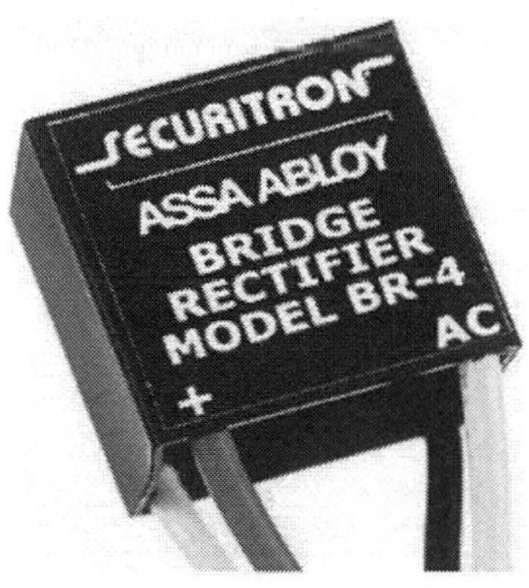

FIGURE 8.6
Full-wave rectifier (https://www.grainger.com/product).

8.1.3.2.1 Full-Wave Center Tapped Rectifier with Resistive Load

Figure 8.7 shows a complete wave rectifier circuit using an adapter with an intermediate junction point. In this circuit, when the sinusoidal wave is in the positive half, the rectifier D_1 is in the front bias position, so the rectifier (D_2) is in reverse bias state.

In the case of a sinusoidal wave in the negative half, the rectifier (D_1) is in reverse bias, and the D_2 is in the front bias state. As a result, it conducts the current and thus the half of the positive and negative wave appears on the output. Figures 8.8a and b show the stages in which the full-wave are obtained.

The average value of the voltage, in this case, shall be calculated from the following formula:

$$V_{av} = 2V_m/\pi \tag{8.19}$$

Note that the average value of the voltage (V_{av}) in this case (full-wave) is twice the value obtained in a half-wave state.

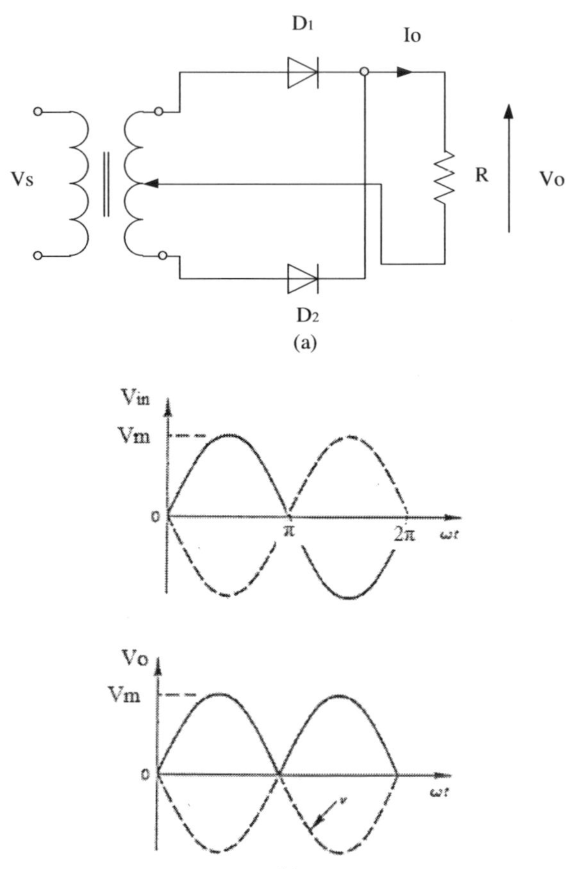

FIGURE 8.7
Full-wave center tapped rectifier. (a) Circuit diagram. (b) Waveforms.

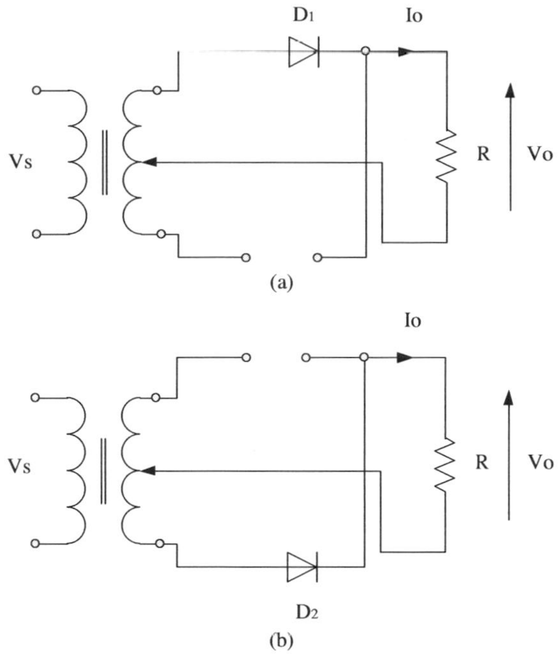

FIGURE 8.8
The stages in which the full-wave is obtained. (a) Diode 1 ON. (b) Diode 2 ON.

Example 8.5

Find the V_{av} of the load voltage a full-wave if the maximum value of the output voltage is 15 V, and create the load current if the load resistance is 3 Ω.

Solution

$$V_{av} = 2\ \text{Vm}/\pi$$

$$= 2 \times 15/3.14$$

$$= 9.55\ \text{v}$$

$$I_{dc} = V_{av}/R$$

$$= 9.55/3$$

$$= 3.18\ \text{A}.$$

Example 8.6

The rectifier shown in Figure 8.9 has a purely resistive load of R. Calculate:

i. The rectifier efficiency.
ii. Form factor.
iii. Ripple factor.
iv. Peak inverse voltage (PIV) for diode D_1.

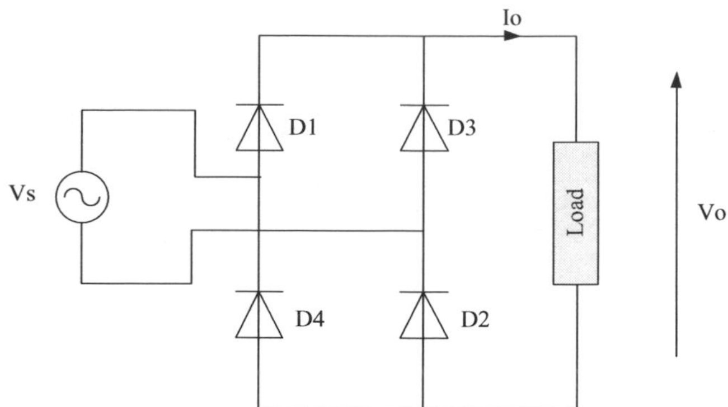

FIGURE 8.9
Single-phase full-wave (bridge) rectifier.

Solution

i. $\eta = \dfrac{V_{dc} \cdot I_{dc}}{V_{rms} \cdot I_{rms}}$

$$= \frac{\dfrac{2V_m}{\pi} \cdot \dfrac{2V_m}{\pi \,.\, R}}{\dfrac{V_m}{\sqrt{2}} \cdot \dfrac{V_m}{\sqrt{2} \,.\, R}} = 81.05\%.$$

ii. $F.F = \dfrac{V_{rms}}{V_{dc}}$

$$= \frac{\dfrac{V_m}{\sqrt{2}}}{\dfrac{2V_m}{\pi}} = 111\%.$$

iii. $R.F = \dfrac{V_{ac}}{V_{dc}} = \sqrt{F.F^2 - 1}$

$$= \sqrt{1.11^2 - 1} = 48\%$$

iv. $PIV = 2.Vm$

8.1.3.2.2 Single-Phase Full-Wave (Bridge) Rectifier

8.1.3.2.2.1 Single-Phase Full-Wave (Bridge) Rectifier with a Resistive Load This rectifier needs four diodes to form the circuit. as shown in Figure 8.9.

The diodes (D_1, D_2) are in a state of ON during the positive part of the input voltage and the diodes (D_3, D_4) are reverse biased. The current follow through the diode (D_1, D_2), and during the negative half of the input voltage wave through D_3 and D_4 were ON and Diodes (D_1, D_2) are in reverse bias. The current follow through the diode in (D_3, D_4) (Figure 8.10).

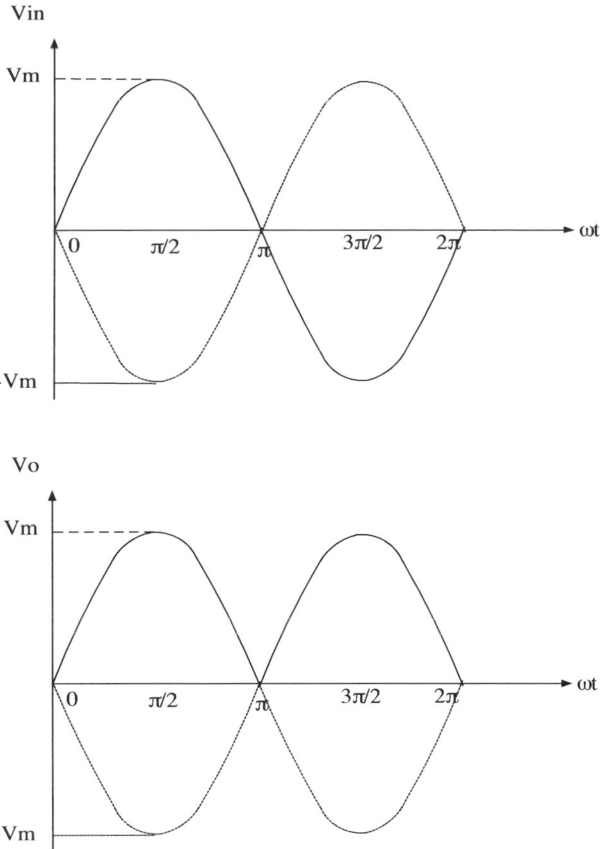

FIGURE 8.10
Input and output voltage of single-phase bridge rectifier.

The voltage and current calculation equations are same used in the full-wave rectifier with the center tap transformer, but should be noted that the diode in the rectifier of the bridge rectifier is subjected to a reverse voltage equal to half of the voltage exerted by the diode in the rectifier with the center tap transformer, which reduces rating of the diode. Figures 8.11a and b show the stages of the rectifier of a full-wave using four diodes.

Formula Derivative:

$$V_{lmean} = \frac{1}{2\pi}\int_0^{\pi} V_{s\,max} \sin\theta d\theta \times 2$$

$$= \frac{V_{s\,max}}{\pi}\int_0^{\pi} sin\theta d\theta$$

$$= \frac{V_{s\,max}}{\pi}\left[-cos\theta\right]_0^{\pi}$$

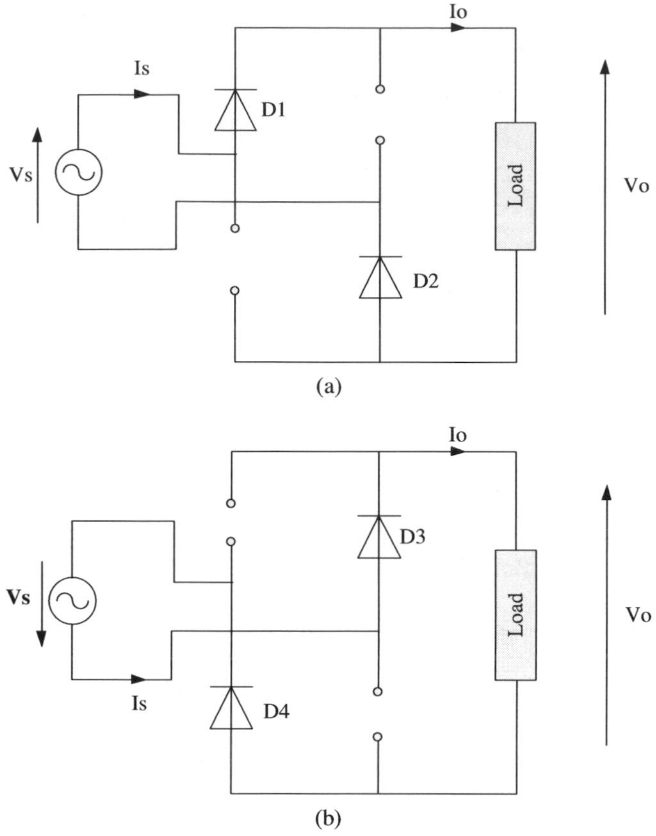

FIGURE 8.11
The stages of operation using a single-phase bridge rectifier. (a) D_1 and D_2 are ON. (b) D_3 and D_4 are ON.

$$= -\frac{V_{s\,max}}{\pi}[cos\pi - cos0]$$

$$\therefore V_{lmean} = \frac{2V_{s\,max}}{\pi}$$

$$\therefore I_{lmean} = \frac{V_{lmean}}{R}$$

$$\text{or } I_{lmean} = \frac{1}{2\pi}\int_0^{\pi} I_{s\,max}\, sin\theta d\theta \times 2$$

$$v_{s\,rms} = \frac{V_{s\,max}}{\sqrt{2}}$$

$$v_{0\,rms} = \sqrt{\frac{1}{2\pi}\int_0^{\pi} V_{smax}{}^2 sin^2\theta d\theta \times 2}$$

$$= V_{smax}\sqrt{\frac{1}{\pi}\int_0^{\pi}\frac{1}{2}(1-cos2\theta)d\theta}$$

$$= V_{smax}\sqrt{\frac{1}{2\pi}\int_0^{\pi}(1-cos2\theta)d\theta} = V_{smax}\sqrt{\frac{1}{2\pi}\int_0^{\pi}1d\theta-\frac{1}{2}\int_0^{\pi}cos2\theta d\theta}$$

$$= V_{smax}\sqrt{\frac{1}{2\pi}\left[\pi-\frac{sin2\pi}{2}-0+\frac{sin2\times 0}{2}\right]}$$

$$= V_{smax}\sqrt{\frac{1}{2\pi}[\pi]}$$

$$v_{o\,rms} = \frac{V_{smax}}{\sqrt{2}}$$

$$\therefore i_{o\,rms} = \frac{v_{o\,rms}}{R} \text{ or } i_{o\,rms} = \sqrt{\frac{1}{2\pi}\int_0^{\pi}I_{smax}{}^2\sin^2\theta d\theta\times 2}.$$

Example 8.7

For the full-wave uncontrolled rectifier (bridge) circuit shown in Figure 8.12 below, $R = 10\Omega$, Calculate V_{Lmean}, I_{Lmean}, $v_{S_{r.m.s}}.v_{o_{r.m.s}}$, and $i_{o_{r.m.s}}$.

Solution

$$V_{L_{mean}} = \frac{1}{2\pi}\int_0^{\pi}V_{max}\sin\theta d\theta\times 2 = \frac{2V_{max}}{\pi} = 127.3\,V$$

$$I_{L_{mean}} = \frac{V_{L_{mean}}}{R} = \frac{127.3}{10} = 12.73\,A$$

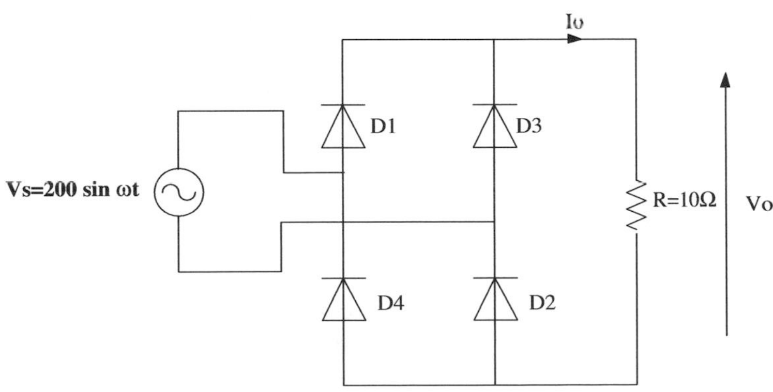

FIGURE 8.12
Circuit of Example 8.7.

$$v_{S_{rms}} = \frac{V_{smax}}{\sqrt{2}} = \frac{200}{\sqrt{2}} = 141.42V$$

$$v_{o\,rms} = \sqrt{\frac{1}{2\pi}\int_0^{\pi} V_{smax}{}^2 sin^2\theta d\theta \times 2} = \frac{V_{max}}{\sqrt{2}} = \frac{200}{\sqrt{2}} = 141.42V$$

$$i_{0\,rms} = i_{s\,rms} = \frac{v_{o\,rms}}{R} = \frac{141.42}{10} = 14.14A.$$

8.1.3.2.2.2 Single-Phase Full-Wave (Bridge) Rectifier with Inductive Load For an RL series-connected load in Figure 8.13, the method of analysis is like that for the half-wave rectifier with the freewheeling diode discussed previously. After a transient that occurs during start-up, the load current $i_{o(t)}$ reaches a periodic steady-state condition. For the bridge circuit, current is transferred from one pair of diodes to the other pair when the source changes polarity. The voltage across the R-L load is a full-wave rectified sinusoid, as it was for the resistive load. The full-wave rectified sinusoidal voltage across the load can be expressed as a Fourier series consisting of a DC term and the even harmonics.

$$V_{dc} = \frac{2\,V_m}{\pi};\quad V_{rms} = \frac{V_m}{\sqrt{2}}$$

$$RF = \sqrt{\left(\frac{V_{rms}}{V_{dc}}\right)^2 - 1} = \sqrt{\left(\frac{\pi}{2\sqrt{2}}\right) - 1} = 0.483.$$

This is significantly less than the half-wave rectifier. With a highly inductive load, which is the usual practical application, virtually constant load current flows. The bridge diode currents are then square blocks of current with magnitude I_{dc}, in this application:

$$\text{Average current in each diode} = I_{D,av} = \frac{I_{dc}}{2}.$$

$$\text{R.M.S. current in each diode} = I_{D,rms=} \frac{I_{dc}}{\sqrt{2}}.$$

$$\text{Diode current, } FF = \frac{I_{D,rms}}{I_{D,av}} = \sqrt{2}.$$

Example 8.8

Repeat Example 8.7 with a heavily inductive load, also find RF and efficiency.

Solution

$$V_{L_{mean}} = \frac{2V_{S_{max}}}{\pi} = \frac{2 \times 200}{\pi} = 127.3\,V$$

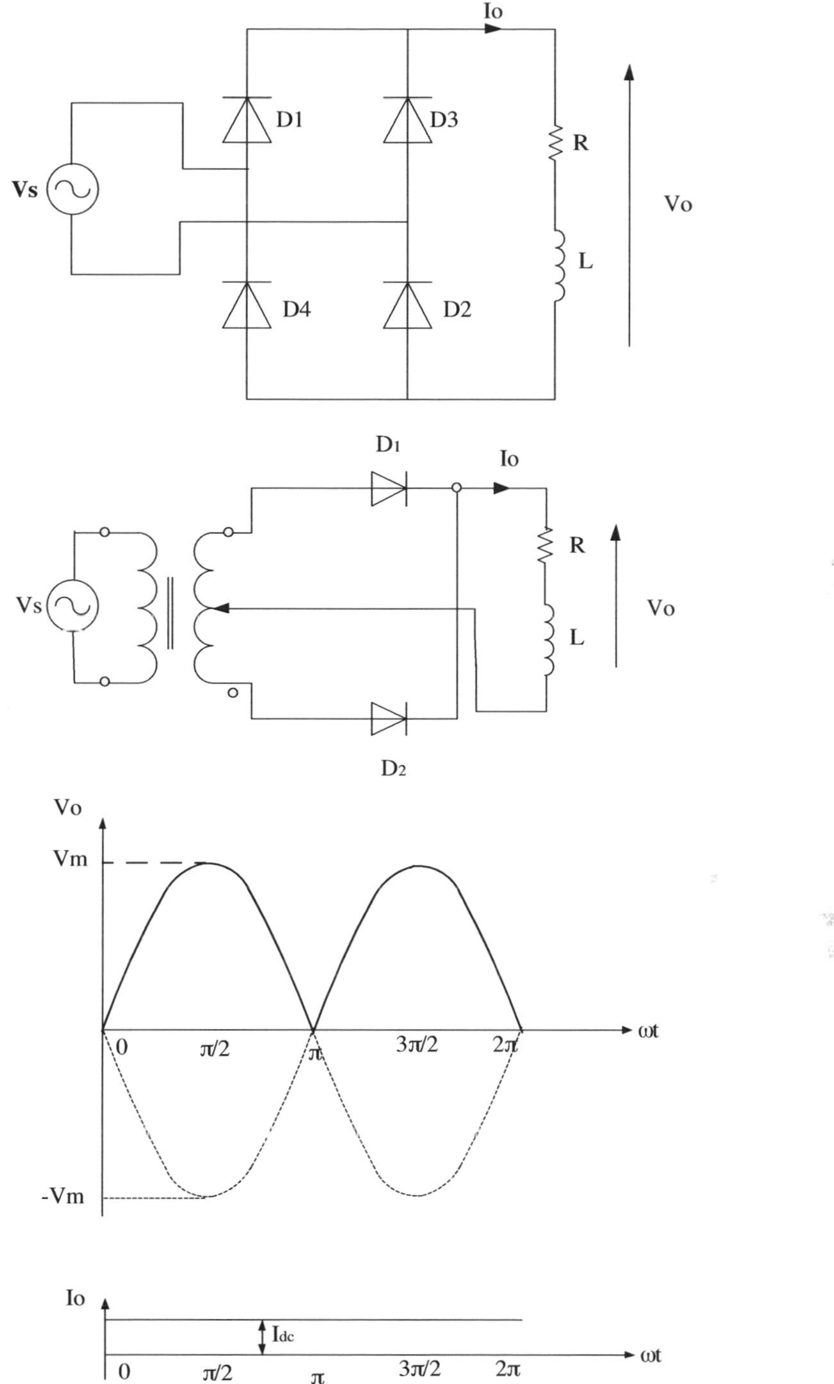

FIGURE 8.13
Single-phase full-wave (bridge and center tap) rectifier with an inductive load and its related waveforms.

$$I_{L_{mean}} = \frac{V_{L_{mean}}}{R} = \frac{127.3}{10} = 12.73\,A$$

$$v_{s_{rms}} = v_{o_{rms}} = \frac{V_{smax}}{\sqrt{2}} = \frac{200}{\sqrt{2}} = 141.42\,V$$

$$i_{s\,rms} = i_{o\,rms} = \sqrt{\frac{1}{2\pi}\int_0^{\pi} I_{lmean}{}^2\theta d\theta \times 2} = I_{lmean} = 12.73\,A$$

$$Form\ factor\ (FF) = \frac{v_{o\,rms}}{V_{lmean}} = \frac{141.42}{127.3} = 1.11$$

$$Ripple\,Factor\left(RF\right) = \sqrt{FF^2 - 1} = \sqrt{(1.11)^2 - 1} = 0.48$$

$$efficiency\left(\eta\right) = \frac{P_{av}}{P_{ac}} = \frac{V_{lmean} \times I_{lmean}}{v_{o\,rms} \times i_{o\,rms}} = \frac{127.3}{141.42} = 90\%.$$

8.1.4 Rectifiers with Filter Circuits

The previous rectifier circuits that the output is a rectifier voltage, the direction of the variable value in the form of pulses, and to reduce the value of the ripples in the voltage, we should be using some types of filters that apply to the output of the rectifier circuits. Either capacitor (C) or inductance and capacitance (LC).

For a given load, a larger capacitor will reduce ripple, but will cost more and will create higher peak currents in the transformer secondary and in the supply feeding it. In extreme cases where many rectifiers are loaded onto a power distribution circuit, it may prove difficult for the power distribution authority to maintain a correctly shaped sinusoidal voltage curve.

For a given tolerable ripple, the required capacitor size is proportional to the load current and inversely proportional to the supply frequency and the number of output peaks of the rectifier per input cycle. The load current and the supply frequency are generally outside the control of the designer of the rectifier system, but the number of peaks per input cycle can be affected by the choice of rectifier design.

A half-wave rectifier will only give one peak per cycle and for this and other reasons are only used in very small power supplies. A full-wave rectifier achieves two peaks per cycle and this is the best that can be done with single-phase input. For three-phase inputs, a three-phase bridge will give six peaks per cycle and even higher numbers of peaks can be achieved by using transformer networks placed before the rectifier to convert to a higher phase order.

To further reduce this ripple, a capacitor-input filter can be used. This complements the reservoir capacitor with an inductor and a second filter capacitor so that a steadier DC output can be obtained across the terminals of the filter capacitor. The inductor presents a high impedance to the ripple current. http://en.wikipedia.org/wiki/Rectifier - cite_note-dcp-0#cite_note-dcp-0 Figure 8.14 shows the filtration process.

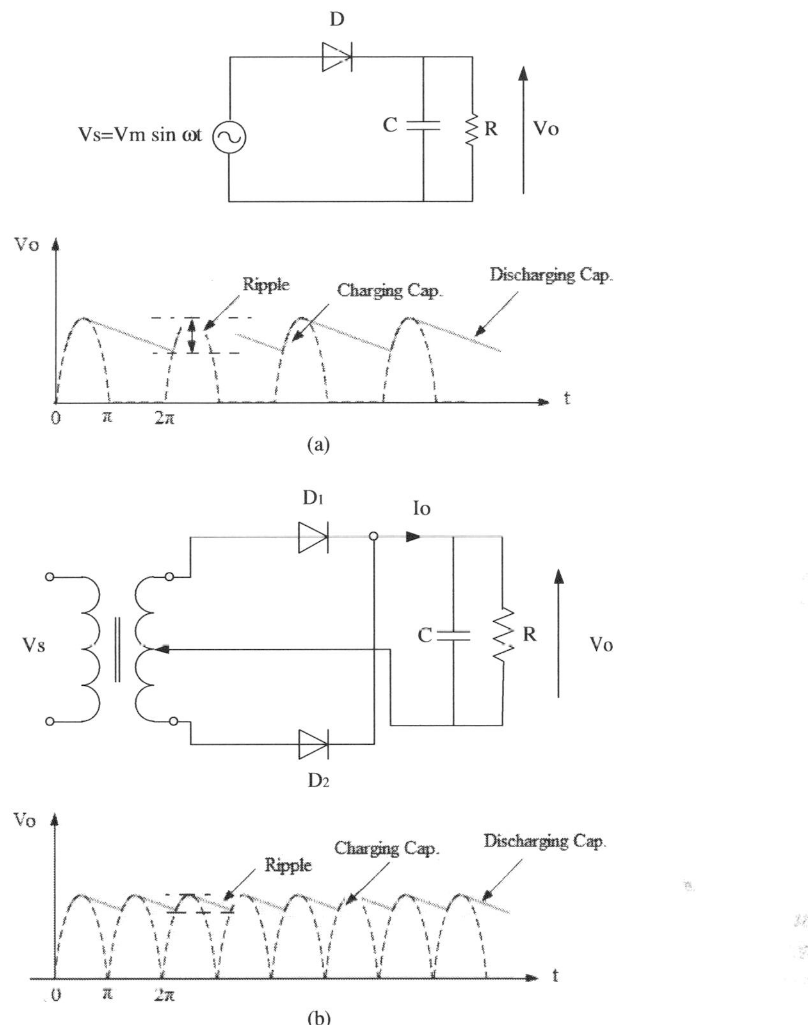

FIGURE 8.14
Filtration process. (a) Half-wave rectifier. (b) Full-wave rectifier.

8.1.5 Controlled Rectifiers

These rectifiers use the thyristors and the delay angle to control the output voltage and current.

8.1.5.1 Thyristor Firing Circuits

The thyristor becomes a conductor if a positive voltage is found between the anode and the cathode, as well as a pulse trigger on the gate. The trigger is a signal on the gate in the form of a pulse that takes a certain amount of time to operate the thyristor.

Shockley diodes are curious devices, but rather limited in application. Their usefulness may be expanded, however, by equipping them with another means of latching. In doing

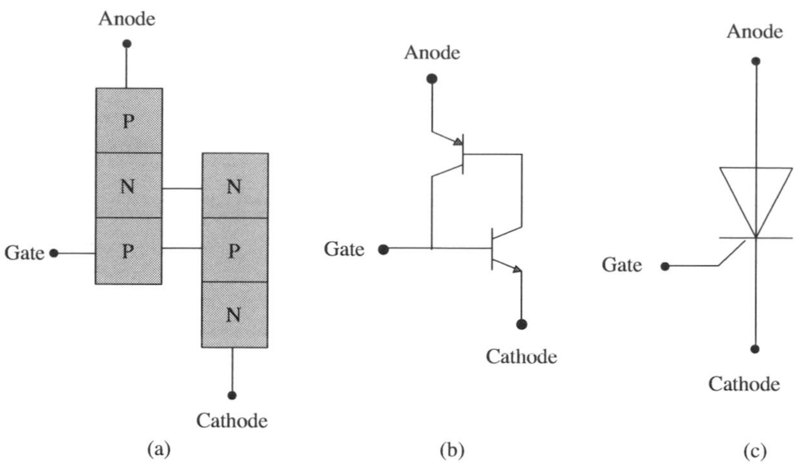

FIGURE 8.15
The silicon controlled rectifier. (a) Physical diagram. (b) Equivalent schematic. (c) Schematic symbol.

so, they become true amplifying devices (if only in an on/off mode), and we refer to them as silicon-controlled rectifiers or SCRs.

The progression from Shockley diode to SCR is achieved with one small addition, nothing more than a third wire connecting to the existing PNPN structure (Figure 8.15).

If SCR's gate is left floating (disconnected), it behaves exactly as a Shockley diode. It may be latched by brake overvoltage or by exceeding the critical rate of voltage rise between anode and cathode, just as with the Shockley diode. Dropout is accomplished by reducing current until one or both internal transistors fall into the cutoff mode, also like the Shockley diode. However, because the gate terminal connects directly to the base of the lower transistor, it may be used as an alternative means to latch the SCR. By applying a small voltage between gate and cathode, the lower transistor will be forced on by the resulting base current, which will cause the upper transistor to conduct, which then supplies the lower transistor's base with the current so that it no longer needs to be activated by a gate voltage. The necessary gate current to initiate latch-up, of course, will be much lower than the current through the SCR from the cathode to anode, so the SCR does achieve a measure of amplification.

This method of securing SCR conduction is called triggering, and it is by far the most common way that SCRs are latched in actual practice. In fact, SCRs are usually chosen so that their brake overvoltage is far beyond the greatest voltage expected to be experienced from the power source so that it can be turned on only by an intentional voltage pulse applied to the gate.

8.1.5.2 The Use of a Thyristor in the Controlled Rectifier Circuits

The thyristors are used in the circuits of controlled rectifier to convert the alternating current to a constant current and voltage that can be controlled by the angle of the trigger of the thyristor (α) and called (trigger angle). The control rectifiers are characterized by simplicity and high efficiency, as well as the low cost of manufacturing, which requires a voltage change to control the speed of the motors.

8.1.5.3 Single-Phase Half-Wave Controlled Rectifier with Resistive Load

The circuit in Figure 8.16 is one of the simplest applications on the controlled rectifier circuits, consisting of alternating current source and thyristor and resistance load.

During the positive half of the wave, the anode voltage is higher than the cathode voltage, and the thyristor is in the forward bias state. When the thyristor is triggered by the current of the gate at the firing angle, the input voltage will appear on the load. When the current follow in the load to be zero the thyristor becomes off, and when the negative half of the input wave will be the anode voltage is lower than the cathode voltage, and the thyristor is in reverse bias and repeated with each cycle (2 π). Figure 8.17 shows the input

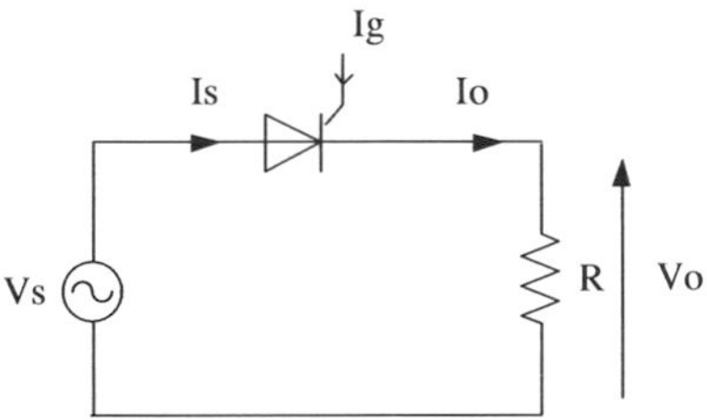

FIGURE 8.16
Half-wave controlled rectifier.

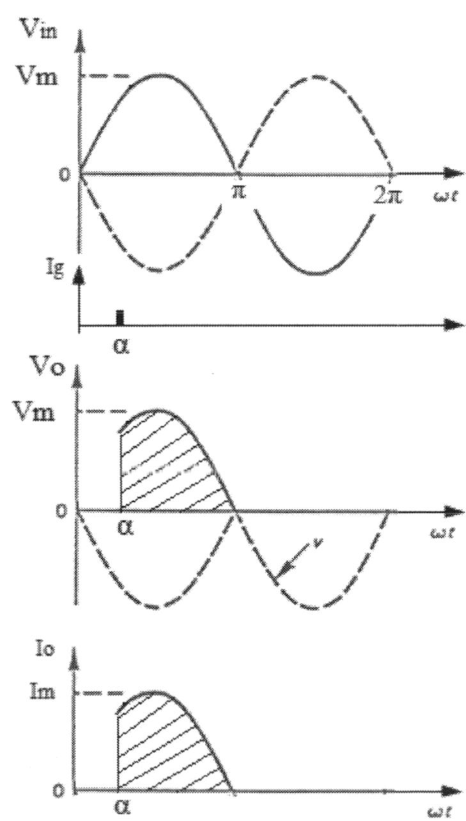

FIGURE 8.17
Input, output voltage, and output current waveforms.

wave and the gate current, load current and voltage. The average value of the mean output voltage can be found in the following equation:

$$V_o = \frac{V_m}{2\pi}(1+\cos \alpha) \tag{8.20}$$

8.1.5.4 Single-Phase Half-Wave Control Rectifier with R-L Load

The importance of this circuit lies in the fact that most industrial loads are an inductive load containing resistance and inductive.

Figure 8.18 shows a circuit with a resistive and inductive load, and Figure 8.19 showing the input and output currents of the voltages and current of this circuit.

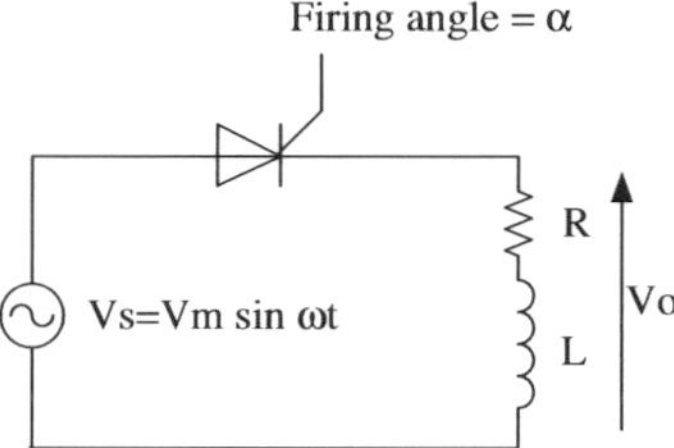

FIGURE 8.18
Single-phase half-wave control rectifier with R-L load circuit.

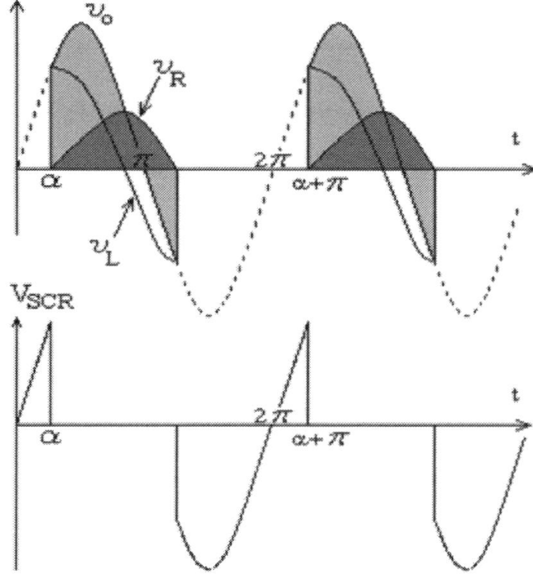

FIGURE 8.19
Output voltage and current and SCR voltage of single-phase half-wave control rectifier with the R-L load.

The current will start to flow in the thyristor when it is triggered, and this current is lagging the negative voltage, due to the transient state of the current caused by the inductor and sometimes freewheeling diode to get rid of the negative part of the load voltages.

The average value of the output voltage without the freewheeling diode can be found in the following formula:

$$V_o = \frac{V_m}{2\pi}(\cos \propto - \cos \beta) \tag{8.21}$$

where β = is the angle at which the thyristor turns OFF.

Example 8.9

A single-phase half-wave controlled rectifier is used to feed a pure resistance load of R = 10 Ω, and AC voltage of 220 V at frequency f = 60 Hz. Calculate:

a. Output voltages at delay angle $\alpha = 90°$.
b. The output current and the thyristor current at the angle of the $\alpha = 90°$.

Solution

(a)

$$V_o = \frac{Vm}{2\pi}(1 + \cos \propto)$$

$$= \frac{220}{2\pi}\sqrt{2}\left(1 + \cos\ 90°\right)$$

$$V_o = 49.51\ \text{V}.$$

(b)

$$I_o = V_o/R = 49.51/10 = 4.951\ \text{A}$$

$$I_{av} = I_o$$

$$= 4.951\ \text{A}.$$

Example 8.10

A single-phase half-wave controlled rectifier with a resistive load and high inductance with a supply voltage of 120 V and frequency of 60 Hz.

a. Calculate the output voltage at $\alpha = 60°$ and $\beta = 200°$.
b. Calculate the output current at the two angles above.

Solution

(a)

$$V_o = (\mathrm{Vm}/2\pi)\cdot(\cos\alpha - \cos\beta)$$

$$= ((\sqrt{2} * 120)/2\pi)\cdot(\cos 60^\circ - \cos 200^\circ)$$

$$V_o = 38.88\ \mathrm{V}.$$

(b)

$$I_o = V_o/R = 38.88/5$$

$$= 7.77\ \mathrm{A}.$$

Example 8.11

A single-phase half-wave controlled rectifier with a resistance R = 2 Ω in series with a very high inductance. The supply voltage of 200 V and f = 60 Hz. Firing angle for each thyristor $\alpha = 60^\circ$. Calculate:

a. The output voltage V_o.
b. The output current I_o.

Solution

(a)

$$V_o = (2.\mathrm{Vm}/\pi)\cos\alpha$$

$$= ((2\sqrt{2} \times 200)/\pi)\cdot\cos 60^\circ$$

$$= 90\ \mathrm{V}.$$

(b)

$$I_o = V_o/R = 90/2 = 45\ \mathrm{A}.$$

Example 8.12

For the single-phase half-wave-controlled rectifier circuit shown in Figure 8.20 has a purely resistive load of *R* and firing angle α = 50% of total power. Determine:

i. Derive an expression for load voltage in term of α.
ii. Efficiency.
iii. FF.
iv. RF.

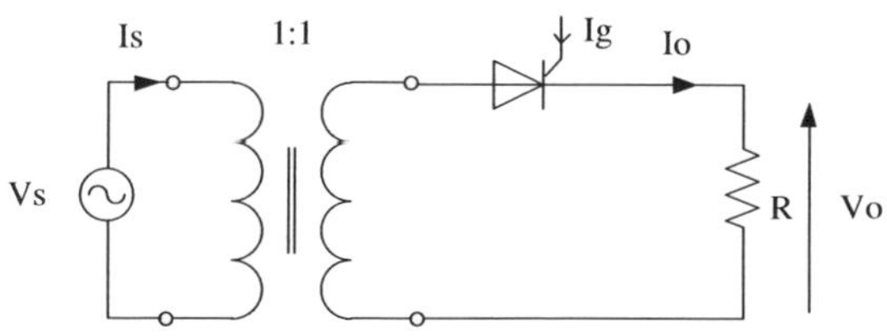

FIGURE 8.20
Circuit of Example 8.12.

Solution

i.

$$V_{lmean} = \frac{1}{2\pi}\int_{\alpha}^{\pi} V_{s\,max} sin\theta d\theta$$

$$= \frac{V_{s\,max}}{2\pi}\int_{\alpha}^{\pi} sin\theta d\theta$$

$$= \frac{V_{s\,max}}{2\pi}\left[-\cos\theta\right]_{\alpha}^{\pi}$$

$$= -\frac{V_{s\,max}}{2\pi}[\cos\pi - \cos\alpha]$$

$$\therefore V_{lmean} = \frac{V_{s\,max}}{2\pi}[1+\cos\alpha].$$

ii.

$$V_{lmean} = \frac{V_{s\,max}}{2\pi}\left[1+\cos 90\right] = \frac{V_{s\,max}}{2\pi} = 0.1592\,V_{s\,max}$$

$$\text{Average power } (P_{av}) = \frac{V_{lmean}{}^{2}}{R} = \frac{(0.1592V_{s\,max})^{2}}{R}$$

$$v_{o\,rms} = \sqrt{\frac{1}{2\pi}\int_{\alpha}^{\pi} V_{smax}{}^{2} sin^{2}\theta d\theta}$$

$$= V_{smax}\sqrt{\frac{1}{2\pi}\int_{\alpha}^{\pi}\frac{1}{2}(1-cos2\theta)d\theta}$$

$$= V_{smax}\sqrt{\frac{1}{4\pi}\int_{\alpha}^{\pi}(1-cos2\theta)d\theta}$$

$$= V_{smax}\sqrt{\frac{1}{4\pi}\int_{\alpha}^{\pi}1d\theta - \frac{1}{2}\int_{\alpha}^{\pi}cos2\theta d\theta}$$

$$= V_{smax}\sqrt{\frac{1}{4\pi}\left[\theta + \frac{sin2\theta}{2}\right]_{\alpha}^{\pi}}$$

$$= V_{smax}\sqrt{\frac{1}{4\pi}\left[\pi - \frac{sin2\pi}{2} - \alpha + \frac{sin2\alpha}{2}\right]}$$

$$= V_{smax}\sqrt{\frac{1}{4\pi}\left[\pi - \frac{sin2\pi}{2} - \frac{\pi}{2} + \frac{sin2\times\frac{\pi}{2}}{2}\right]}$$

$$= V_{smax}\sqrt{\frac{1}{4\pi}\left[\frac{\pi}{2}\right]} = V_{smax}\sqrt{\frac{1}{8}} = 0.3536\,V_{smax}$$

$$\text{AC power}\,(P_{ac}) = \frac{v_{0\,rms}{}^2}{R} = \frac{(0.3536V_{s\,max})^2}{R}$$

$$efficiency(\eta) = \frac{P_{av}}{P_{ac}} = \frac{\dfrac{\left(0.1592V_{s\,max}\right)^2}{R}}{\dfrac{(0.3536V_{s\,max})^2}{R}} = 20.27\%$$

iii.

$$Form\,factor(FF) = \frac{v_{o\,rms}}{V_{lmean}} = \frac{0.3536\,V_{s\,max}}{0.1592\,V_{s\,max}} = 2.221.$$

iv.

$$Ripple\,Factor(RF) = \sqrt{FF^2 - 1} = \sqrt{(2.221)^2 - 1} = 1.9831.$$

8.1.5.5 Single-Phase Full-Wave Control Rectifier with R-L Load

The fully controlled rectifier contains four thyristors. We will assume that the load here contains resistance and high inductance to filter the output current.

The thyristors (T_1, T_2) are connected to the positive part of input voltage after being triggered at angle α, and the work continues to the next firing angle for (T_3, T_4) at $\alpha + \pi$. The occurrence of the thyristors (T_3, T_4) in the negative part of the input wave due to the transient condition of the presence of inductance in the load, as shown in Figure 8.21. Figure 8.22 shows input and output voltages and source current. The average output voltage can be found by the following equation (Figure 8.23):

$$V_o = \frac{2V_m}{\pi}\cos\alpha \tag{8.22}$$

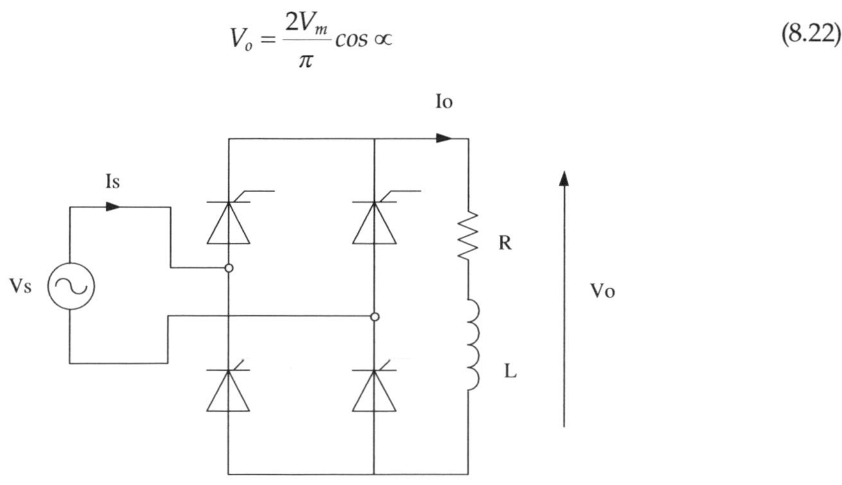

FIGURE 8.21
Single-phase full-wave control rectifier with R-L load circuit.

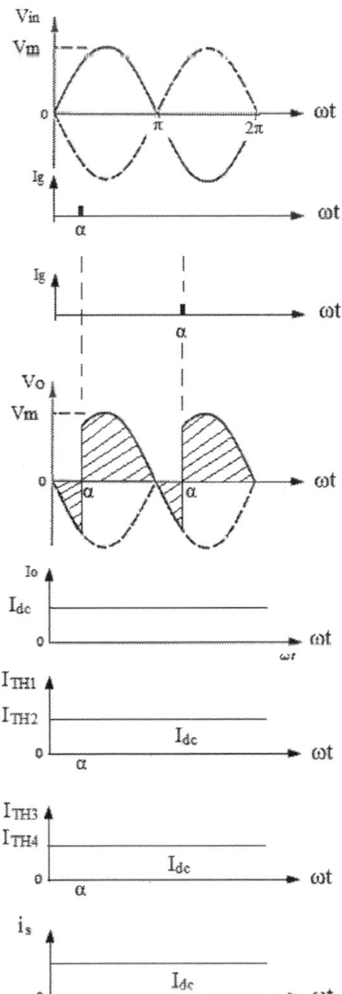

FIGURE 8.22
Input, output voltages, load, SCR, and source currents.

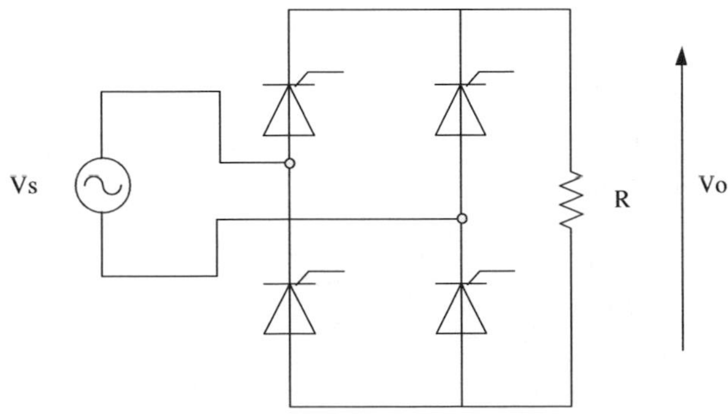

FIGURE 8.23
Single-phase full-wave control rectifier with R load circuit.

Formula Derivative:

a. resistive load:

$$V_{lmean} = \frac{1}{2\pi}\int_{\alpha}^{\pi} V_{s\,max} sin\theta d\theta \times 2$$

$$= \frac{V_{s\,max}}{\pi}\int_{\pi}^{\pi} sin\theta d\theta$$

$$= \frac{V_{s\,max}}{\pi}\left[-cos\theta\right]_{\alpha}^{\pi}$$

$$= -\frac{V_{s\,max}}{\pi}\left[cos\pi - cos\alpha\right]$$

$$\therefore\ V_{lmean} = \frac{V_{s\,max}}{\pi}\left[1+\cos\alpha\right]$$

$$\therefore I_{lmean} = \frac{V_{lmean}}{R}$$

$$\text{or}\quad I_{lmean} = \frac{1}{2\pi}\int_{\alpha}^{\pi} I_{s\,max}\sin\theta d\theta \times 2$$

$$v_{s\,rms} = \frac{V_{s\,max}}{\sqrt{2}}$$

$$v_{0\,rms} = \sqrt{\frac{1}{2\pi}\int_{\alpha}^{\pi} V_{smax}{}^{2} sin^{2}\theta d\theta \times 2}$$

$$= V_{smax}\sqrt{\frac{1}{\pi}\int_{\alpha}^{\pi}\frac{1}{2}(1-cos2\theta)d\theta}$$

$$= V_{smax}\sqrt{\frac{1}{2\pi}\int_{\alpha}^{\pi}(1-cos2\theta)d\theta} = V_{smax}\sqrt{\frac{1}{2\pi}\int_{\alpha}^{\pi}1d\theta - \frac{1}{2}\int_{\alpha}^{\pi}cos2\theta d\theta}$$

$$v_{0rms} = V_{smax}\sqrt{\frac{1}{2\pi}\left[\pi - \frac{sin2\pi}{2} - \alpha + \frac{sin2\alpha}{2}\right]}$$

$$\therefore i_{0rms} = \frac{v_{orms}}{R}$$

Example 8.13

For the fully controlled full-wave rectifier circuit shown in Figure 8.24, firing angle $\alpha = 90^{\circ}$, calculate the efficiency.

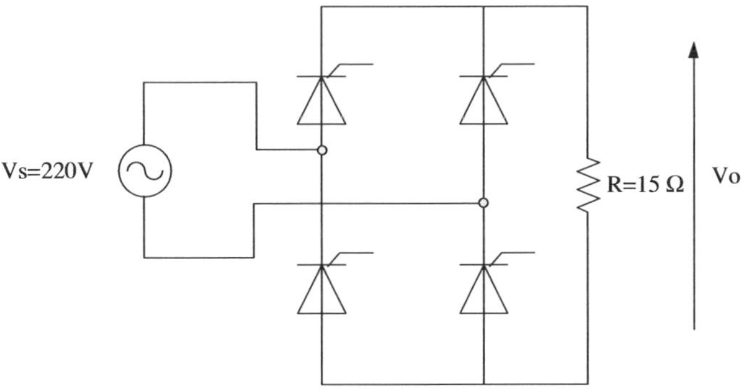

FIGURE 8.24
Circuit of Example 8.13.

Solution

$$V_{lmean} = \frac{V_{s\,max}}{\pi}[1+\cos\alpha] = \frac{220\times\sqrt{2}}{\pi}[1+cos\,90°] = 99.08\,V$$

$$I_{lmean} = \frac{V_{lmean}}{R} = \frac{99.08}{15} = 6.6\,A$$

$$v_{0\,rms} = \sqrt{\frac{1}{2\pi}\int_{\alpha}^{\pi} V_{smax}{}^{2}\sin^{2}\theta d\theta \times 2} = V_{smax}\sqrt{\frac{1}{2\pi}\left[\pi - \frac{sin2\pi}{2} - \alpha + \frac{sin2\alpha}{2}\right]} = 155.5V$$

$$i_{orms} = \frac{v_{orms}}{R} = \frac{155.5}{15} = 10.36\,A$$

$$efficiency(\eta) = \frac{P_{av}}{P_{ac}} = \frac{V_{lmean}\times I_{lmean}}{v_{o\,rms}\times i_{o\,rms}} = \frac{99.08\times 6.6}{155.5\times 10.36} = 40.59\%.$$

8.1.5.6 Single-Phase Full-Wave Half Control Rectifier with R-L Load

When a converter contains both diode and thyristors, it is called *half-controlled*. These three circuits produce identical load waveforms, neglecting any differences in the number type of semiconductor voltage drops. The power to the load is varied by controlling the angle α, at which the load voltage from going negative, extend the conduction period, and reduce the AC ripple (Figure 8.25).

$$Average\ output\ voltage = V_{dc} = \frac{1}{\pi}\int_{\alpha}^{\pi} V_m \sin\omega t\, d\omega t$$

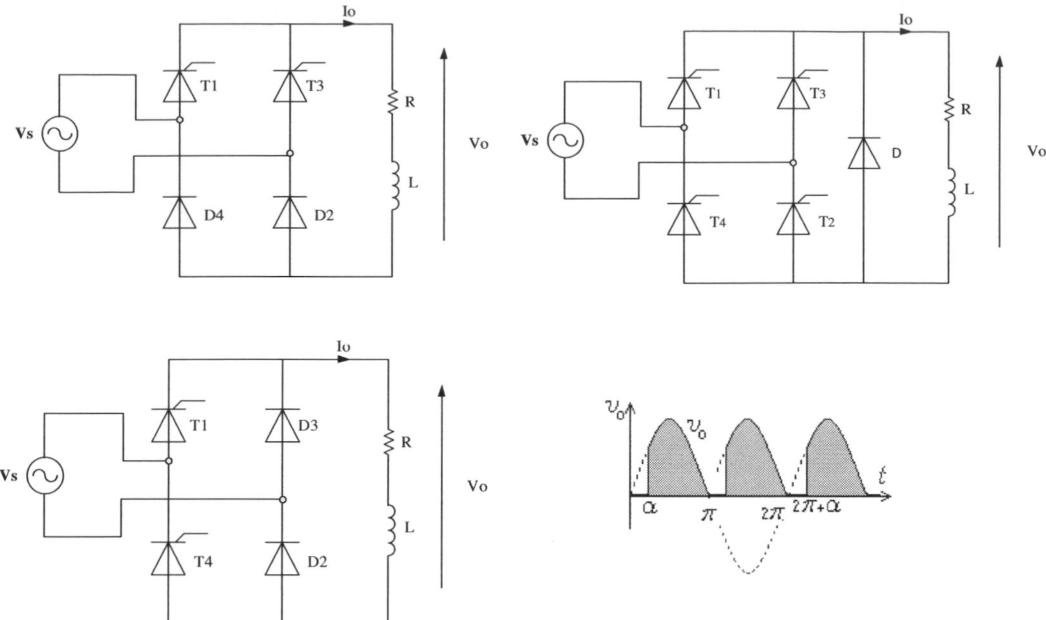

FIGURE 8.25
Half-controlled rectifiers and output voltage.

$$= \frac{V_m}{\pi}(1 + cos\alpha)$$

$$RMS\ output\ voltage = V_{rms} = \sqrt{\frac{1}{\pi}\int_{\alpha}^{\pi}(V_m \sin\omega t)^2 d\omega t}$$

$$= V_m\sqrt{\frac{\pi - \alpha + \frac{1}{2}\sin 2\alpha}{2\pi}}.$$

Inversion is not possible.

Example 8.14

For the fully controlled full-wave rectifier circuit shown in Figure 8.26, firing angle $\alpha = 45$. Calculate efficiency.

Solution

$$\varphi = \tan^{-1}\frac{\omega L}{R} = \tan^{-1} 1\frac{2\pi \times 50 \times 5 \times 10^{-3}}{10} = 8.9°$$

$$V_{lmean} = \frac{V_{s\,max}}{\pi}\left[cos\ \varphi + cos\ \alpha\right] = \frac{150}{3.14}\left[cos\ 8.9° + cos\ 45°\right] = 81V$$

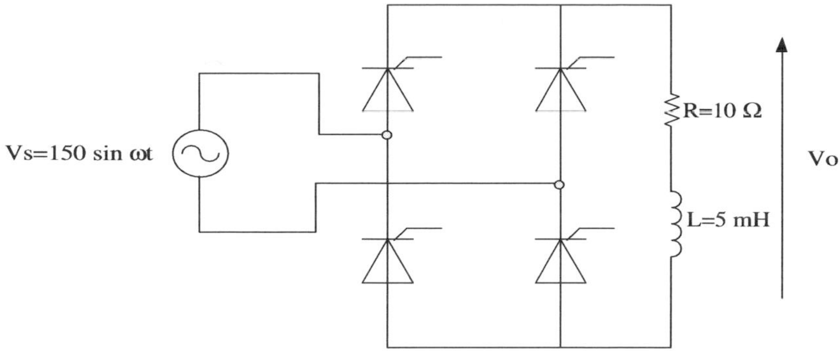

FIGURE 8.26
Fully controlled full-wave rectifier circuit of Example 8.14.

$$I_{lmean} = \frac{V_{lmean}}{R} = \frac{81}{10} = 8.1A$$

$$v_{0\,rms} = \sqrt{\frac{1}{2\pi}\int_{\alpha}^{\pi+\varphi} V_{smax}{}^{2} sin^{2}\theta d\theta \times 2} \approx 101.1V$$

$$i_{orms} = \frac{v_{orms}}{R} = \frac{101.1}{10} = 10.11A$$

$$efficiency(\eta) = \frac{P_{av}}{P_{ac}} = \frac{V_{lmean} \times I_{lmean}}{v_{o\,rms} \times i_{o\,rms}} = \frac{81 \times 8.1}{101.1 \times 10.11} = 64.25\%$$

Example 8.15
For the fully controlled full-wave rectifier circuit shown in Figure 8.27, firing angle $(\alpha) = 45°$. Calculate the efficiency.

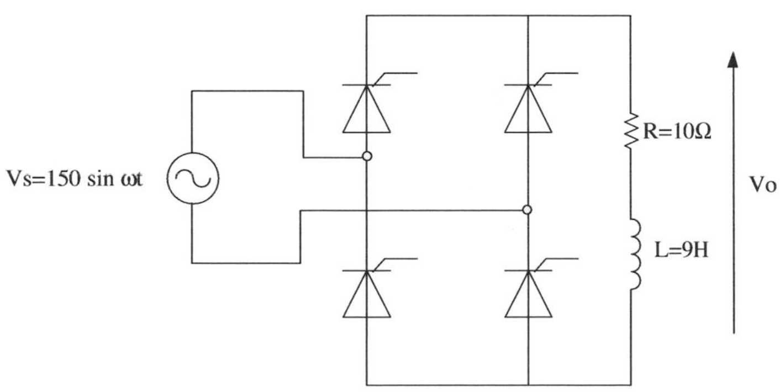

FIGURE 8.27
Fully controlled full-wave rectifier circuit of Example 8.15.

Solution

$$\varphi = \tan^{-1}\frac{\omega L}{R} = \tan^{-1}\frac{2\pi \times 50 \times 9}{10} = 89.79°$$

$$V_{lmean} = \frac{2V_{s\ max}}{\pi}[\cos\alpha] = \frac{2\ 150}{3.14}[\cos 45°] = 67.5\,\text{V}$$

$$I_{lmean} = \frac{V_{lmean}}{R} = \frac{67.5}{10} = 6.75\,A$$

$$v_{0\,rms} = \sqrt{\frac{1}{2\pi}\int_{\alpha}^{\pi+\alpha} V_{smax}^{2}\sin^{2}\theta d\theta \times 2} \approx 105.8\,V$$

$$i_{0\,rms} = \sqrt{\frac{1}{2\pi}\int_{\alpha}^{\pi+\alpha} I_{lmean}^{2}\ \theta d\theta \times 2} = I_{lmean} = 67.5\,A$$

$$efficiency(\eta) = \frac{P_{av}}{P_{ac}} = \frac{V_{lmean} \times I_{lmean}}{v_{o\,rms} \times i_{o\,rms}} = \frac{67.5 \times 6.75}{105.8 \times 6.75} = 63.8\%.$$

8.2 Power Electronics Circuits with MATLAB Program

8.2.1 MATLAB Simulation of Single-Phase Half-Wave Uncontrolled Rectifier

1. Open MATLAB Program, and open the *Simulink Library Browser*. From which, open new module (Figure 8.28)
2. From the library, Simulink opens the branch ***Sinks***, from which get the block ***Scope***. Set the following: Number of Axis = 2. Also get the block ***Display***
3. From the library, Simulink opens the branch ***Signal Routing*** and choose **Mux** block

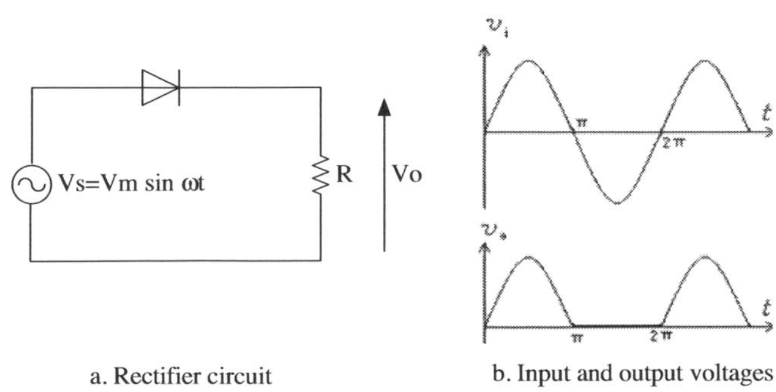

FIGURE 8.28
Single-phase half-wave uncontrolled rectifier. (a) Rectifier circuit (b) Input and output voltages.

4. Open the library ***Sim Power Systems***, and then open the branch *Electrical Sources*. From which get the block ***AC Voltage Source***. Set the following: Peak Amplitude Voltage = 100 V, Frequency = 50, Sample Time = 0.0001
5. From the same library open the *Elements* and get the block ***Ground (Input)***
6. Open the branch *Elements* and get the Block ***Series RLC branch***. Set the following: Resistance = 10 Ω, Inductance = 0, Capacitance reactance = inf, or choose brunch type on only resistance
7. From the branch, *measurements* get the block ***Voltage Measurement and current measurement***
8. From the *Power, Electronic* branch gets the block ***Diode***. Set the following: Snubber resistance = $1e^9$
9. Connect the circuit in Figure 8.29.

NOTE: a. Add power graphical user interface (GUI) block in Figure 8.29.

b. From the menu *simulation*, choose *configuration parameters*, and from which set the type of the solver to the following: ode23tb (stiff/TR-BDF2)

Now, run the module and draw V_o and V_{lmean}. As shown in Figures 8.30 and 8.31.
Controlled *Half-Wave* Rectifier with a Resistive Load.

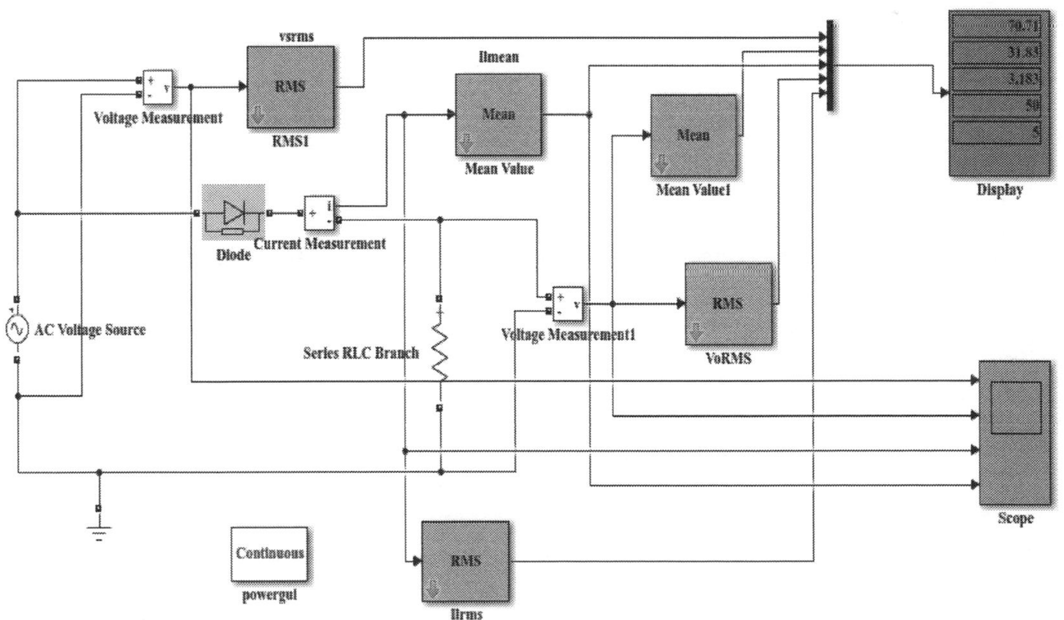

FIGURE 8.29
Single-phase half-wave uncontrolled rectifier Simulink.

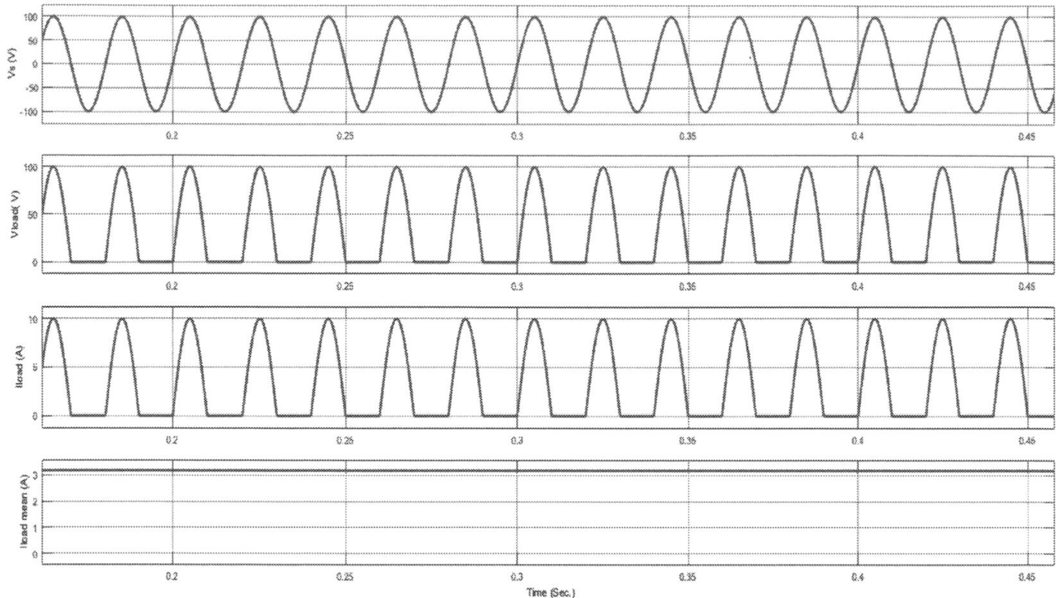

FIGURE 8.30
Waveforms.

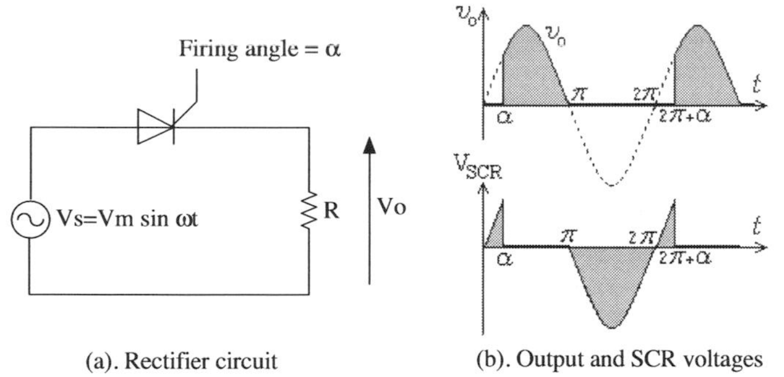

FIGURE 8.31
Single-phase half-wave controlled rectifier. (a) Rectifier circuit (b) Output and SCR voltages.

8.2.2 MATLAB Simulation of Single-Phase Half-Wave Controlled Rectifier

1. Open MATLAB Program, and open the *Simulink Library Browser.* From which, open new module
2. From the *library, Simulink* open the branch *Sinks,* from which get the block ***Scope***. Also get the block ***Display***
3. From the *library, Simulink* open the branch *Sources,* from which get the block ***Pulse Generator***. Set the Period = 0.05, Pulse width % = 1

4. Open the library *Sim Power Systems,* and then open the branch *Electrical Sources*. From which get the block ***AC Voltage Source***. Set the following: Peak Amplitude Voltage = 200 V, Frequency = 50 Hz, Sample Time = 0.0001
5. From the same library open the branch *Connectors* and get the block ***Ground (Input)***
6. Open the branch *Elements* and get the Block ***Series RLC branch***. Set the following: Resistance = 10 Ω, Inductance = 0, Capacitance reactance = inf
7. From the branch *measurements* get the block ***Voltage Measurement***
8. From the *Power, Electronic* branch gets the block ***Detailed Thyristor***. Set the following: Snubber resistance = 1e9
9. Connect the circuit in Figures 8.32 and 8.33
10. Now, from the menu *simulation*, choose *configuration parameters,* and from which set the type of the solver to the following: ode23tb (stiff/TR-BDF2)
11. Open the block ***Pulse Generator*** and change the **Phase delay** to the values **0.001666, 0.0025, 0.00333,** and **0.005**. These values are equivalent to 30°, 45, 60°, and 90°, respectively

$$\text{Firing angle} = \frac{\alpha}{360} \times \frac{1}{\text{freq}}.$$

12. Draw the waveform of *Vo* and V_{lmean}.

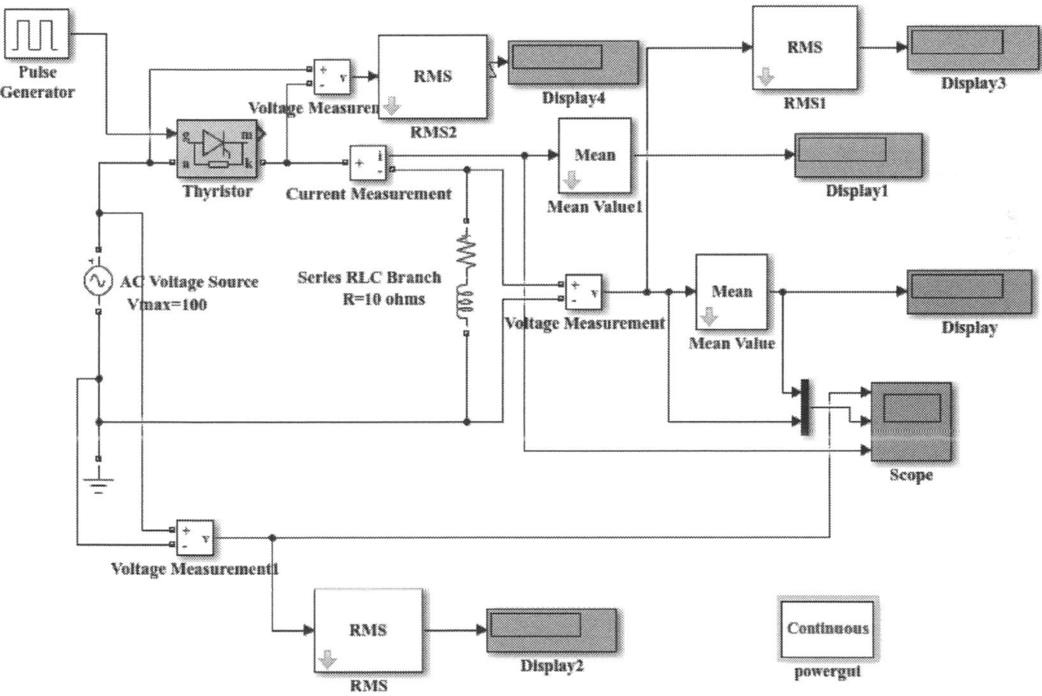

FIGURE 8.32
Single-phase half-wave uncontrolled rectifier Simulink.

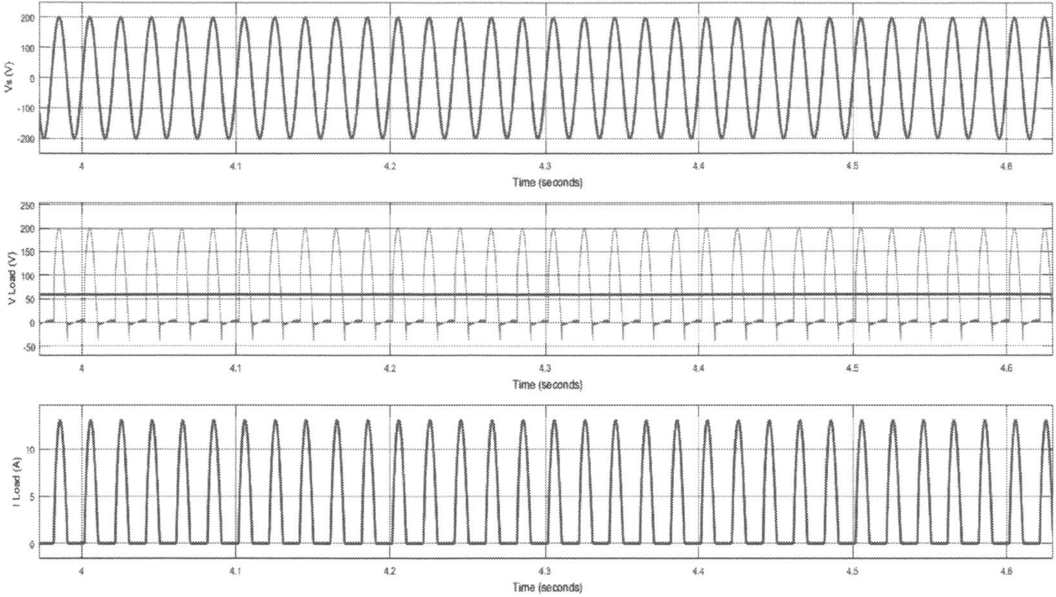

FIGURE 8.33
Waveforms.

8.2.3 MATLAB Simulation of Single-Phase Half-Wave Controlled Rectifier with an Inductive Load

To implement the diagram in Figure 8.34 using MATLAB/Simulink

1. Copy the branch *Elements* and get the Block ***Series RLC branch***. Set the following: Resistance = 10 Ω, Inductance = 9mH, Capacitance reactance = inf
2. Connect the circuit as in Figure 8.35 and run the module
3. Use the Simulink library browser as Figure 8.36
4. Set the parameters as shown in Figures 8.36 through 8.43
5. Draw the waveforms of V_o, V_R, and V_L.

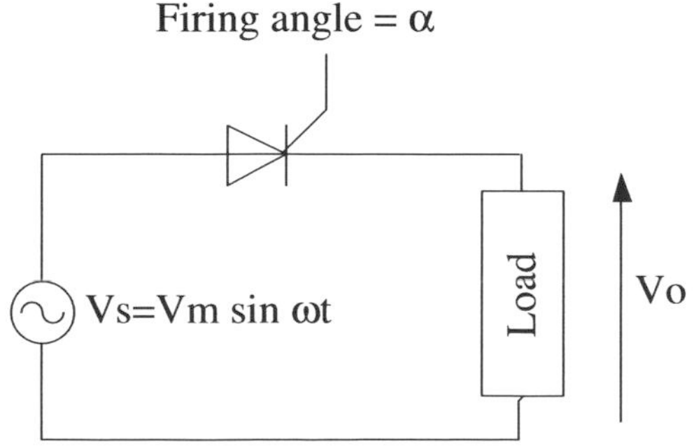

FIGURE 8.34
Single phase half wave-controlled rectifier.

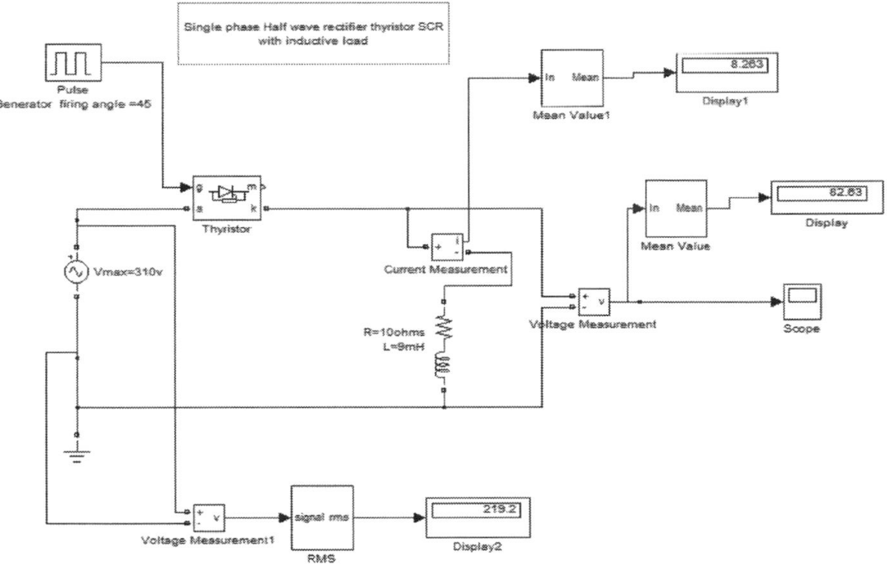

FIGURE 8.35
Single-phase half-wave controlled rectifier with an inductive load Simulink.

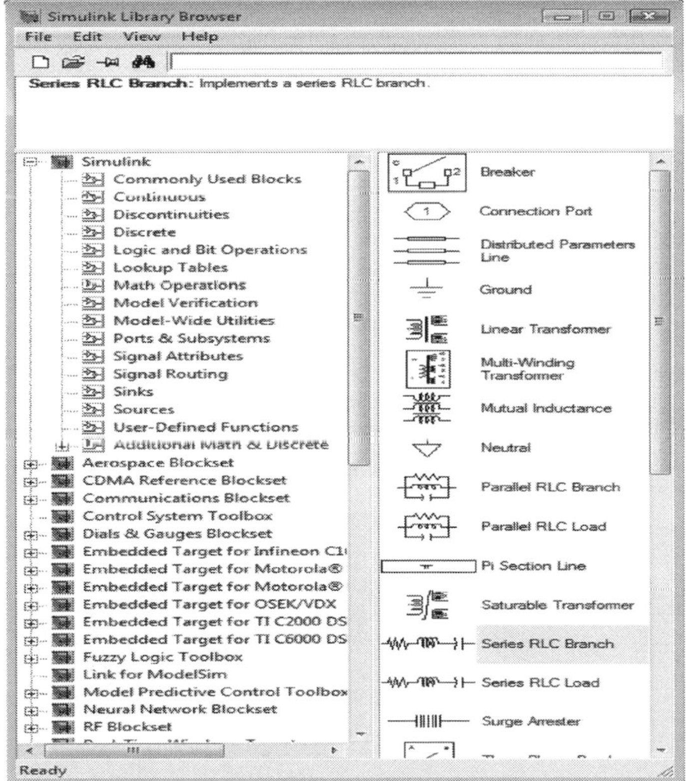

FIGURE 8.36
Simulink library browser.

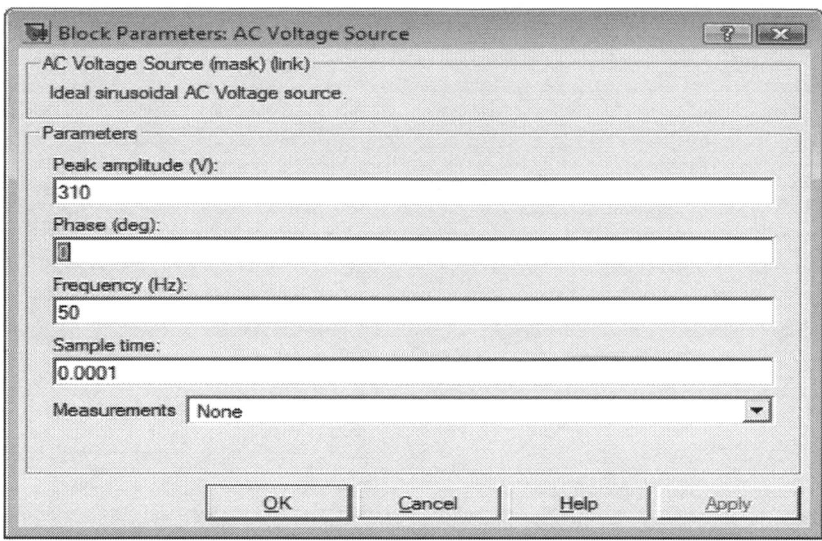

FIGURE 8.37
Block parameters of AC voltage source.

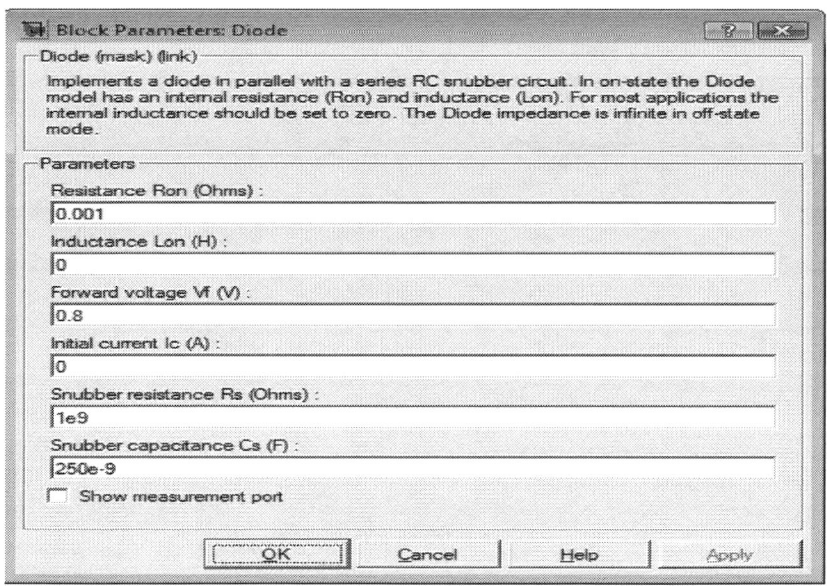

FIGURE 8.38
Block parameters of the diode.

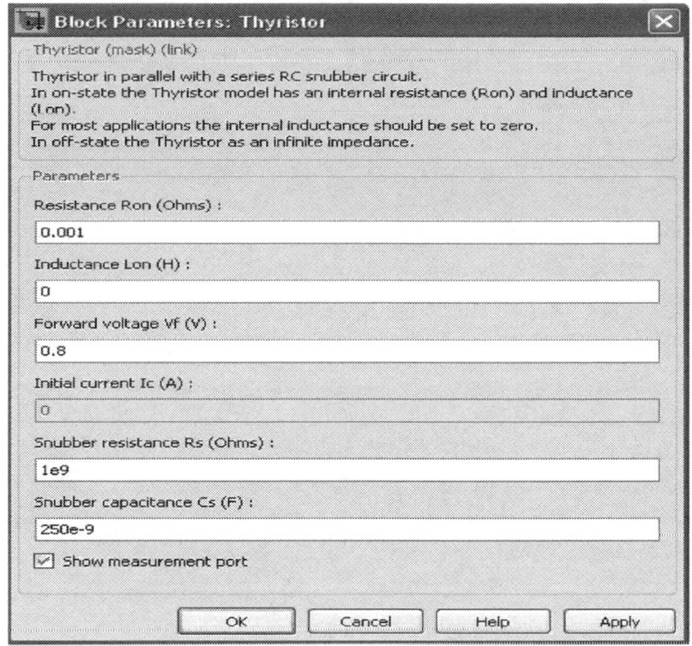

FIGURE 8.39
Block parameters of SCR.

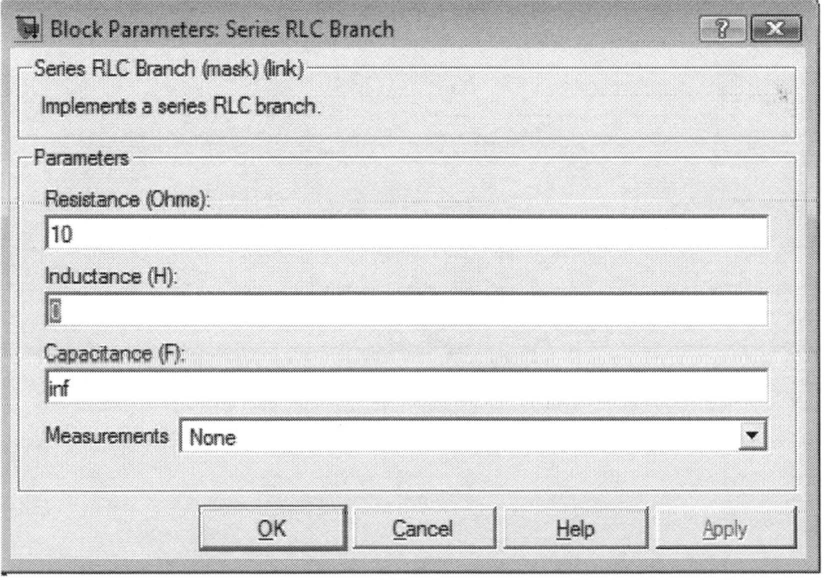

FIGURE 8.40
Block parameters of series RLC branch.

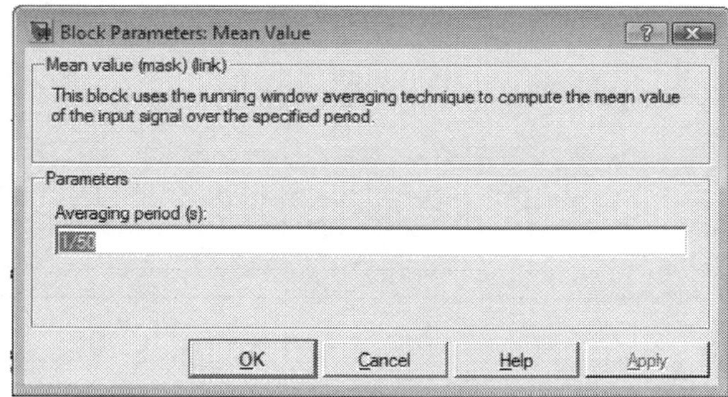

FIGURE 8.41
Mean value block parameters.

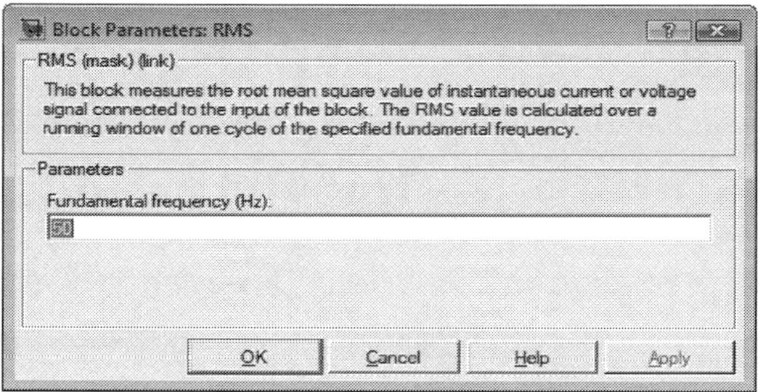

FIGURE 8.42
RMS block parameters.

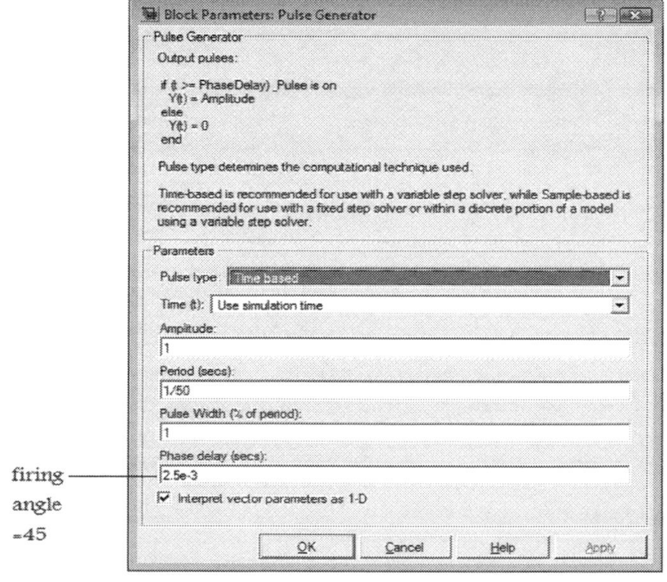

FIGURE 8.43
Pulse generator block parameters.

Example 8.16

For the full-wave uncontrolled rectifier circuit shown in Figure 8.26, $R = 10\Omega$, $V_{max} = 200\,V$ and firing angle $(\alpha) = 45°$. Trace and calculate V_{Lmean}, I_{Lmean}, and $v_{s_{r.m.s}}$ (Figures 8.44 and 8.45).

Solution

$$V_{L_{mean}} = \frac{1}{2\pi}\int_0^{\pi} V_{max} sin\theta d\theta \times 2 = \frac{2V_{max}}{\pi} = 127.3\,volts$$

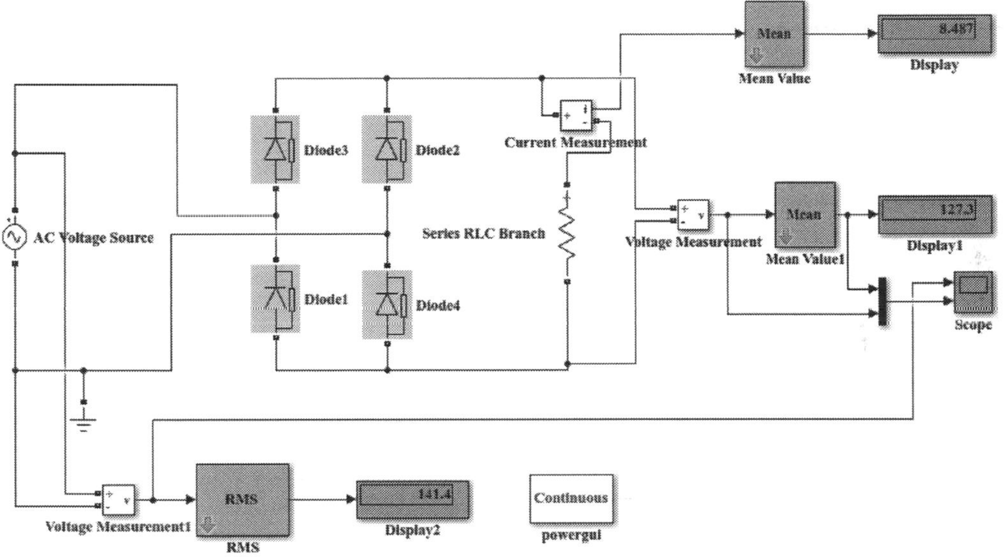

FIGURE 8.44
Full-wave uncontrolled rectifier circuit Simulink.

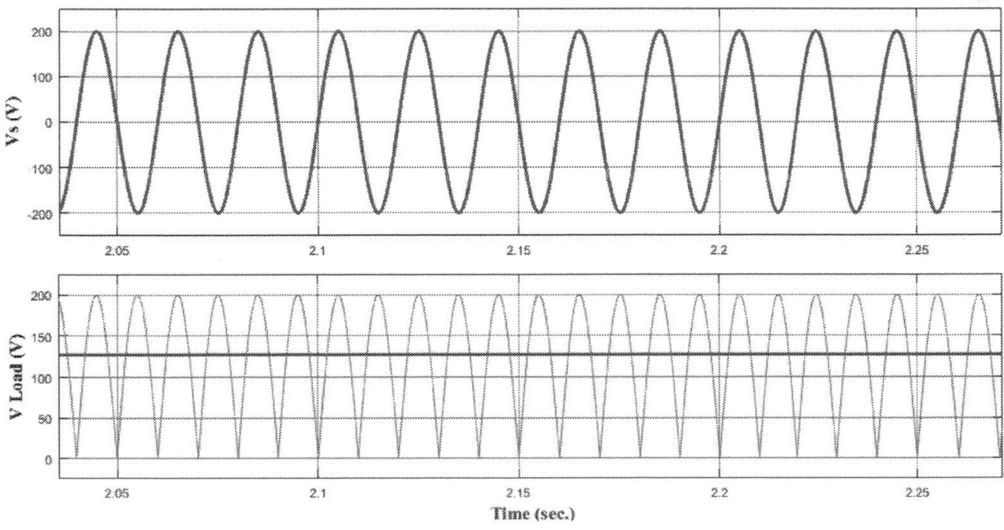

FIGURE 8.45
Waveforms.

$$I_{L_{mean}} = \frac{V_{L_{mean}}}{R} = \frac{127.3}{15} = 8.49\,Amper$$

$$V_{S_{rms}} = \frac{V_{max}}{\sqrt{2}} = \frac{200}{\sqrt{2}} = 141.42\,volts$$

Example 8.17

For the full-wave controlled rectifier circuit shown in Figure 8.46, $R = 10\Omega$, $L = 5mH$, $V_{max} = 150$ volt, and firing angle $(\alpha) = 30°$. Trace and calculate V_{Lmean}, I_{Lmean}, and $v_{S_{r.m.s}}$ (Figure 8.47).

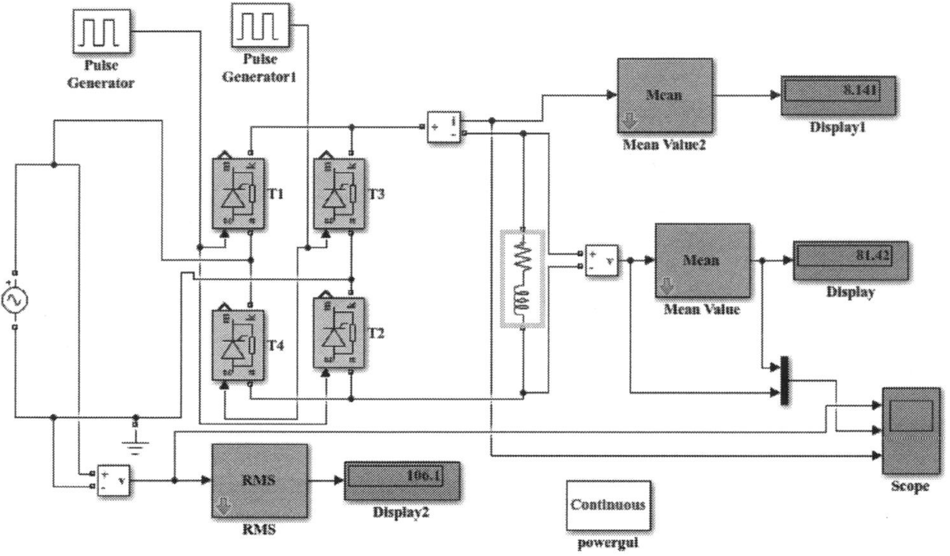

FIGURE 8.46
Full-wave controlled rectifier circuit Simulink.

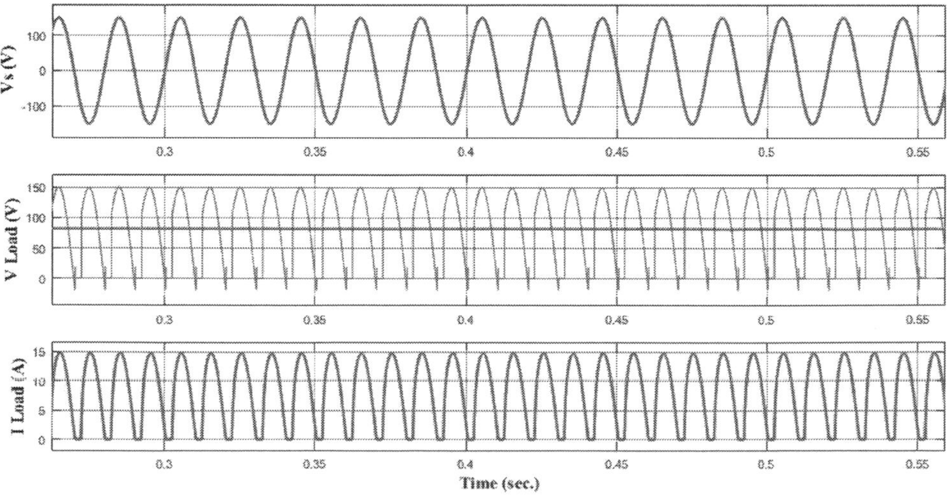

FIGURE 8.47
Waveforms.

Solution

$$\phi = tan^{-1}\frac{\omega L}{R} = 8.9°$$

$$V_{L_{mean}} = \frac{1}{2\pi}\int_{\alpha}^{\pi+f} V_{max} sin\theta d\theta \times 2 = \frac{V_{max}}{\pi}\left[cos\phi + cos\alpha\right] = 88.5\, volts$$

$$I_{L_{mean}} = \frac{V_{L_{mean}}}{R} = \frac{88.5}{10} = 8.85\, Amper$$

$$V_{SRMS} = \frac{V_{max}}{\sqrt{2}} = \frac{150}{\sqrt{2}} = 106.01\, volts.$$

8.3 DC-DC Converter Basics

A DC-DC converter is a device that accepts a DC input voltage and produces a DC output voltage. Typically, the output produced is at a different voltage level than the input. In addition, DC-DC converters are used to provide noise isolation, power bus regulation, etc. This section considers a summary of some of the popular DC-DC converter topologies.

8.3.1 Step-Down (Buck) Converter

The circuit in Figure 8.48, the transistor turning ON will put voltage V_{in} on one end of the inductor. This voltage will tend to cause the inductor current to rise. When the transistor is OFF, the current will continue flowing through the inductor, but now flowing through the diode. We initially assume that the current through the inductor does not reach zero, thus the voltage at V_x will now be only the voltage across the conducting diode during the full OFF time. The average voltage at V_x will depend on the average ON time of the transistor provided the inductor current is continuous as in Figure 8.49.

To analyze the voltages of this circuit let us consider the changes in the inductor current over one cycle. From the relation:

$$V_x - V_0 = L\frac{di}{dt} \tag{8.23}$$

The change of current satisfies:

$$di = \int_{ON} (V_x - V_0)dt + \int_{OFF} (V_x - V_0)dt \tag{8.24}$$

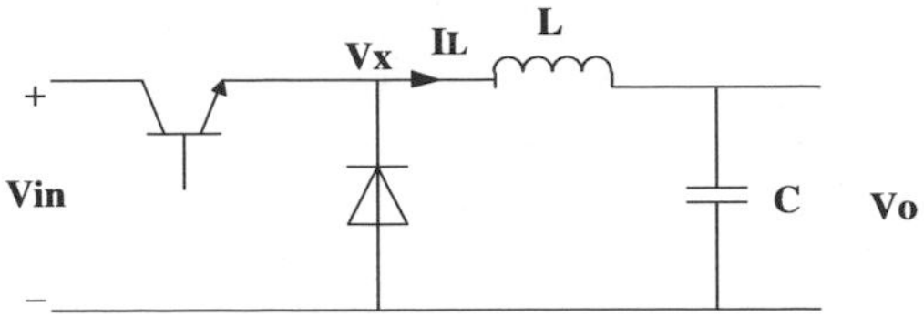

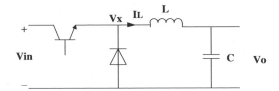

FIGURE 8.48
Buck converter.

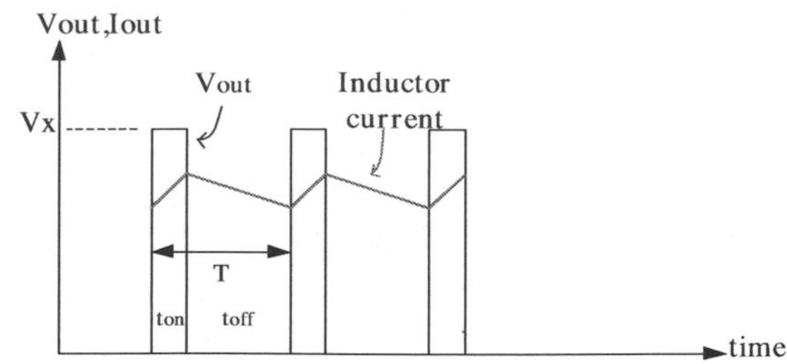

FIGURE 8.49
Voltage and current changes.

For steady-state operation, the current at the start and end of a period T will not change. To get a simple relation between voltages, we assume no voltage drop across transistor or diode while ON and a perfect switch change. Thus, during the ON time $V_x = V_{in}$ and in the OFF $V_x = 0$.

So

$$0 = di = \int_0^{t_{on}} (V_{in} - V_0)dt + \int_{t_{on}}^{T} (-V_0)dt \tag{8.25}$$

which simplifies to:

$$(V_{in} - V_0).t_{on} + (-V_0).t_{off} = 0$$

or

$$\frac{V_o}{V_{in}} = \frac{t_{on}}{T} \tag{8.26}$$

and defining "duty ratio" as:

$$D = \frac{t_{on}}{T} \tag{8.27}$$

The voltage relationship becomes $V_o = D\,V_{in}$ since the circuit is lossless and the input and output powers must match on the average $V_o \times I_o = V_{in} \times I_{in}$. Thus, the average input and output current must satisfy $I_{in} = D\,I_o$. These relations are based on the assumption that the inductor current does not reach zero.

8.3.1.1 Transition between Continuous and Discontinuous

When the current in the inductor L remains always positive then either the transistor T_1 or the diode D_1 must be conducting. For continuous conduction, the voltage V_x is either V_{in} or 0. If the inductor current ever goes to zero then the output voltage will not be forced to either of these conditions. At this transition point, the current just reaches zero as seen in Figure 8.50. During the ON time V_{in}–V_{out} is across the inductor thus:

$$I_L(peak) = (V_{in} - V_{out}).\frac{t_{on}}{L} \tag{8.28}$$

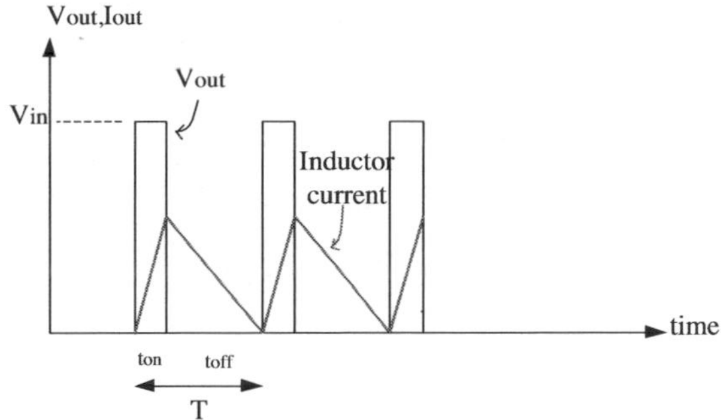

FIGURE 8.50
Buck converter at the boundary.

The average current which must match the output current satisfies:

$$I_L(av) = \frac{I_L(peak)}{2} = (V_{in} - V_{out}).\frac{dT}{2L} \tag{8.29}$$

If the input voltage is constant, the output current at the transition point satisfies:

$$I_L(av) = I_{out} = V_{in}.\frac{(1-d)}{2L}.T \tag{8.30}$$

8.3.1.2 Voltage Ratio of Buck Converter (Discontinuous Mode)

As for the continuous conduction analysis, we use the fact that the integral of the voltage across the inductor is zero over a cycle of switching T. The transistor OFF time is now divided into segments of diode conduction δ_dT and zero conduction δ_oT as shown in Figure 8.51. The inductor average voltage thus gives:

$$(V_{in} - V_o)\, DT + (-V_o)\, \delta.dT = 0 \tag{8.31}$$

$$\frac{V_{out}}{V_{in}} = \frac{d}{d + d_\delta} \tag{8.32}$$

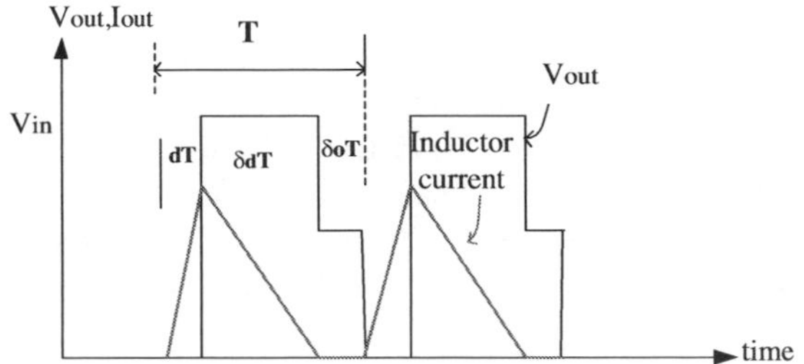

FIGURE 8.51
Buck converter—discontinuous conduction.

for the case $d + \delta_d < 1$. To resolve the value of δ_d, consider the output current which is half the peak when averaged over the conduction times $d + \delta_d$:

$$I_{out} = \frac{I_{L(peak)}}{2}.d + \delta_d \tag{8.33}$$

Considering the change of current during the diode conduction time:

$$I_{L(peak)} = \frac{V_o(\delta_d T)}{L} \tag{8.34}$$

Thus, from equations 8.33 and 8.34, we can get:

$$I_{out} = \frac{V_o \delta_d T(d + d_\delta)}{2L} \tag{8.35}$$

using the relationship in equation 8.32:

$$I_{out} = \frac{V_{in} d T d_\delta}{2L} \tag{8.36}$$

And solving for the diode conduction:

$$\delta_d = \frac{2LI_{out}}{V_{in} dT} \tag{3.37}$$

The output voltage is thus given as:

$$\frac{V_{out}}{V_{in}} = \frac{d^2}{d^2 + \left(\frac{2LI_{out}}{V_{in}T}\right)} \tag{3.38}$$

Defining $k^* = 2L/(V_{in} T)$, we can see the effect of discontinuous current on the voltage ratio of the converter. As seen in Figure 8.52, once the output current is high enough, the voltage ratio depends only on the duty ratio "d." At low currents, the discontinuous operation tends to increase the output voltage of the converter toward V_{in}.

8.3.2 Step-Up (Boost) Converter

The schematic in Figure 8.53 shows the basic boost converter. This circuit is used when a higher output voltage than input is required.

While the transistor is ON $V_x = V_{in}$ and the OFF state the inductor current flows through the diode giving $V_x = V_o$. For this analysis, it is assumed that the inductor current always remains flowing (continuous conduction). The voltage across the inductor is shown in Figure 8.54, and the average must be zero for the average current to remain in a steady-state:

$$V_{in}.t_{on} + (V_{in} - V_o)\, t_{off} = 0$$

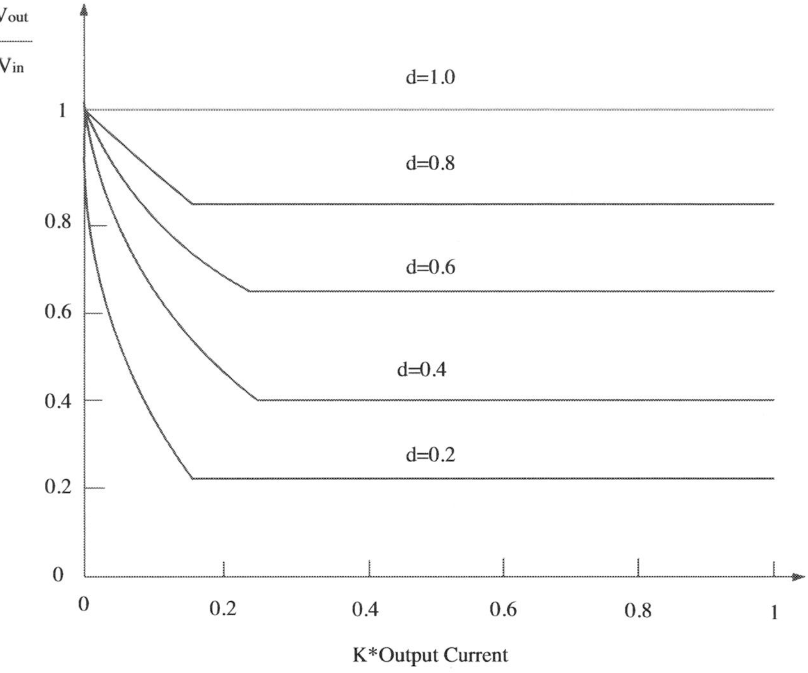

FIGURE 8.52
Output voltage versus current.

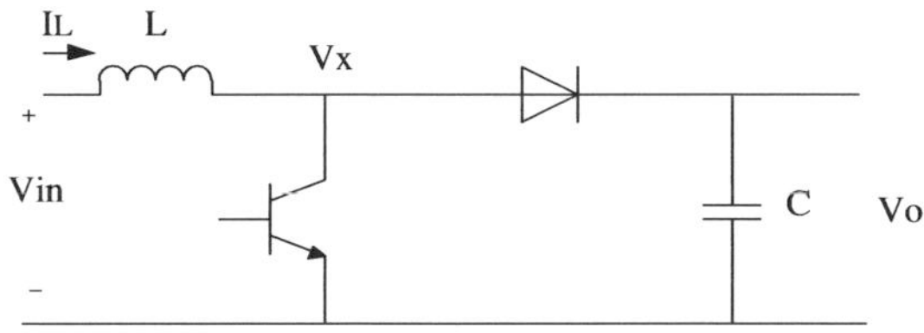

FIGURE 8.53
Boost converter circuit.

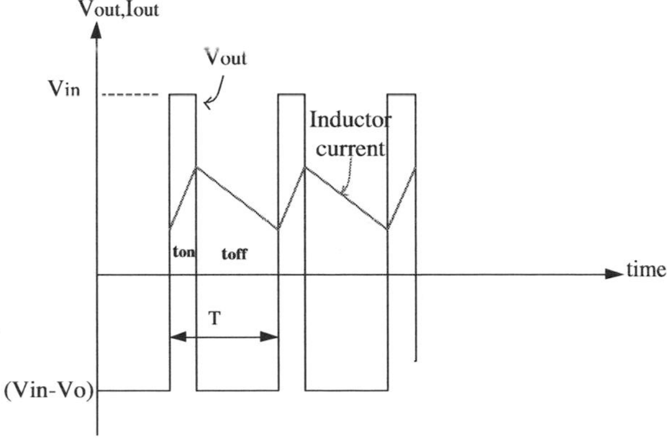

FIGURE 8.54
Voltage and current waveforms (boost converter).

This can be rearranged as:

$$\frac{V_{out}}{V_{in}} = \frac{T}{t_{off}} = \frac{1}{(1-D)} \tag{8.39}$$

And for a lossless circuit, the power balance ensures:

$$\frac{I_o}{I_{in}} = (1-D) \tag{8.40}$$

Since the duty ratio "D" is between 0 and 1, the output voltage must always be higher than the input voltage in magnitude. The negative sign indicates a reversal of sense of the output voltage.

8.3.3 Buck-Boost Converter

With continuous conduction for the buck-boost converter in Figure 8.55, $V_x = V_{in}$ when the transistor is ON and $V_x = V_o$ when the transistor is OFF. For zero net current change over a period, the average voltage across the inductor is zero.

From Figure 8.56:

$$V_{in} \cdot t_{on} + V_o \cdot t_{off} = 0,$$

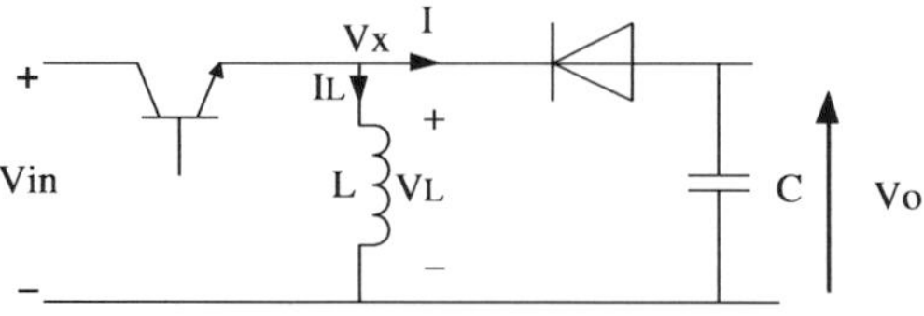

FIGURE 8.55
Schematic for the buck-boost converter.

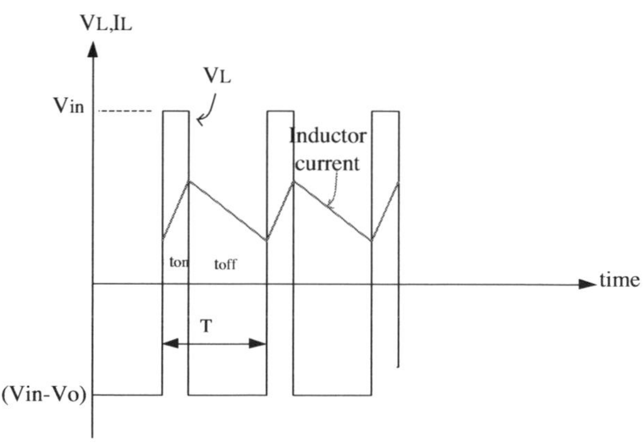

FIGURE 8.56
Waveforms for the buck-boost converter.

which gives the voltage ratio:

$$\frac{V_{out}}{V_{in}} = \frac{T}{t_{off}} = \frac{D}{(1-D)} \tag{8.41}$$

and the corresponding current:

$$\frac{I_{out}}{I_{in}} = \frac{(1-D)}{D} \tag{8.42}$$

Since the duty ratio D is between 0 and 1, the output voltage can vary between lower or higher than the input voltage in magnitude. The negative sign indicates a reversal of sense of the output voltage.

8.3.4 Converter Comparison

The voltage ratios achievable by the DC-DC converters are summarized in Figure 8.57. Notice that only the buck converter shows a linear relationship between the control (duty ratio) and output voltage. The buck-boost can reduce or increase the voltage ratio with rectifier gain for a duty ratio of 50%.

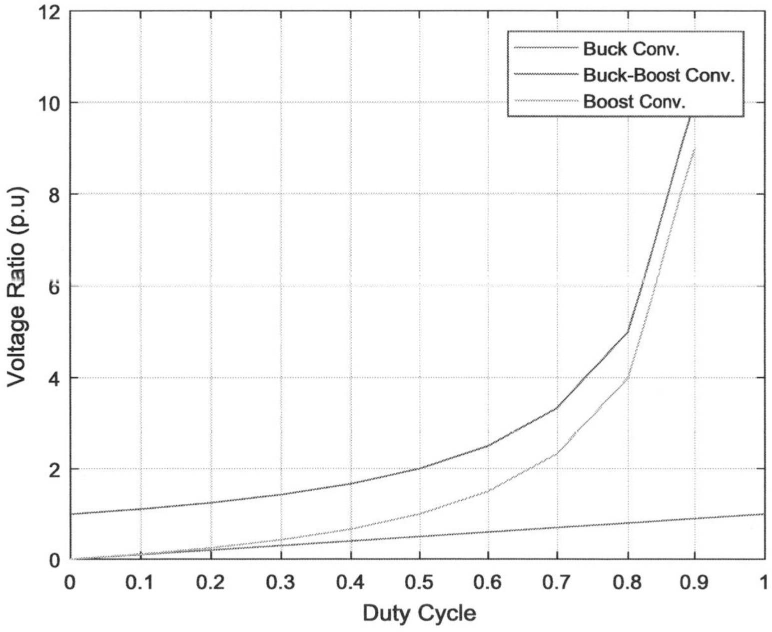

FIGURE 8.57
Comparison of voltage ratio.

MATLAB program

```
clc
clear all
D=0:0.1:1;
X1=D; %X1=Vo/Vin Buck converter
X2 = 1./(1-D); %X2=Vo/Vin Boost converter
X3=D./(1-D); %X3=Vo/Vin Buck-Boost converter
figure
plot(D, X1,D, X2,D, X3)
title('Comparison of voltage ratio')
xlabel('Duty Cycle')
ylabel('Voltage Ratio (p.u)')
grid
```

Example 8.18

A step-down (buck) converter is used to supply an inductive load of R = 3Ω and L = 9 mH, with a DC voltage of 60V as shown in Figures 8.58 and 8.59. If D = 0.4, F_s = 2 KH_z, calculate:

a. I_{Lmean} and V_{Lmean}.
b. Trace the output voltage and current.

a. $Ts = 1/2000 = 5\times10^{-4}$

$$D = \frac{a}{Ts}$$

$$a = D\times Ts = 0.4\times5\times10^{-4} = 2\times10^{-4}$$

$$Ts = a + b$$

$$b = Ts - a = 5\times10^{-4} - 2\times10^{-4} = 3\times10^{-4}$$

$$I_{max} = \frac{Ed}{R}\left[\frac{e^{\frac{RTs}{L}} - e^{\frac{Rb}{L}}}{e^{\frac{RTs}{L}} - 1}\right] = 8.4A$$

$$I_{min} = I_{max}e^{\frac{-Rb}{L}} = 7.6A$$

$$I_{Lmean} = \frac{I_{max} + I_{min}}{2} = 7.8A$$

b. $V_{max} = I_{max} \times R = 25.2V$

$$V_{min} = I_{min} \times R = 22.8V$$

$$V_{Lmean} = \frac{V_{max} + V_{min}}{2} = 24V$$

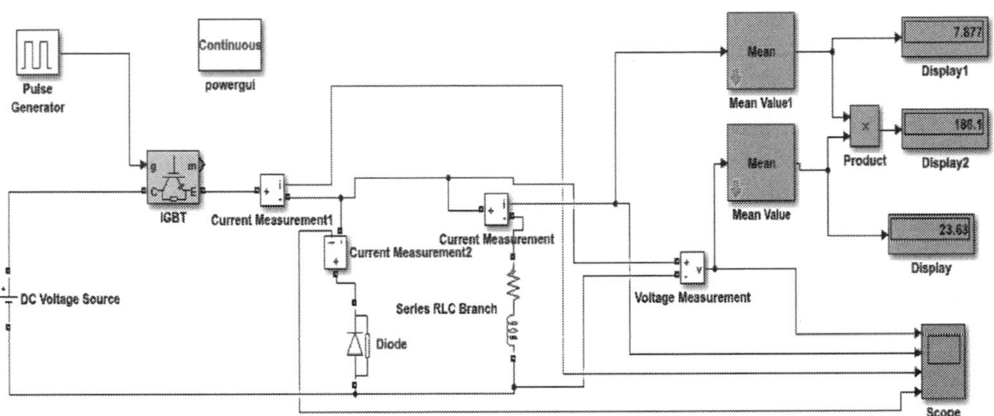

FIGURE 8.58
Step-down (buck) converter of Example 8.18.

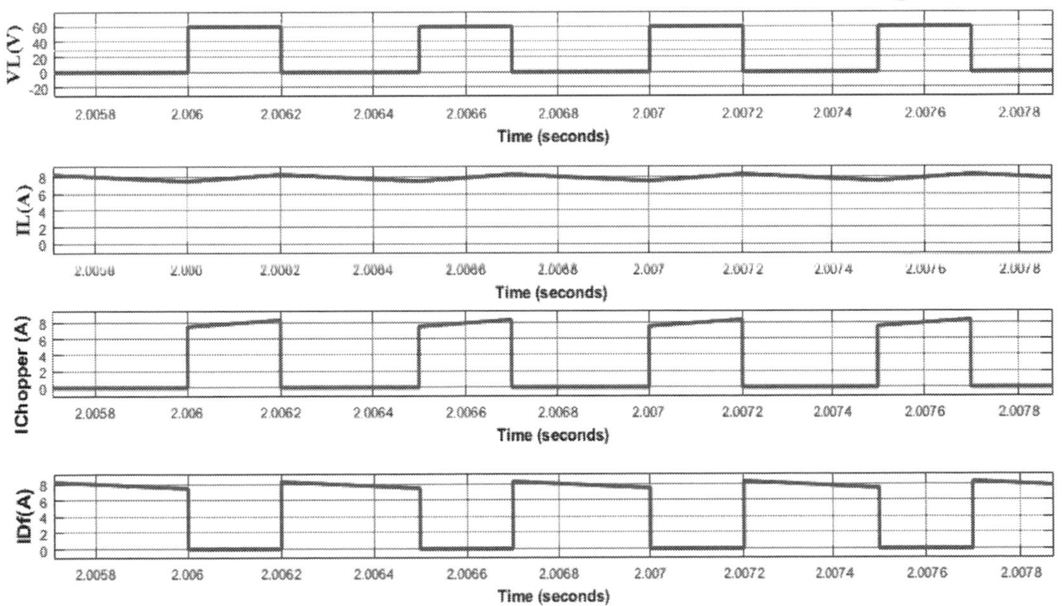

FIGURE 8.59
Waveforms.

Problems

8.1 A single-phase half-wave controlled rectifier the thyristor turn ON at an angle $\alpha = 90°$. If the load resistance of R = 10 Ω and supply voltage of 120 V.

Required:

1. Draw input voltage, output voltage, and thyristor current
2. Draw the input voltage, output voltage, and thyristor current if $\alpha = 0$.

8.2 A single-phase full-wave control rectifier, all the thyristor works at delay angle $\alpha = 90°$ supplied bus AC voltage of 120 V. Draw the input voltage, output voltage output current if load:

1. Pure resistance
2. Resistance with high-value inductance.

8.3 A single-phase full-wave bridge rectifier operate at an angle of $\alpha = 90°$, load A pure resistance of R = 20 Ω is supplied by an AC source of V = 20 sin ωt at frequency f = 50 Hz. Calculate:

1. Output voltage
2. Output current.

8.4 The circuit is shown in Figure 8.60.

A. Draw the waveform of the load voltages of the thyristor at $\alpha = 60°$

B. Draw the voltages of the load pure resistance at $\alpha = 90°$

C. Draw the output current I_o for A and B above.

8.5 The circuit is shown in Figure 8.61. If the rated load is 10 Ω and the source voltage is 120 volt. Required:

a. Derive an expression for load and thyristor current

b. Use an equation for load voltage. Calculate the load current at $\alpha = 90°$.

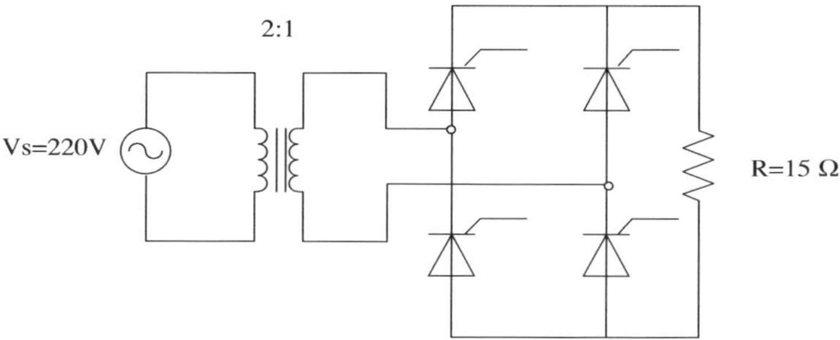

FIGURE 8.60
Circuit diagram for Problem 8.4.

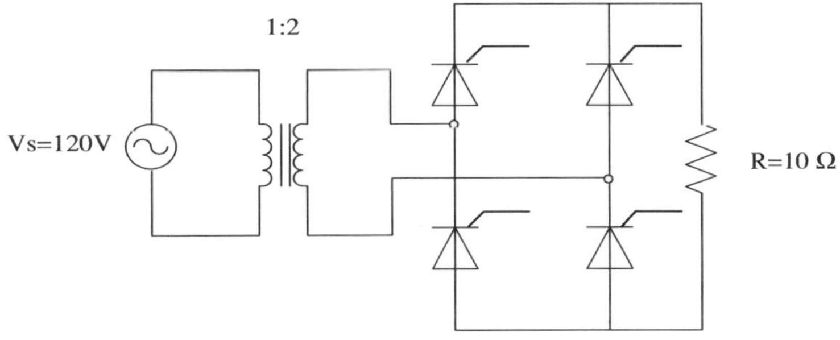

FIGURE 8.61
Circuit diagram for Problem 8.5.

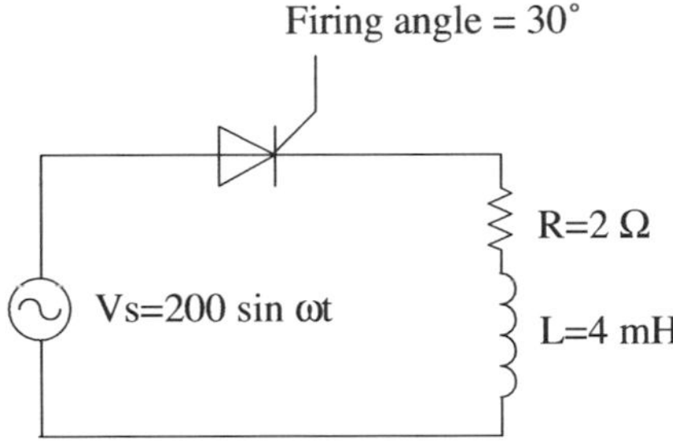

FIGURE 8.62
Circuit diagram for Problem 8.7.

8.6 A step-down (buck) converter is used to supply an inductive load of $R = 10\Omega$ and $L = 5mH$, with a D.C voltage of 80V, if D = 0.6, $F_s = 20\ KH_z$, and input DC voltage, calculate:

1. I_{Lmean} and V_{Lmean}
2. Trace the output voltage and current.

8.7 For the single-phase half-wave controlled rectifier circuit shown in Figure 8.62, $R = 2\Omega$, L = 4mH, $V_s = 200\sin\omega t$ volt, and firing angle $(\alpha) = 30°$.

i. Calculate V_{Lmean}, I_{Lmean}, and $v_{S_{r.m.s}}$
ii. Repeat the circuit at placed free-wheeling diode (FWD) and calculate V_{Lmean}, I_{Lmean}, $v_{S_{r.m.s}}$ and RF.

8.8 For the full-wave half-controlled rectifier circuit shown in Figure 8.63, $R = 17\Omega, L = 8H$, $V_{max} = 150$ volt and firing angle $(\alpha) = 30°$. Calculate and trace V_{Lmean}, I_{Lmean}, and $v_{S_{r.m.s}}$.

8.9 The DC chopper in Figure 8.64 has an inductive load of R = 0.5 Ω and L = 0.5 mH. The input voltage value is Vs = 120 V, f = 5 KHz, E = 0 V, and duty cycle = 0.4. Calculate the I_{min} and I_{max}, the average load current I_{av}, the average value of the

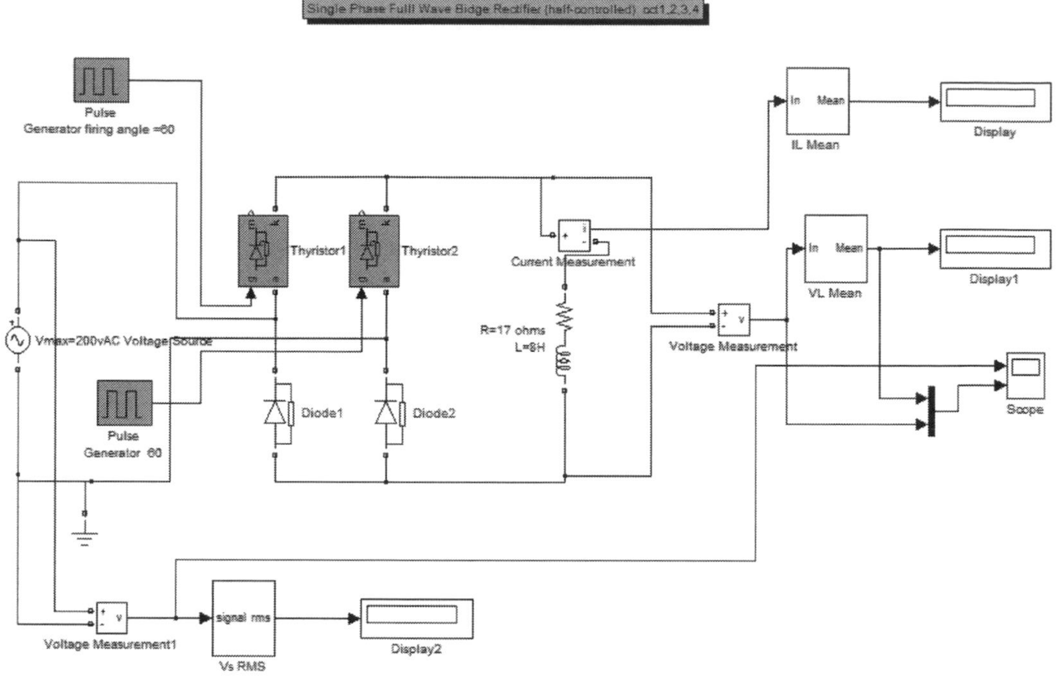

FIGURE 8.63
Full-wave half-controlled rectifier circuit Simulink.

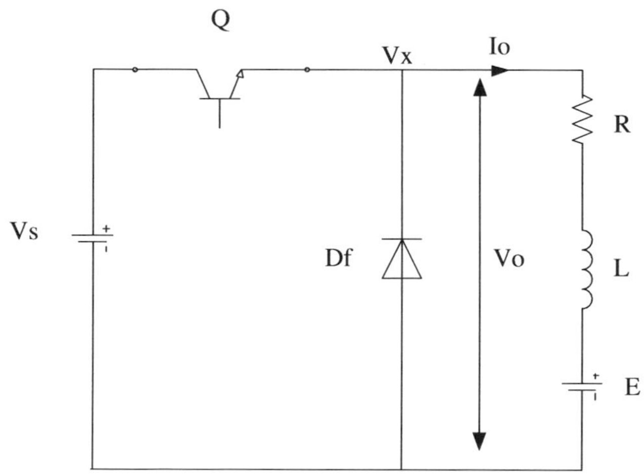

FIGURE 8.64
Circuit diagram for Problem 8.9.

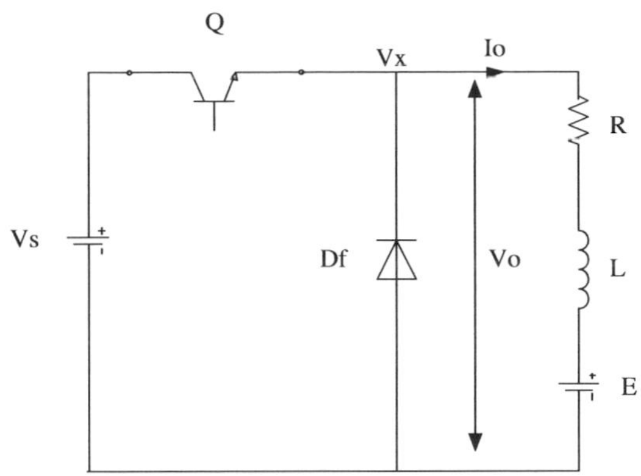

FIGURE 8.65
Circuit diagram for Problem 8.10.

diode current I_D, and the effective input resistance R_i. Draw the voltage across the chopper, the output voltage, and the current in the freewheeling diode.

8.10 The DC chopper in Figure 8.65 has an inductive load of R = 1.2 Ω and L = 0.5 mH. The input voltage value is Vs = 140 V, f = 5 KHz, E = 10 V, and duty cycle = 0.75. Calculate the I_{min} and I_{max}, the average load current I_{av}, the average value of the diode current I_D, and the effective input resistance R_i. Draw the voltage across the chopper, the output voltage, and the current in the freewheeling diode.

9

Concept of DC Drive

In imagination, there's no limitation.

Mark Victor Hansen

A DC motor is used to drive a mechanical load. Variable DC drives have been used to control DC motors longer than variable frequency drives have been used to control AC motors. The first motor-speed control used DC motors because of the simplicity of controlling the voltage to the armature and field of a DC motor.

9.1 DC Motors Drive

The brushed DC motor is one of the earliest motor designs. Today, it is the motor of choice in most of the variable speed and torque control applications.

9.1.1 Advantages

- Easy to understand the design
- Easy to control the speed
- Easy to control torque
- Simple, cheap drive design.

9.1.1.1 Easy to Understand the Design

The design of the brushed DC motor is quite simple. A permanent magnetic field is created in the stator by either of two means:

- Permanent magnets
- Electromagnetic windings.

If the field is created by permanent magnets, the motor is said to be a "permanent-magnet DC motor." (PMDC). If created by electromagnetic windings, the motor is often said to be a "shunt wound DC motor." (SWDC). Today, because of cost-effectiveness and reliability, the PMDC is the motor of choice for applications involving fractional horsepower DC motors, as well as most applications up to about 3 horsepower.

At 5 horsepower and greater, various forms of the shunt wound DC motor are most commonly used. This is because the electromagnetic windings are more cost effective than permanent magnets in this power range.

The section of the rotor where the electricity enters the rotor windings is called *the commutator.* The electricity is carried between the rotor and the stator by conductive graphite-copper brushes (mounted on the rotor), which contact rings on the stator. In most DC motors, several sets of windings or permanent magnets are present to smooth out the motion.

9.1.1.2 Easy to Control the Speed

Controlling the speed of a brushed DC motor is simple. The higher the armature voltage, the faster the rotation. This relationship is linear to the motor's maximum speed. The maximum armature voltage which corresponds to a motor's rated speed (these motors are usually given a rated speed and a maximum speed, such as 1750/2000 rpm) is available in certain standard voltages, which roughly increase in conjunction with horsepower. Thus, the smallest industrial motors are rated 90 and 180 V. Larger units are rated at 250 V and sometimes higher. Specialty motors for use in mobile applications are rated 12, 24, or 48 V. Other tiny motors may be rated 5 V.

Most industrial DC motors will operate reliably over a speed range of about 20:1 down to about 5%–7% of base speed. This is a much better performance than the comparable AC motor. This is partly due to the simplicity of control, but is also partly due to the fact that most industrial DC motors are designed with variable speed operation in mind and have added heat dissipation features which allow lower operating speeds.

9.1.1.3 Easy to Control Torque

In a brushed DC motor, torque control is also simple since output torque is proportional to current. If you limit the current, you have just limited the torque which the motor can achieve. This makes this motor ideal for delicate applications such as textile manufacturing.

9.1.1.4 Simple, Cheap Drive Design

The result of this design is that variable speed or variable torque electronics are easy to design and manufacture. Varying the speed of a brushed DC motor requires little more than a large enough potentiometer. In practice, these have been replaced for all but sub-fractional horsepower applications by the Silicon Controlled Rectifier (SCR) and pulse-width modulation (PWM) drives, which offer relatively precisely control voltage and current.

Large DC drives are available up to hundreds of horsepower. However, over about 10 horsepower careful consideration should be given to the price/performance trade-offs with AC inverter systems since the AC systems show a price advantage in the larger systems. (But they may not be capable of the application's performance requirements.)

9.1.2 Disadvantages

DC motors have the following disadvantages:

1. Expensive to produce
2. Not can reliably control at lowest speeds
3. Physically larger
4. High maintenance.

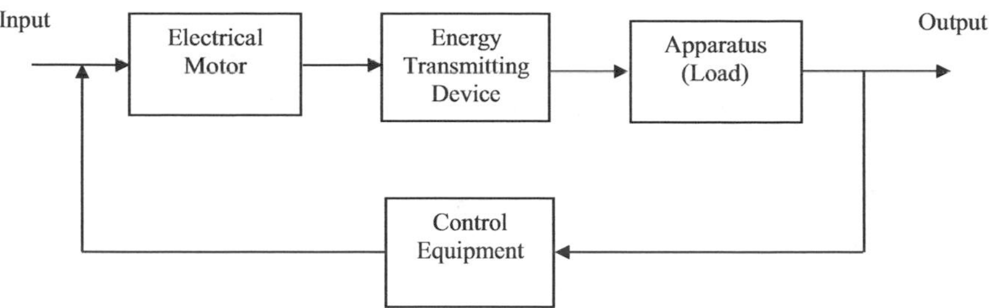

FIGURE 9.1
A block diagram of an electrical drive.

A drive consists of three main parts: prime mover; energy transmitting device; and actual apparatus (load), which perform the desired job. The function of the first two parts is to impart motion and operate the third one. In electrical drives, the prime mover is an electric motor or an electromagnet. A block diagram of an electrical drive shown in Figure 9.1.

The electromagnetic forces or torque developed by the driving motor tend to propagate motion of the drive system. This motion may be uniform translational or rotational motion or non-uniform as in case of starting, braking, or changing the load.

In the uniform motion, the motor torque (Figure 9.2) must be in a direction opposite to that of the load and equal to it, taken on the same shaft (motor or load shaft) the steady-state condition has a uniform motion at ω_m = Constant:

$$T_m = T_L$$

$$T_m - T_L = 0 \tag{9.1}$$

For a uniform motion $T_m \neq T_L$, therefore the system will either in acceleration if $T_m > T_L$ or deceleration when $T_m < T_L$ and the dynamic equation relating them is given by:

$$T_m - T_L = J\frac{d\omega}{dt} \tag{9.2}$$

where J = moment of inertia of all rotating part transformed to the motor shaft.

Types of the load: There are two main loads type:

Active load: which provides active torque (gravitational), deformation in elastic bodies, springs forces, compressed air, …), these load may cause motion of the system.

Passive loads: are that loads which have a torque all the times opposing the motion such as frictional load and shearing loads.

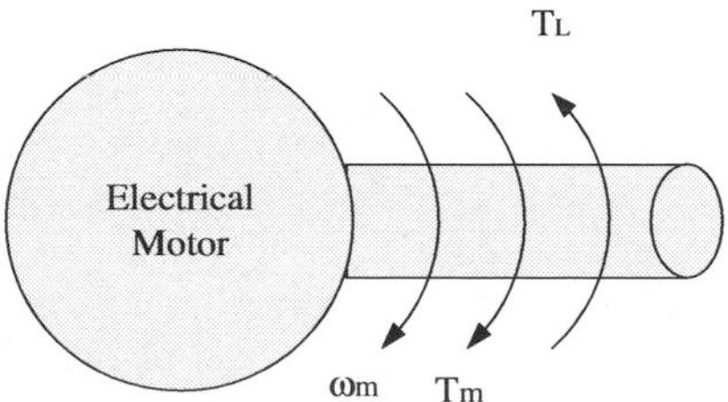

FIGURE 9.2
Motion of the motor drive and torque direction.

Also, loads may be sub-divided into:

1. *Constant loads*: which are unchanged with time or with a variation of speed
2. *Linear varying loads*:
   a. First order, $T_L = a + b\omega$ where a, b are constant
   b. Second order (second varying load), $T_L = a + b\,\omega + c\,\omega^2$, where a, b, c are constant and ω is the speed.
3. *The frictional load*: may be considered constant, which pump load consider linear, and compressor loads may consider the second order.

The electrical drive may operate in one of three main modes of operation:

1. *Continuous mode:* by which the motor is started and operate for enough time to reach steady-state temperature then when stopped (Figure 9.3), it must be left for enough to reach the initial temperature (room temperature).

   In such an operation and to avoid thermal stresses:

$$T_L \leq T_{rated} \text{ (torque of the motor).}\theta$$

$$T_s > T_L \text{ for the normal motor operation,}$$

   where T_s = starting torque.
2. *Interruptive operation:* In such operation, the motor operates and stops for approximately equal intervals by which at any interval, the motor will not reach steady-state temperature when started and not return to initial temperature when stopped as shown in Figure 9.4. In such operation and according to the ON and OFF intervals $T_L < T_{rated}$ of the motor (depending on the number of ON and OFF intervals).

   For each interval $T_s > T_L$ to ensure starting. The number of intervals is limited by the temperature reached which must not exceed the designed temperature limit θ_{max}. Such operation of drives may be found in the control system

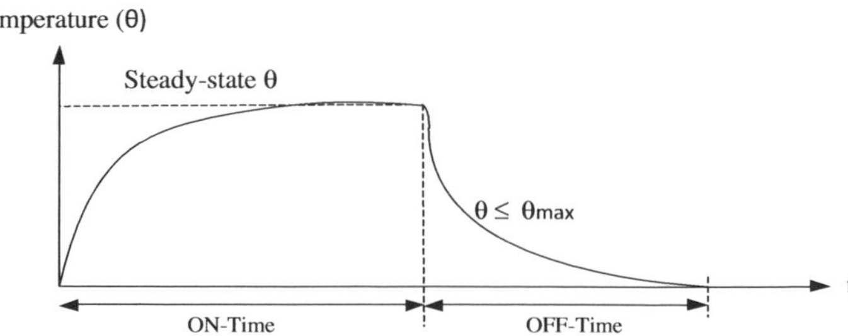

FIGURE 9.3
Continuous mode.

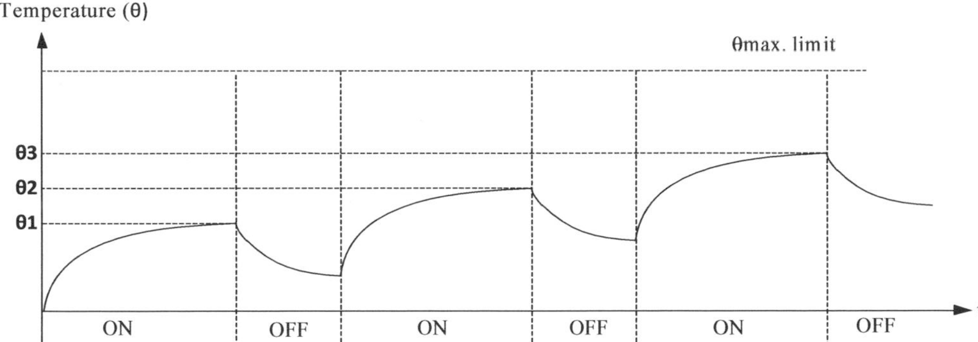

FIGURE 9.4
Interruptive operation.

3. *Short time operation*: in such operation, the motor is started and operate for a small interval of time, then switch-OFF for a very long period as shown in Figure 9.5.

   Θ may reach θ_{Max}, but the long OFF time will be sufficient to return the temperature to its initial value $t_{ON} << t_{OFF}$.

   The motor may be overloaded during the ON time so the temperature θ may reach θ_{Max} at small times, but the long OFF period will return it back to its initial value.

In such drive, $T_L > T_{rated}$ motor so multi-successive ON (starting) may damage the motor. Such derive may be found in cars starting system.

For quadrants operation of drives: By the effect of active and passive loads, and according to the relationship and directions of w, T_m, and T_L the electrical drive may operate in any quadrant of 4-quadrants diagram whose axis is $\pm T$ and $\pm\omega$ as shown in Figure 9.6.

The first quadrant (I) represents motor operation by which the motor torque and speed are in the same direction (+ve) and opposing the load torque T_L.

The speed of the motor and its torque is positive (counter clockwise direction). The speed may be changed from 0 to ω_o, where ω_o = no-load angular speed and the motor torque

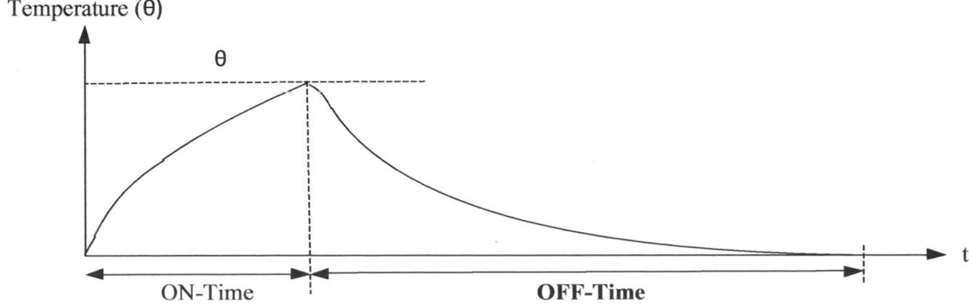

FIGURE 9.5
Short time operation.

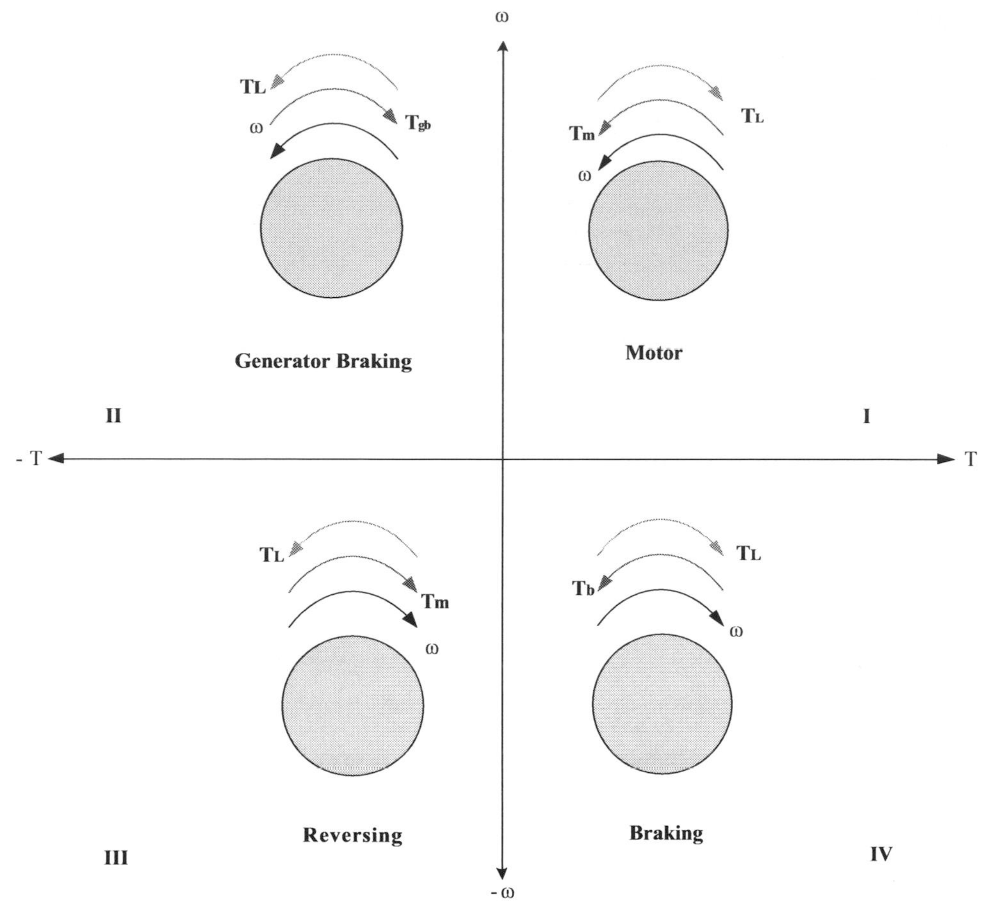

FIGURE 9.6
Quadrants operation of drives.

change from 0 at no-load to T_s at $\omega = 0$ (starting torque). For normal motor operation, the following relation must be satisfied:

$T_r \geq T_L \geq 0$ for the normal thermal operating condition

$0 < \omega_L \leq \omega_o$ according to the selective method of speed variation

$Ts > T_L$ for all motor operation to ensure starting

T_r = rated torque of the motor (N·m)

T_L = load torque of the motor shaft (N·m)

T_s = starting torque of the motor (N·m).

The normal steady-state operation is when $T_m = T_L$ which make the dynamic torque $(T_m - T_L) = 0$. This means that:

$$T_m - T_L = J\frac{d\omega}{dt} = 0$$

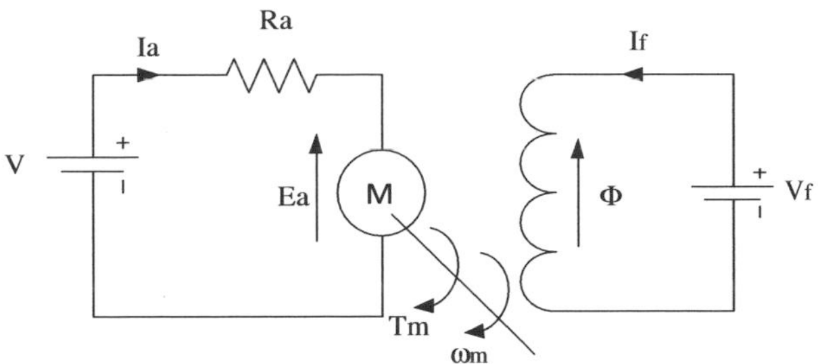

FIGURE 9.7
Equivalent circuit for separate excited DC motor.

or ω = constant = operating speed.

If for any reason the motor speed exceeds ω_o, then the motor back EMF I will be generator then the supply voltage (V). From equivalent circuit for separately excited DC motor in Figure 9.7, the armature current:

$$I_a = \frac{V - E}{R_a} \tag{9.3}$$

will be negative (return to the supply) and produce a torque opposing the speed (brake).

Therefore, moving the second quadrant will change the operation from motor to generator braking. At generator braking, ω is positive, but greater than ω_o and the motor torque opposes the motion (the load torque in the same direction as the motion).

At no-load speed ω_o, $E = V$, $I_a = 0$, $T_m = 0$ (Y-axis) the 3rd quadrant represent also motor operation, but in the reverse direction with respect to that in the first quadrant (i.e., rotation in the clockwise direction and T_m also in the clockwise direction and opposing T_L).

The 4th quadrant represents brake operation by which the motor takes current from the supply and produce positive torque, but opposing the motion.

The motion is started by the effect of the load which is greater than T_s (active) and hence the motor move opposite to its electrical direction causing E to change its direction. So,

$$I_a = \frac{V - E}{R_a} \rightarrow I_a = \frac{V + E}{R_a}$$

Since E is negative.

And hence T_m increase as ω increase until at certain ω:

$T_m = T_L$, and the load will brake at speed $= \omega\ \omega_o > \omega > 0$.

9.2 Torque-Speed Characteristics

In many applications, the rotational speed of the motor may be much higher than the required speed or vice versa. In other applications, multi-speeds are required. To change the load speed, there are mechanical and electrical methods as well as their combination. In mechanical method, the use of pullies, the gear may make the load speed equal to the required speed from given motor rotational speed.

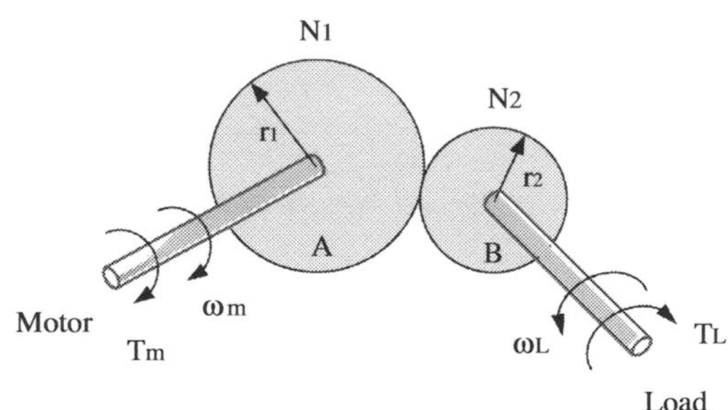

FIGURE 9.8
A simple two gears.

A simple two gears may have radius r_1 and r_2 and number of tooth N_1 and N_2 as shown in Figure 9.8, correspondingly, may give the following relationship. Since teeth pitch is the same for both gears:

So when gear A moves one revolution, ($2\pi r_1$), gear B must move the same length ($2\pi r_2$) according to the radius.

i.e.

$$1 \times 2\pi r_1 = i(2\pi r_2) \tag{9.4}$$

where:

$$i = \frac{Load\ \ speed}{Motor\ \ speed} = \frac{\omega_L}{\omega_m} \tag{9.5}$$

$$i = \frac{I_1'}{I_2'} = \frac{\mathrm{N}_1}{\mathrm{N}_2} \tag{9.6}$$

$$\omega_L = I{\cdot}\omega_m \tag{9.7}$$

$I > 1$ if $r_1 > r_2$ or $N_1 > N_2$.

If the gears have the efficiency of 100%:

So,

$$P_{in} = P_{out}.$$

Or

$$T_m{\cdot}\omega_m = T_L{\cdot}\omega_L,$$

$$\therefore T_L = T_m \cdot \frac{\omega_M}{\omega_L} = \frac{T_m}{i} \tag{9.8}$$

And if the efficiency $\eta \neq 100\%$ (losses) $1 > \eta > 0$:

So,

$$P_{in} = P_{out}/\eta \tag{9.9}$$

Or

$$T_m{\cdot}\omega_m = T_L{\cdot}\omega_L/\eta \tag{9.10}$$

$$\therefore T_L = \eta \times T_m \cdot \frac{\omega_M}{\omega_L} = \eta \times \frac{T_m}{i} \tag{9.11}$$

If there are multi-stages of gears with gear ratio = i_1, i_2, i_3, …

So,

$$\omega_L = (i_1 \times i_2 \times i_3 \times \ldots)\omega_m \tag{9.12}$$

For ideal case:

$$T_L = \frac{T_m}{i_1 \times i_2 \times i_3 \times \ldots\ldots} \tag{9.13}$$

For actual case:

$$T_L = \frac{(\eta_1 \times \eta_2 \times \eta_3 \times \ldots.)T_m}{i_1 \times i_2 \times i_3 \times \ldots\ldots} \tag{9.14}$$

where η_n represents the efficiency of stage n.

When the load is transferred to the motor shaft, and hence its speed will be ω_m, so the load torque at the motor shaft will be:

$$P_L = \omega_L \times T_L = \omega_m \times T'_L. \tag{9.15}$$

For ideal case:

$$T'_L = \frac{\omega_L}{\omega_m} \cdot T_L = i \cdot T_L \tag{9.16}$$

For actual case:

$$T'_L = \frac{\omega_L}{\omega_m \cdot \eta} \cdot T_L = \frac{i \cdot T_L}{\eta} \tag{9.17}$$

The equivalent system will be as shown (Figure 9.9):

So steady-state operation when:

$$T'_L = \frac{i \cdot T_L}{\eta} \tag{9.18}$$

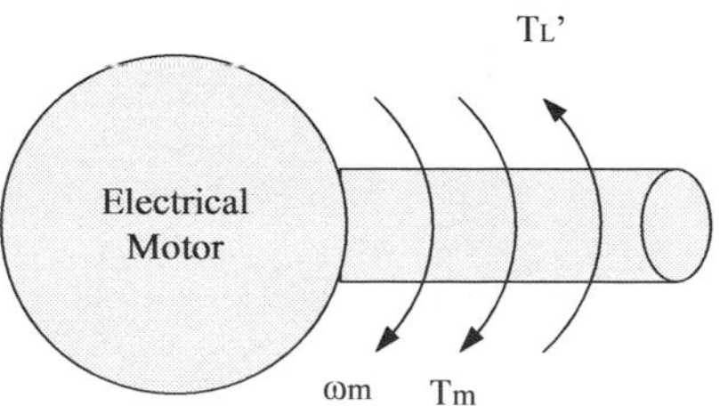

FIGURE 9.9
Equivalent system for a motor drive with the load.

Electrical methods: there are many electrical methods which are used to make the load operation at the required speed by changing ω_m. To select any method, the following points are taken into consideration:

1. The direction of speed variation: UP; DOWN; and UP and DOWN with respect to operating speed at natural characteristics
2. The dynamic range of speed variation, i.e., the possible ratio of maximum to minimum speed achieved:

$$D = \text{Dynamic range of speed} = \frac{\omega_m}{\omega_{min}}.$$

ω_{min} is taken to be at least 10% of rated speed

3. The stiffness coefficient of mechanical characteristics (β) which is equal to:

$$\beta = -\frac{dt}{d\omega} \tag{9.19}$$

This factor represents the sensitivity of the motor to the load variation. The ideal motor must keep constant speed when its load changes, i.e., $\beta = \infty$, and as this factor decrease, it means the motor will be much sense to load variation

4. The value of the load torque
5. The efficiency, simplicity, cost, ….

Different motors have a different electrical method of speed variation; therefore DC and AC motors will take each alone with their possible speed variation methods.

9.3 DC Motors Parametric Methods

9.3.1 Separate Excited and Shunt Motor

The main equations for the separately excited DC motor are:

$$I_a = \frac{V - E}{R_a} \tag{9.20}$$

$$E = k\,\phi\,\omega \tag{9.21}$$

$$T = k\,\phi\,Ia \tag{9.22}$$

$$K = constant = \frac{Z.P}{2\pi \cdot a} \tag{9.23}$$

where:

Z = conductors/armature

P = no. of poles

A = no. of parallel paths

A = 2 for wave connected, and a = P for lap connected.

The mechanical characteristics ($\omega = f(T)$) can be found from Equations 9.20 through 9.22 as:

$$\omega = \frac{V}{K\Phi} - \frac{R_a}{(K\Phi)^2} \cdot T \tag{9.24}$$

For separate or shunt motor, inter poles and compensating windings ϕ can be considered = constant independent of load.

So for natural characteristics ϕ = constant = ϕ_{nom}.

The mechanical characteristics will be a straight line since V, k, ϕ, R_a are constants.

If $V = V_{nom}$, $\phi = \phi_{nom}$, and no additional elements are inserted, the mechanical characteristics are natural characteristics. From Figure 9.10:

At no-load ($T = I_a = 0$)

$$\omega_o = \frac{V}{K\Phi} \quad (\text{no load speed}; E = V) \tag{9.25}$$

At $\omega = 0$, $T = T_{so}$ = starting condition is given by:

$$T_{so} = K\Phi \cdot \frac{V}{R_a} \quad (\text{when } E = 0) \tag{9.26}$$

If the motor is operating directly with the load at its natural characteristics:

So, $T_r \geq T_L \geq 0$, and $\omega_0 \geq \omega_L \geq \omega_r$.

So for any operating load torque, the operating point will lay on the mechanical characteristics between points A and B, for example at T_L, ω_L (point C).

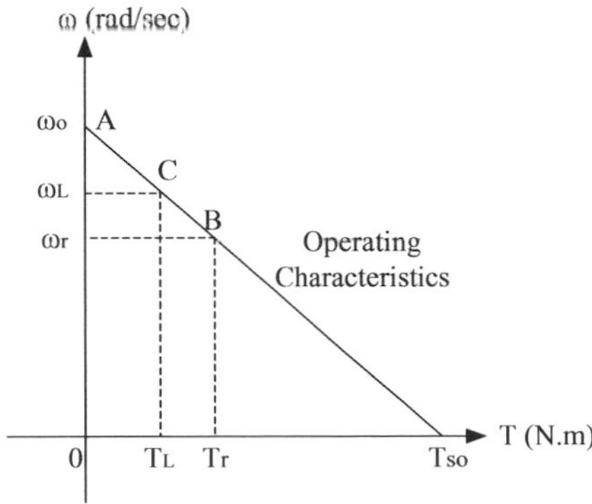

FIGURE 9.10
Torque-speed characteristics.

The stiffness coefficient:

$$\beta = -\frac{dT}{d\omega} = \frac{(K\Phi)^2}{R_a} \tag{9.27}$$

9.3.1.1 Adding Resistance to the Armature

From the equivalent circuit for adding resistance to the armature shown in Figure 9.11, the mechanical characteristics will be represented by:

$$\omega_i = \frac{V_{nom}}{K\Phi_{nom}} - \frac{R_a + R_d}{(K\Phi_{nom})^2} \cdot T \tag{9.28}$$

which gives different speed ω_i.

At different added R_d for given load torque T (Figure 9.12).

At no-load ($T = I_a = 0$):

$$\omega_i = \frac{V_{nom}}{K\Phi_{nom}} = \omega_o = \text{constant independent of } R_d.$$

At $\omega = 0$, $T = T_{si}$ = starting condition, and $E = 0$.

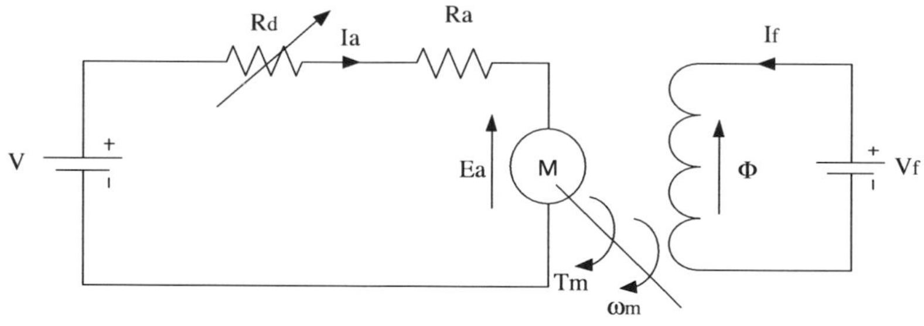

FIGURE 9.11
Adding resistance to the armature.

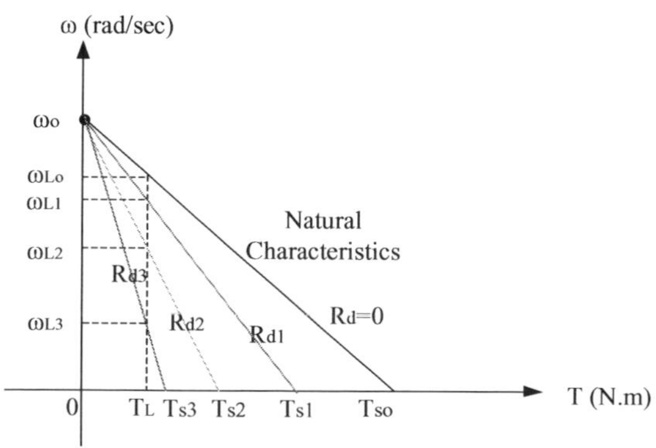

FIGURE 9.12
Torque-speed characteristics when adding resistance to the armature.

$$T_{si} = K\Phi_{nom} \cdot \frac{V_{nom}}{R_a + R_d}$$

$$T_{si} \downarrow as\, R_d \uparrow$$

$$\beta = -\frac{dT}{d\omega} = \frac{(K\Phi_{nom})^2}{R_a + R_d} \downarrow as\, R_d \uparrow.$$

If $$0 < R_{d1} < R_{d2} < R_{d3}.$$

So $$T_{so} > T_{s1} > T_{s2} > T_{s3}$$

$$\omega_{Lo} > \omega_{L1} > \omega_{L2} > \omega_{L3}$$

$$\beta_o > \beta_1 > \beta_2 < \beta_3.$$

The method can decrease the speed only to ω_{Lo} and the stiffness coefficient will be as R_d increase. The method is simple with low efficiency (high losses) due to I_a^2 (R_d + R), therefore it can be used only for small motors. Any operating condition can be achieved by a certain R_d which make the developed characteristics pass through that point. For example, it is required to operate a load of T_L at speed ω_L. Given natural characteristics, such motor can drive such load as shown in Figure 9.13.

If

$$T_L \le T_m \quad at \ \ rated$$

$$T_r = \frac{P_r}{\frac{2\pi N_r}{60}} \tag{9.29}$$

If the speed is given in rpm, the natural characteristics are not suitable if:

$$\omega = \frac{V_{nom}}{K\Phi_{nom}} - \frac{R_a}{(K\Phi_{nom})^2} \cdot T_L \tag{9.30}$$

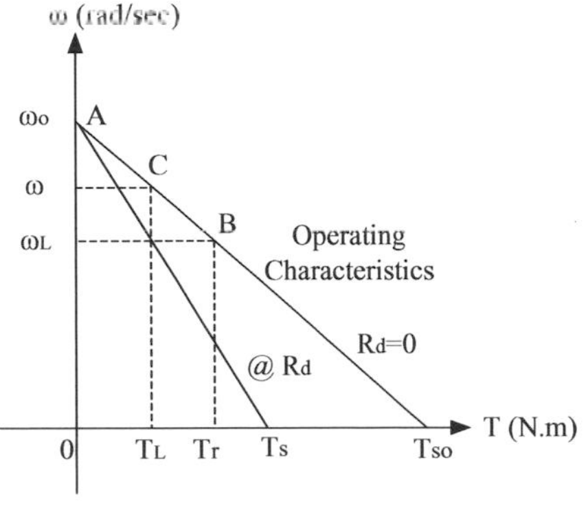

FIGURE 9.13
Torque-speed characteristics when adding resistance to the armature for two values of resistance.

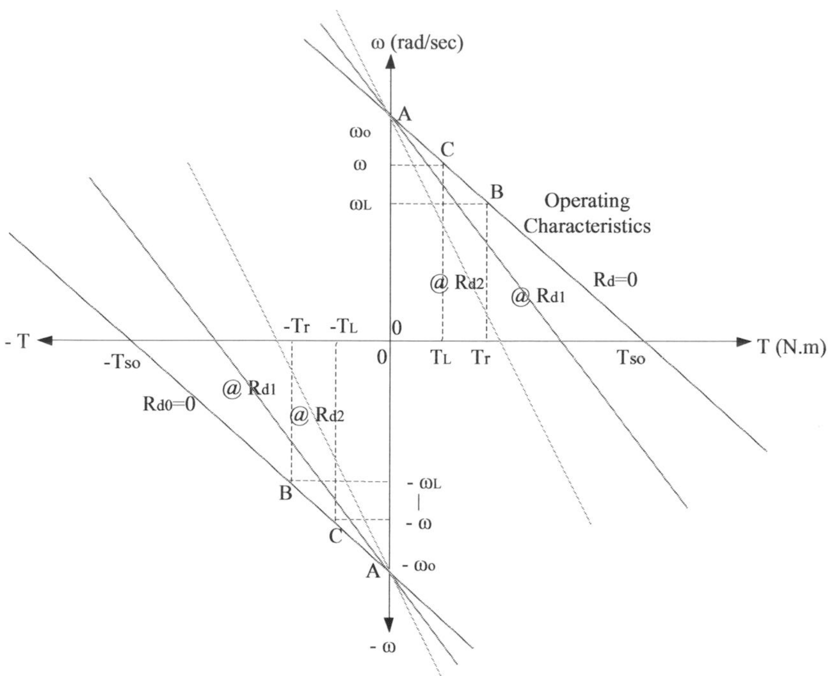

FIGURE 9.14
Torque-speed characteristics when adding resistance to the armature in two operation mode.

It is greater than ω_L required.

To find R_d such that the load T_L operates at ω_L, the characteristics must have R_d so that it passes through the point (T_L, ω_L) (Figure 9.14). R_d can be found if the point in the 1st quadrant (motor operation) either by using the equation:

$$\omega_L = \frac{V_{nom}}{K\Phi_{nom}} - \frac{R_a + R_d}{\left(K\Phi_{nom}\right)^2} \cdot T_L \tag{9.31}$$

So R_d can be found if the other parameters are known, or by similarity:

$$\frac{\omega_o}{T_s} = \frac{\omega_0 - \omega_L}{T_L} \tag{9.32}$$

So,

$$T_s = \frac{\omega_o}{\omega_0 - \omega_L} T_L \tag{9.33}$$

$$T_s - K\Phi_{nom} \cdot \frac{V_{nom}}{R_a + R_d} \tag{9.34}$$

For the motor to operate in the fourth quadrant, each developed characteristics can be extended to the other quadrant.

So, for a brake operation, ω_L will be negative will T_L is positive. For generator braking, ω_L is positive $>\omega_o$, but T_L is negative.

For reverse operation:

$$-\omega_L = \frac{V}{K\Phi} - \frac{R_a - R_d}{(K\Phi)^2} \cdot (-T_L) \tag{9.35}$$

9.3.1.2 Changing the Armature Supply Voltage

If the armature supply voltage can vary,
so $V_{nom} \geq V > 0$ will give a set of mechanical characteristics:

$$\omega = \frac{V}{K\Phi_{nom}} - \frac{R_a}{(K\Phi_{nom})^2} \cdot T_L \tag{9.36}$$

At any selected voltage, the characteristics are a straight line parallel to each other given by:

$$\omega_{0i} = \frac{V_{noim}}{K\Phi_{nom}} \tag{9.37}$$

$$T_{si} = K\Phi_{nom} \cdot \frac{V_i}{R_a} \tag{9.38}$$

As $V_i \downarrow$, $\omega_{oi} \downarrow$ *and* $T_{si} \downarrow$

$$\beta = -\frac{dT}{d\omega} = \frac{(K\Phi)^2}{R_a} = \text{constant.}$$

So the mechanical characteristics are parallel to each other as shown in Figure 9.15.

$$V_{nom} > V_1 > V_2$$

$$\omega_o > \omega_{o1} > \omega_{o2}$$

$$T_{so} > T_{s1} > T_{s2}.$$

At constant load T_L:

$$\omega_L > \omega_{L1} > \omega_{L2}.$$

So the speed variation only down. By the use of power electronics, the armature supply voltage can be changed. The efficiency of this method is better than that with R_d and the stiffness coefficient remains constant.

9.3.1.3 Changing the Field Flux

By adding resistance to the field circuit or by inserting series winding, the flux of the motor can be changed over and under ϕ_{nom} of the single winding as shown in Figure 9.16.

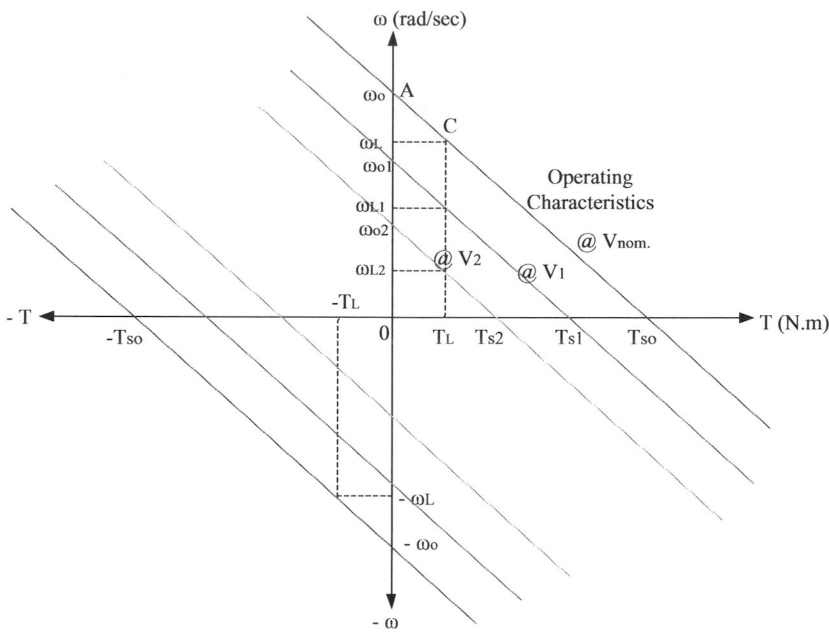

FIGURE 9.15
Torque-speed characteristics changing the armature supply voltage in two operation mode.

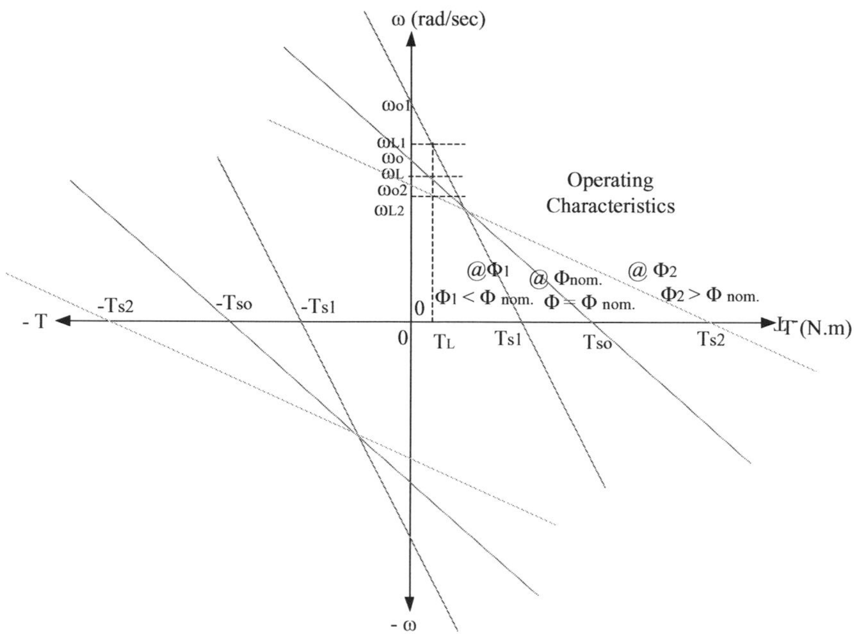

FIGURE 9.16
Torque-speed characteristics changing the field flux in two operation mode.

$$\omega_i = \frac{V_{nom}}{K\Phi_i} - \frac{R_a}{(K\Phi_i)^2} \cdot T_L.$$

$$\text{As } \Phi_i \uparrow, \omega_{oi} = \frac{V_{nom}}{K\Phi_i} \downarrow \quad \text{and} \quad T_{si} = \frac{V_{nom}}{R_a} \uparrow.$$

So for $\phi_1 < \phi_{nom}$

$$\omega_1 > \omega_o$$

$$T_{s1} > T_{so}$$

And for $\phi_2 > \phi_{nom}$

$$\omega_{o2} < \omega_o$$

$$T_{s2} > T_{so}.$$

According to the value of T_L, increasing or decreasing of the flux may give different results of speed (uncertainty), for example, decreasing the flux from ϕ_1 to ϕ_2 will give increasing of speed if the load torque is T_1 from ω_1 to ω_2. If the load torque is T_2 the change from the flux from ϕ_1 to ϕ_2 will not change the speed.

At load torque T_3 the changing from the flux from ϕ_1 to ϕ_2 will decrease the speed from ω_3 to ω_4 as shown in Figure 9.17.

The method affects the stiffness coefficient:

$$\beta = -\frac{dT}{d\omega} = \frac{(K\Phi)^2}{R_a} \downarrow$$

When $\phi_i \downarrow$, but the method is more efficient than that with R_d in the armature.

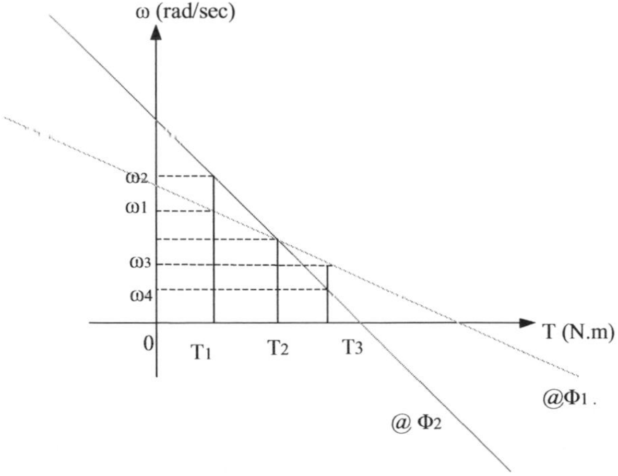

FIGURE 9.17
Torque-speed characteristics changing the field flux for two values of flux.

Example 9.1

A separate excited, 3 kW, 150 Volt, 1610 rpm, DC motor has an armature resistance = 1 Ω and is to be used in the drive through a reducer of Nm/NL = 10. If the load torque is 150 N·m, calculate the required R_d added to the armature to operate the load at 1 rad/sec? What voltage reduction is used for such operation if no R_d is added?

Solution

$$\omega_r = \frac{2\pi \cdot n_r}{60} = \frac{2\pi \times 1610}{60} = 168.6\,\text{rad/sec}$$

$$T_L' = \frac{150}{10} = 15\,\text{N}\cdot\text{m}$$

$$T_{mr} = \frac{P_r}{\omega_r} = \frac{3\times 10^3}{168.6} = 17.79\,\text{N}\cdot\text{m}$$

(such motor can drive the load 15 N·m).

For natural characteristics,

$$\omega_r = \frac{V}{K\Phi} - \frac{R_a}{(K\Phi)^2}\cdot T_r$$

$$168.6 = \frac{150}{K\Phi} - \frac{1}{(K\Phi)^2}\times 17.79$$

Solving:

$$168.6(K\Phi)^2 - 150\,(K\Phi) + 17.79 = 0$$

$$K\Phi = 0.7487.\ Or\ K\Phi = 0.1409$$

$$\omega_o = \frac{V}{K\Phi}$$

$$\omega_o = \frac{150}{0.7487} = 200.34\,rad/sec$$

Or:

$$\omega_o = \frac{150}{0.1409} = 1064.58\ \ rad/sec$$

By neglecting the high value of ω_o.

So $k\Phi = 0.7487$.

To find R_d required:

$$10 = 200.34 - \frac{1+R_d}{(0.7487)^2}\cdot 15$$

$$R_d = 6.11\ \Omega$$

$$\text{For } R_d = 0\ \Omega$$

$$10 = \frac{V}{0.7487} - \frac{1}{(0.7487)^2} \cdot 15$$

$$V = 27.52\ \text{V}.$$

So the reduction of supply voltage = 150 – 27.52 = 122.478 Volt.

9.3.2 Series Motor

For the series motor, the value of R_f (field resistance) is added to armature total resistance and the change of load (torque) will change also ϕ.

So ϕ is no longer constant, but the function of the armature current, for example, (Figure 9.18):

$$K\Phi = \frac{A \cdot I_a^2 + B \cdot I_a + C}{D \cdot I_a} \tag{9.39}$$

where A, B, C, and D are constants whose values give a linear relationship at small I_a and constant kϕ at very high I_a (saturation). Figure 9.19 shows torque-speed characteristics for the series motor.

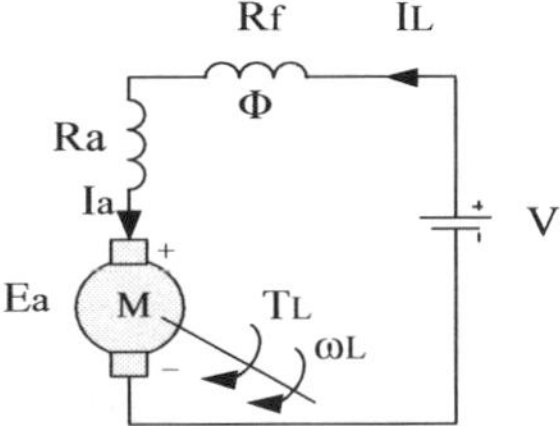

FIGURE 9.18
Equivalent circuit for series motor.

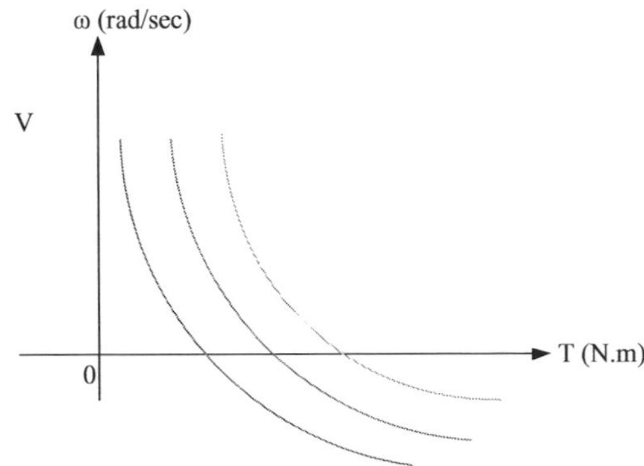

FIGURE 9.19
Torque-speed characteristics for the series motor.

At no-load ideally $\phi = 0$ and $\omega_o = \infty$, therefore no-load or high load operation of the series motor is not allowed since the resultant speed is dangerously high. For the same reason, a movement from 1st quadrant (generator braking) is not allowed also. The mechanical characteristics are given by the same equation:

$$\omega = \frac{V}{K\Phi} - \frac{R_a}{(K\Phi)^2} \cdot T \text{ with } \Phi \text{ changing with T.}$$

Adding the resistance to the armature or changing the supply voltage will change the ϕ also. The characteristics are a no-longer straight line.

Example 9.2

A series motor 120 Volt, 2.2 KW, 850 rpm, has $R_a = 0.5\ \Omega$ and $R_F = 0.3\ \Omega$. The field is linearly proportional to armature current. Find the required additional armature resistance to drive a load of 20 N·m at speed of 250 rpm.

Solution

Let $k\phi = A \cdot I_a$,
where A = constant.

$$\omega_r = \frac{2\pi \cdot n_r}{60} = \frac{2\pi \times 850}{60} = 89.011 \text{rad/sec}$$

$$T_{mr} = \frac{P_r}{\omega_r} = \frac{2.2 \times 10^3}{89.011} = 24.716 \text{N.M}$$

$$T = K\Phi \cdot I_a = A \cdot I_a^2$$

$$\omega_r = \frac{V}{K\Phi} - \frac{R_a + R_F}{(K\Phi)^2} \cdot T_r$$

$$89.011 = \frac{120}{K\Phi} - \frac{0.5 + 0.3}{(K\Phi)^2} \cdot 24.716.$$

Solving:

$$89.011(K\Phi)^2 - 120(K\Phi) + 19.7728 = 0.$$

$$\text{Either } K\Phi = 1.156$$

$$\text{or} = 0.192.$$

$$\omega_o = \frac{V}{K\Phi}$$

$$\omega_o = \frac{120}{1.156} = 103.8 \text{ rad/sec}$$

or

$$\omega_o = \frac{120}{0.192} = 625 \text{ rad/sec.}$$

By neglecting the high value of ω_o,

so

$$\therefore k\phi = 1.156 = A{\cdot}I_a$$

$$T_r = A{\cdot}I_a^2 = 24.716$$

$$\therefore \frac{T_r}{K\Phi} = I_a = \frac{24.716}{1.156} = 21.38\,A$$

$$1.156 = A{\cdot}21.38,$$

$$\text{so } A = 0.054$$

To find R_d required:

$$10 = 200.34 - \frac{1 + R_d}{(0.7487)^2} \cdot 15,$$

$$R_d = 6.11\ \Omega.$$

$$\text{For } R_d = 0\ \Omega$$

$$T_L = A{\cdot}I_a^2$$

$$20 = 0.054\ I_a^2$$

$$I_a = 19.24 \text{ A.}$$

$$\text{New } k\phi = A{\cdot}I_a = 0.054 \times 19.24 = 1.039$$

$$\omega_L = \frac{2\pi n_L}{60}$$

$$\omega_L = \frac{2\pi \times 250}{60} = 26.18\, rad/sec$$

$$\omega_L = \frac{V}{K\Phi} - \frac{R_a + R_F + R_d}{(K\Phi)^2} \cdot T_L$$

$$26.18 = \frac{120}{1.039} - \frac{0.5 + 0.3 + R_d}{(1.039)^2} \cdot 20,$$

$$R_d = 4.021\ \Omega.$$

9.4 DC Drive Circuits

The DC drives are classified under:

1. Single-phase DC drives
2. Three-phase DC drives
3. Chopper drives.

First, the basic operating characteristics of DC motor are presented and then power electronics circuits strategies as mentioned in previous chapters are described. In this section, the single-phase rectifier drive and DC chopper drive will discuss.

9.4.1 DC Drive Rectifier Circuits

9.4.1.1 Single-Phase Half-Wave Converter Drives a Separately Excited DC Motor

Assume that this circuit fed through single-phase half-wave converter is shown in Figure 9.20. Motor field circuit is fed through a single-phase semi converter in order to reduce the ripple content in the field circuit. Single-phase half-wave converter feeding a DC motor offers one-quadrant drive, Figure 9.21a. The waveforms for source voltage V_s, armature terminal voltage V_t, armature current i_a, source current i_s, and freewheeling diode current i_{fd}. Note that thyristor current $i_T = i_s$: The armature current is assumed ripple free.

For single-phase half-wave converter, the average output voltage of the converter, V_o = armature terminal voltage, V_t is given by:

$$V_0 = V_t = \frac{V_m}{2\pi}(1+\cos\alpha) \qquad For\, 0<\alpha<\pi, \tag{9.40}$$

where V_m = maximum value of source voltage. For single-phase semi converter in the field circuit, the average output voltage is given by:

$$V_f = \frac{V_m}{\pi}(1+\cos\alpha_1) \qquad For\, 0<\alpha_1<\pi. \tag{9.41}$$

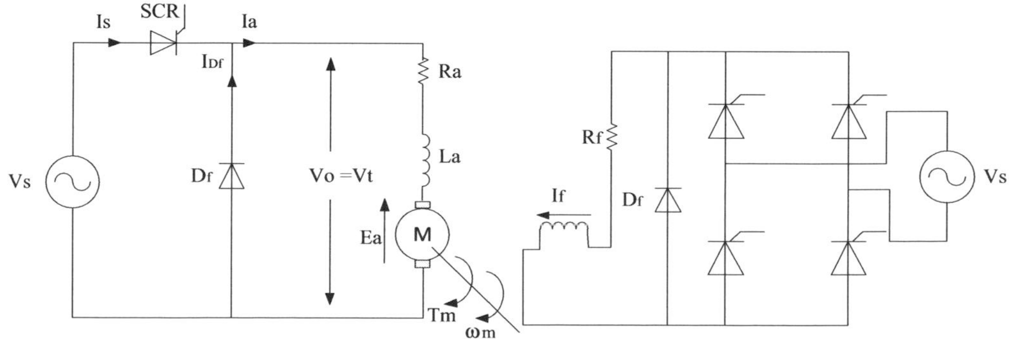

FIGURE 9.20
Circuit diagram of single-phase half-wave converter drive.

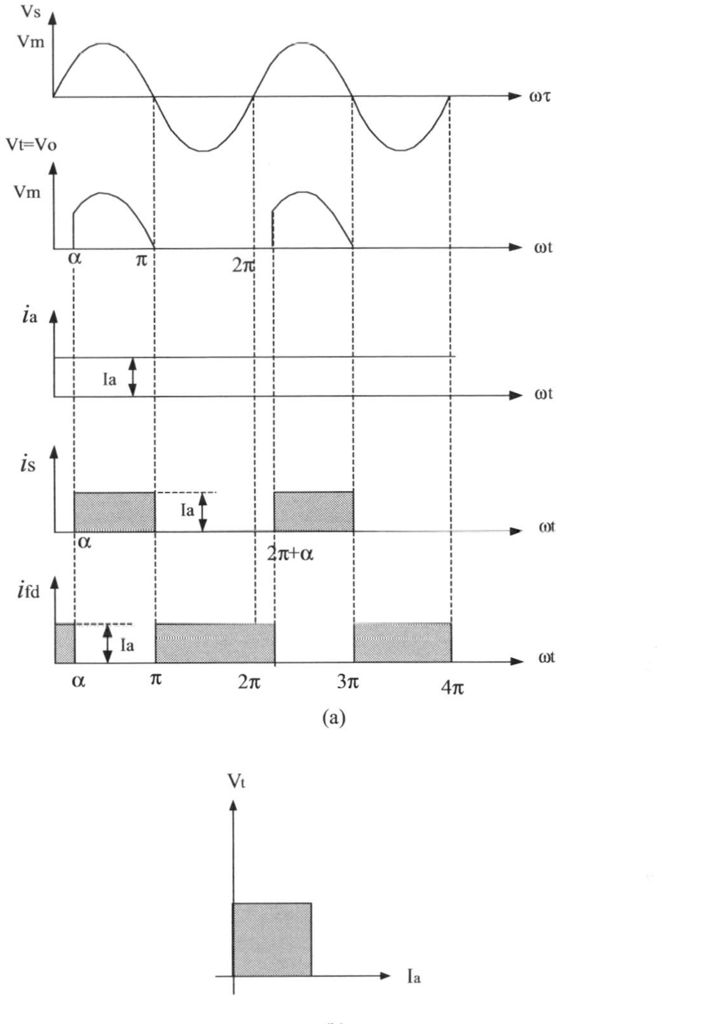

FIGURE 9.21
Waveforms single-phase half-wave converter drive (a) Voltages and currents waveforms (b) V-I characteristics.

It is seen from the waveforms of Figure 9.21a that RMS value of armature current, $I_{ar} = I_{aRMS}$ value of source or Thyristor current, and the relation between armature current and terminal voltage given in Figure 9.21b.

$$I_{sr} = \sqrt{I_a^2 \frac{\pi - \alpha}{2\pi}} = I_a \left(\frac{\pi - \alpha}{2\pi} \right)^{1/2}. \tag{9.42}$$

The RMS value of freewheeling-diode current,

$$I_{sr} = \sqrt{I_a^2 \frac{\pi + \alpha}{2\pi}} = I_a \left(\frac{\pi + \alpha}{2\pi} \right)^{1/2}. \tag{9.43}$$

$$Apparent\ input\ power = (rms\ source\ voltage) \cdot (rms\ source\ current)$$
$$= V_s \cdot I_{sr}. \tag{9.44}$$

$$\text{Power delivered to motor} = E_a I_a + I_a^2 \cdot r_a = (E_a + I_a r_a)\, I_a = V_t \cdot I_a \tag{9.45}$$

Input supply power factor:

$$pf = \frac{E_a I_a + I_a^2 r_a}{V_s \cdot I_{sr}} \tag{9.46}$$

Example 9.3

A separately excited DC motor has $k\varphi = 1.44$, $R_a = 0.86\ \Omega$. When it is operating at 150 rad/sec, the armature current is $I_a = 40A$. The terminal voltage V_t is held constant under all conditions.

1. Compute V_t
2. Compute the no-load speed in rad/sec.

Solution

1.

$$E_a = k\varphi w_m = (1.44) \cdot (150) = 216V$$

$$V_t = E_a + I_a r_a = 216 + (40) \cdot (0.86).$$
$$= 250.4V$$

2. "No-load" speed occurs when T = 0. Therefore,

$$\omega = \frac{V_t}{k\varphi} - \frac{R_a T}{(k\varphi)^2} = \frac{V_t}{k\varphi}. \text{ When T} = 0$$

$$\omega_0 = \frac{250.4}{1.44} = 173.9\,\text{rad/sec.}$$

Example 9.4

A separately excited DC motor is supplied from 220 V, 60 Hz source through a single-phase half-wave controlled converter. Its field is fed through i-phase semi converter with zero degrees firing-angle delay. Motor resistance R_a = 0.1 Ω, and motor constant = 0.6 V-sec/rad. For rated load torque of 15 N·m at 1200 rpm and for continuous ripple free currents, determine:

1. The firing angle delay of the armature converter
2. The RMS value of thyristor and freewheeling diode currents
3. The input power factor of the armature converter.

Solution

1. The average current = 15/0.5 = 30 A

$$\omega_m = \frac{2\pi \cdot n_m}{60}$$

$$\omega_m = \frac{2\pi \cdot 1200}{60} = 125.66\, rad\,/\,sec.$$

Motor EMF $E_a = K_m \cdot \omega_m = 0.6 \times 125.66 = 75.4$ V.

For 1-phase half-wave converter feeding a DC motor,

$$V_0 = V_t = \frac{V_m}{2\pi}(1 + \cos\alpha) = E_a + I_a r_a$$

$$V_0 = V_t = \frac{220 \cdot \sqrt{2}}{2\pi}(1 + \cos\alpha) = 75.4 + 30 \times 0.1.$$

Solving for α to get α = 54.32° = 0.948 rad.

2. RMS value of thyristor current is:

$$I_{sr} = \sqrt{I_a^2 \frac{\pi - \alpha}{2\pi}} = I_a\left(\frac{\pi - \alpha}{2\pi}\right)^{1/2}$$

$$I_{sr} = 30\left(\frac{\pi - 0.948}{2\pi}\right)^{1/2} = 17.725\,A,$$

the RMS value of freewheeling diode current is:

$$I_{sr} = \sqrt{I_a^2 \frac{\pi + \alpha}{2\pi}} = I_a\left(\frac{\pi + \alpha}{2\pi}\right)^{1/2}$$

$$I_{sr} = 30\left(\frac{\pi + 0.948}{2\pi}\right)^{1/2} = 24.2\,A.$$

3. The input power factor of armature converter:

$$pf = \frac{V_t \cdot I_a}{V_s \cdot I_{sr}}$$

$$pf = \frac{78.4 \times 30}{220 \times 24.2} = 0.441\, lagging.$$

Example 9.5

Draw the speed response for separate excited DC motor drive: A 5 hp, 240 V separately excited DC motor with parameters given as follows in Figure 9.22.

Solution

The Simulink of the circuit as shown in Figure 9.23, and the related waveforms given in Figure 9.24.

Block Parameters: DC_Motor 5 hp; 240V; 16.2 A; 1220rpm

DC machine (mask) (link)

Implements a (wound-field or permanent magnet) DC machine.
For the wound-field DC machine, access is provided to the field connections so that the machine can be used as a separately excited, shunt-connected or a series-connected DC machine.

Configuration | Parameters | Advanced

Armature resistance and inductance [Ra (ohms) La (H)]	[0.78 0.016]
Field resistance and inductance [Rf (ohms) Lf (H)]	[150 112.5]
Field-armature mutual inductance Laf (H) :	1.234
Total inertia J (kg.m^2)	0.05
Viscous friction coefficient Bm (N.m.s)	0.01
Coulomb friction torque Tf (N.m)	0
Initial speed (rad/s) :	1
Initial field current:	1.6

OK | Cancel | Help | Apply

FIGURE 9.22
DC machine mask.

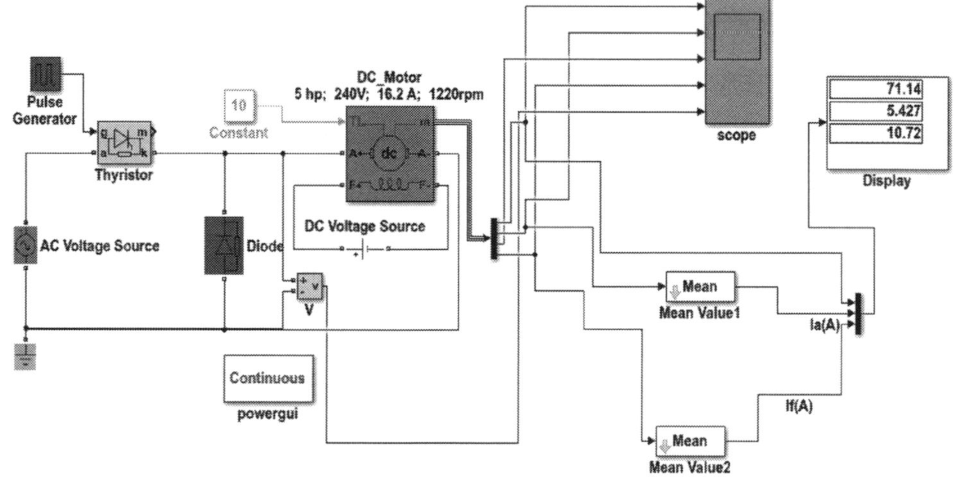

FIGURE 9.23
Separate excited DC motor drive Simulink.

9.4.1.2 Single-Phase Full-Wave Converter Drives a Separately Excited DC Motor

Example 9.6

A separately excited DC motor drives a rated load torque of 40 Nm at 1000 rpm as shown in Figure 9.25. The field circuit resistance is 200 Ω and armature circuit resistance is 0.05 Ω. The field winding, connected to single-phase, 120 V source, is fed through 1-phase full converter with zero degrees firing angle. The armature circuit is also fed through another full converter from the same converter single-phase, 120 V source.

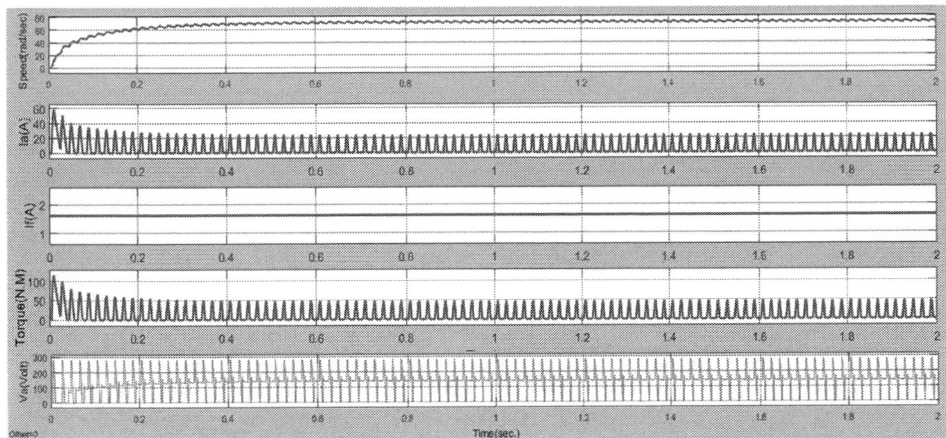

FIGURE 9.24
Waveforms.

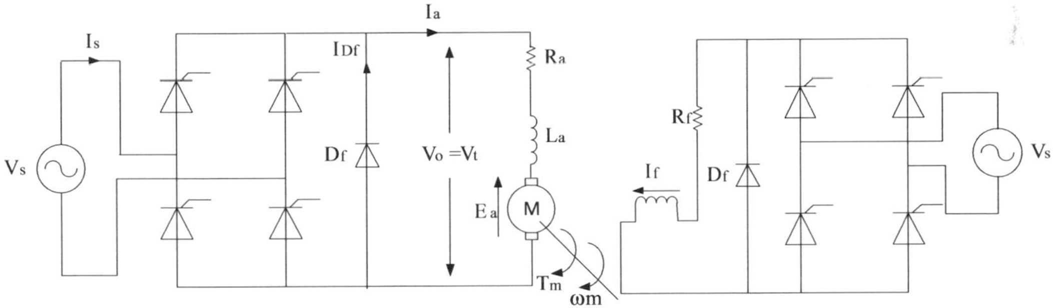

FIGURE 9.25
Separately excited DC motor drives.

With magnetic saturation neglected, the motor constant is 0.8 V-sec/A-rad. For ripple free armature and field currents. Determine:

1. Rated armature current
2. Firing angle delay of armature converter at rated load
3. Speed regulation at full load
4. Input power factor (PF) of the armature converter and the drive at rated load.

Solution

1. For field converter, angle delay $\alpha = 0°$.

The field voltage,

$$V_f = \frac{2V_m}{\pi} = \frac{2\sqrt{2} \times 120}{\pi} = 108.03\,V$$

$$I_f = \frac{V_f}{r_f} = \frac{108.03}{200} = 0.54\,A.$$

With magnetic saturation neglected, $\phi_1 = K_1 I_f$

$$E_a = K_a \phi \omega_m = K_a \cdot K_1 \cdot I_f \cdot \omega_m = K \cdot I_f \cdot \omega_m$$

and the torque $T = K_a \phi I_a = K_a \cdot K_1 \cdot I_f \cdot I_a = K \cdot I_f \cdot I_a$.
The rated armature current, I_a:

$$I_a = \frac{T}{K \cdot I_f}$$

$$I_a = \frac{40}{0.8 \times 0.54} = 92.59\,A$$

2.

$$V_0 = V_t = \frac{2V_m}{\pi}(\cos\alpha) = E_a + I_a r_a$$

$$V_0 = V_t = \frac{2\sqrt{2} \times 120}{\pi}(\cos\alpha) = 0.8 \times 0.54 \times \frac{2\pi \times 1000}{60} + 92.59 \times 0.05$$

$$\alpha = 62.51°.$$

3. At the same firing angle of $\alpha = 62.51°$ motor EMF at no-load,

$$E_a = K_a \phi \omega_m = K_a \cdot K_1 \cdot I_f \cdot \omega_m = K \cdot I_f \cdot \omega_m = V_t = 49.87\ \text{V}.$$

The speed at no-load:

$$\omega_{mo} = \frac{E_a}{K \cdot I_f} = \frac{49.87}{0.8 \times 0.54} = 115.43\, rad/sec.$$

Or 1102.37 rpm.
Speed regulation at full load:

$$N_R\% = \frac{\text{No load speed} - \text{Full load speed}}{\text{Full load speed}}$$

$$N_R\% = \frac{1102.37 - 1000}{1000} = 10.23\%.$$

4. The input power factor of the armature converter

$$pf = \frac{V_t \cdot I_a}{V_s \cdot I_{ar}}$$

$$pf = \frac{49.87 \times 92.59}{120 \times 92.59} = 0.415 \quad lagging$$

RMS value of current in armature converter,

$$I_{ar} = 92.59\ \text{A}.$$

RMS value of current in field circuit,

$$I_{fT} = I_f = 0.54 \text{ A}.$$

Total current taken from the source,

$$I_{sr} = \sqrt{I_{ar}^2 + I_{fr}^2}$$

$$I_{sr} = \sqrt{92.59^2 + 0.54^2} = 92.591 A.$$

Input S = $V_s \cdot I_{sr}$ = 120 × 92.591 = 11.11 kVA.

With no loss in the converters, total power input to motor and field:

$$= V_t \cdot I_a + V_f I_f = 49.87 \times 59.03 + 108.03 \times 0.54 = 3 \text{ KW}.$$

$$Input\ power\ factor = \frac{P}{S} = \frac{3}{11.11} = 0.27\,\text{lagging}.$$

Example 9.7

Draw the speed response for DC shunt motor drive: A 5 hp, 240 V, 1750 rpm, separate excited DC motor with parameters given as follows:

R_a = 2.581 Ω, L_a = 0.028 H, R_f = 281.3 Ω, L_f = 156 H, J = 0.02215 kg·m².

Solution

The Simulink of the circuit as shown in Figure 9.26, and the related waveforms given in Figure 9.27.

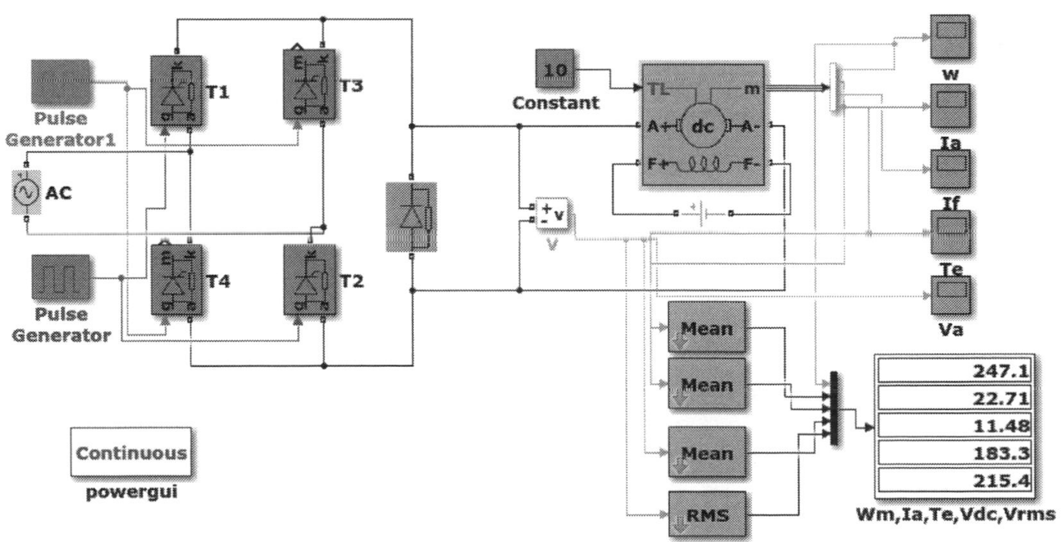

FIGURE 9.26
Circuit diagram using MATLAB/Simulink.

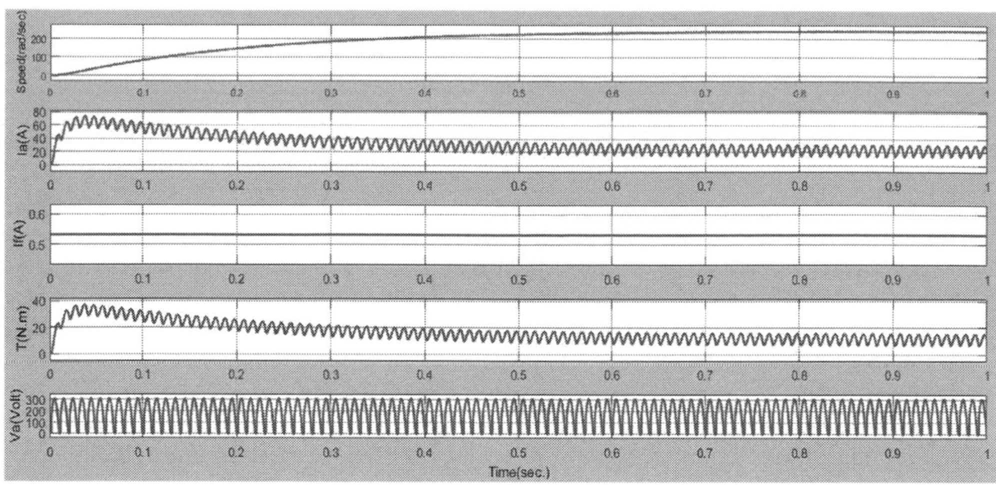

FIGURE 9.27
Waveforms.

9.5 DC Chopper Drive

The speed of a DC Motor is directly proportional to the line voltage applied to it. Given a fixed DC source, VS, and a power Insulated Gate Bi-polar Transistor (IGBT) to act as a switch, it is possible to control the average voltage applied to the motor using a technique called PWM.

In the circuit shown in Figure 9.28 below, the source voltage, V_s, is "chopped" to produce an average voltage somewhere between 0% and 100% of V_s. Thus, the average value of the voltage applied to the motor, V_m, is controlled by closing and opening the "switch," Q1. To close the switch, a firing signal is delivered to the gate of the IGBT, causing it to conduct between source and drain. To open the switch, the firing signal is removed and the IGBT is self-biased to stop conducting. In PWM, the switch is closed and opened every modulation period.

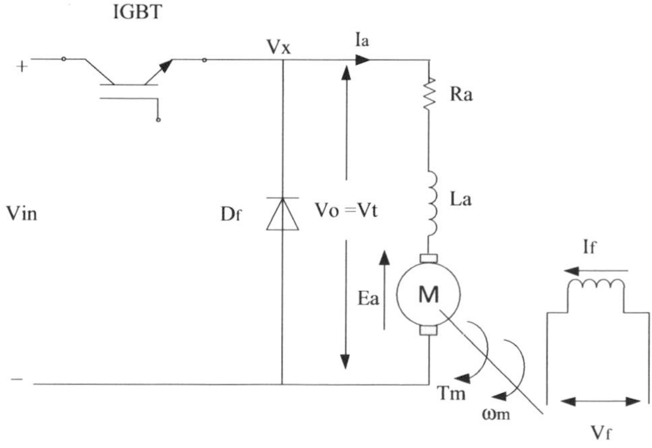

FIGURE 9.28
DC chopper drive circuit.

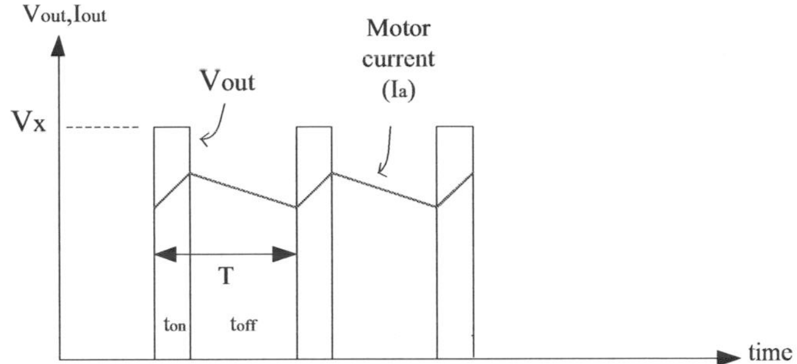

FIGURE 9.29
Waveforms of DC chopper drive circuit.

In discussing the period of modulation, let time be divided into uniform periods of 1 millisecond each and let a period be called T, the modulation period. During T, there is a time, t_0 to t_1, during which the MOSFET Q1 is on, and a time, t_1 to t_2, during which it is off, as indicated in Figure 9.29. This is true for each period and therefore Q1 turns on and off 1000 times every second when T = 1 ms.

When Q1 is on, V_s volts are applied to the motor for t_1 milliseconds. When Q1 is off, zero volts are applied to the motor. However, the armature current, I_a, is still allowed to circulate through the diode. The magnitude of the armature current will diminish between t_1 and t_2 as losses in the motor dissipate energy.

The voltage V_m seen by the motor can be expressed in terms of the source voltage V_s and the "ON" time t_1 and the period of modulation T. The equation is:

$$V_m = \alpha V_s \tag{9.47}$$

where:

$$\alpha = t_1/T \tag{9.48}$$

The symbol α is called *the duty cycle*. As the duty cycle is increased from 0% to 100%, the average voltage applied to the motor increases from 0 to V_s volts and the motor speeds up.

It is sometimes more useful to think of the frequency of modulation as opposed to its period. The modulation frequency is just the inverse of the period of modulation, $f_m = 1/T$. State of the art PWM converters have modulation frequencies as high as 200 k, although some people argue that frequencies above 20 k lead to unnecessary expense.

Example 9.8

Draw speed, armature current, field current, and torque for the DC chopper drive circuit shown in Figure 9.30 using MATLAB/Simulink.

Solution

The Simulink of the circuit as shown in Figure 9.30 and the related waveforms given in Figure 9.31.

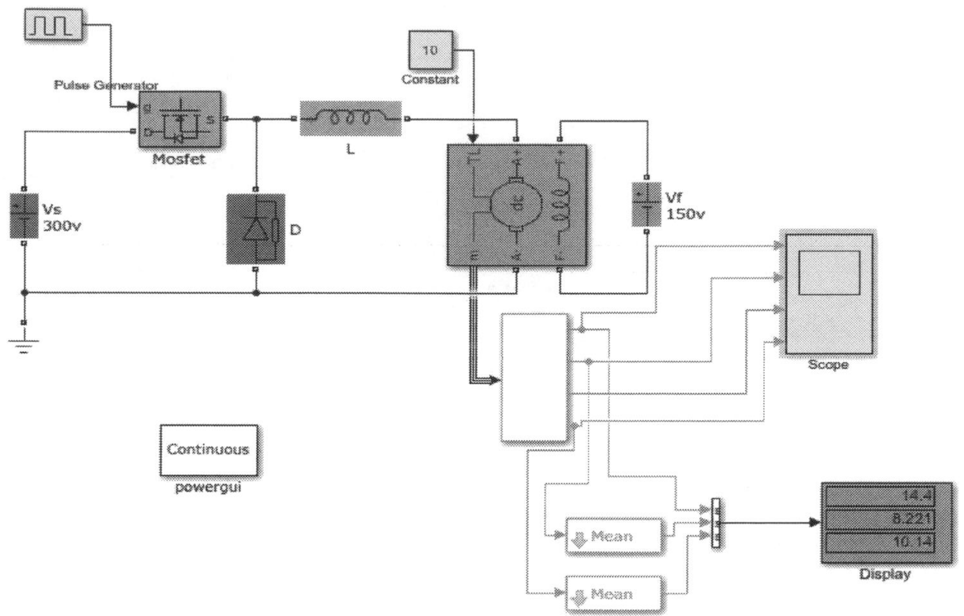

FIGURE 9.30
DC chopper drive circuit using MATLAB/Simulink.

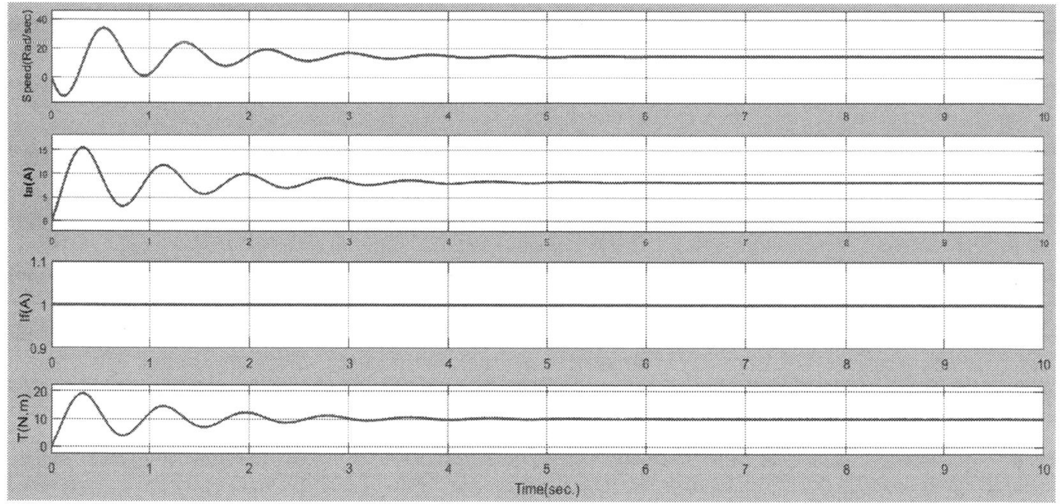

FIGURE 9.31
Waveforms.

Example 9.9

A step-down (buck) converter in Figure 9.32 is used to supply 100 V DC motor and an inductive load of $L = 4mH, R_a = 0$, if $F_s = 5\ KHz$, $T = 50N \cdot m$, and input DC voltage an is 300 V, calculate:

1. V_{Lmean} and speed of duty cycle range 0.4–0.8
2. Trace the output voltage and current of case D = 0.7.

Solution

The Simulink of the circuit as shown in Figure 9.33 and the related waveforms given in Figure 9.34.

1.

Duty Cycle (D)	I_{Lmean} (A)	V_{Lmean} (V)	Speed (Rad/sec)
0.4	67.7	120	104.8
0.5	67.4	150	140.5
0.6	67.7	180.2	176.5
0.7	68.04	210.5	212.6
0.8	68.35	240.4	248.1

2.

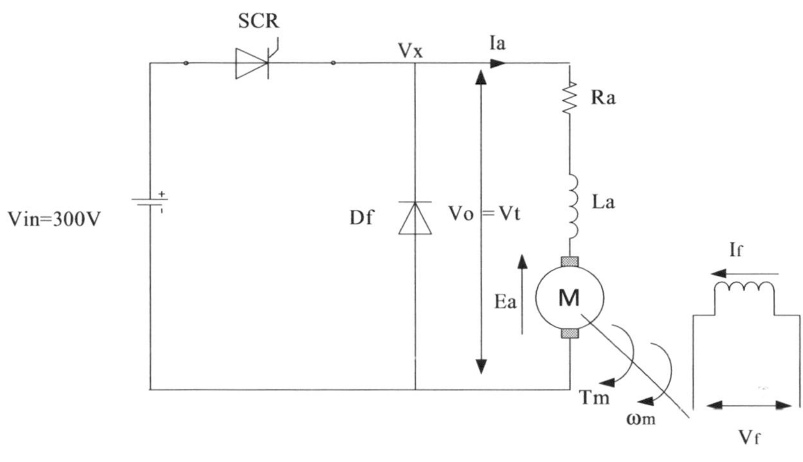

FIGURE 9.32
Step-down (buck) converter.

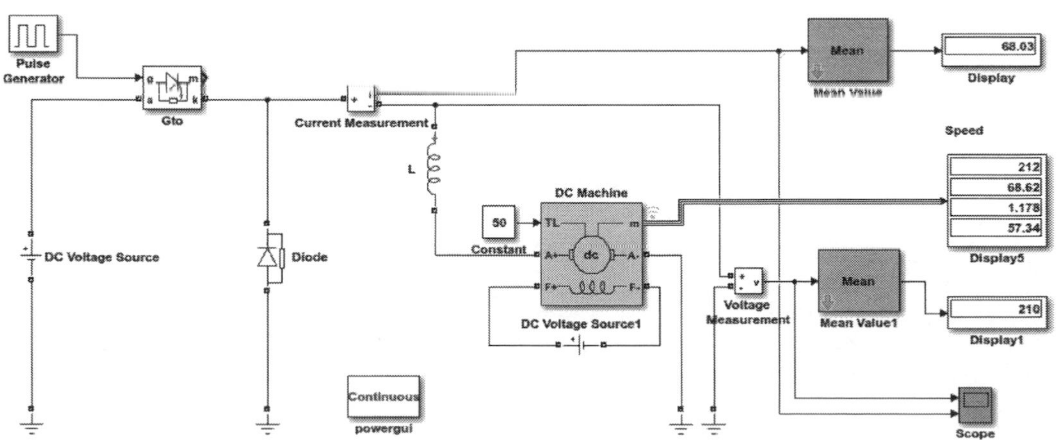

FIGURE 9.33
Step-down (buck) converter MATLAB/Simulink.

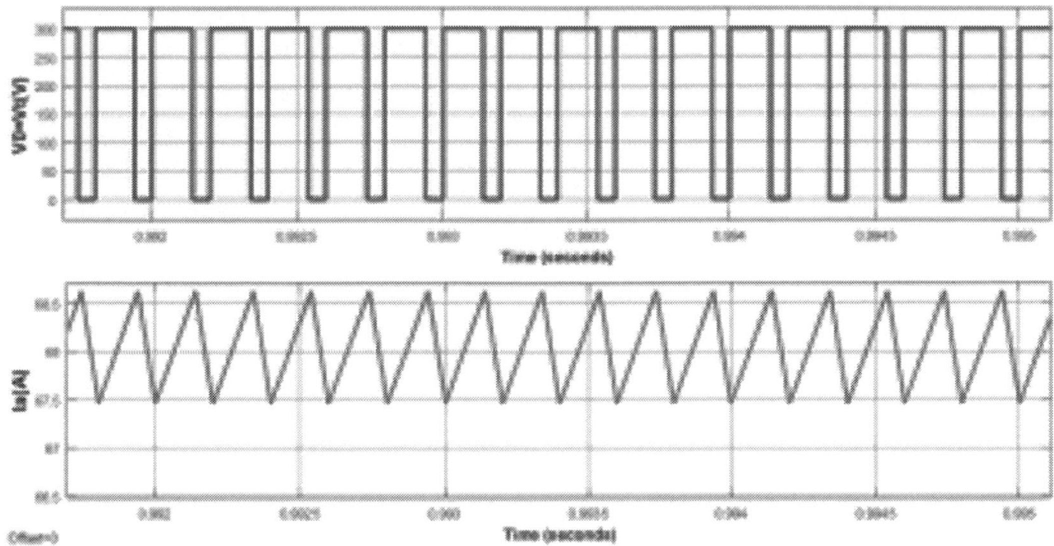

FIGURE 9.34
Waveforms.

9.6 Electrical Braking of Separate Excited DC Motor

The separately excited DC motor can be braking by one of the following methods:

1. Generator braking
2. Supply reverse braking
3. Dynamic braking.

9.6.1 Generator Braking

When the speed of a separately excited motor exceeds its no-load speed, the motor will exert a torque oppose the motion and the direction of the armature current is reversed.

The current will move from the machine to the supply (the machine will operate as a generator), and the produced torque is opposing the motion, hence, it is called generator braking or regenerative braking. Figure 9.35 shows torque-speed characteristics in the case of generator braking.

Suppose a motor drive a load T_L at speed ω_L according to the mechanical characteristics as shown in Figure 9.35. If the voltage is reduced or ϕ is increased such that the new mechanical characteristics have $\omega_o < \omega_L$ as shown.

Then at t = 0 when V is changed, the motor will produce a torque = T_{gb} opposing the motion. As ω reduced from ω_L to ($\omega_{o\ new}$), this torque will change from T_{gb} to zero. For the operating characteristics, $\omega_o > \omega_L$ and the machine operate as a motor with E = $(k\phi \cdot \omega_L)$ is less than V, the armature current is:

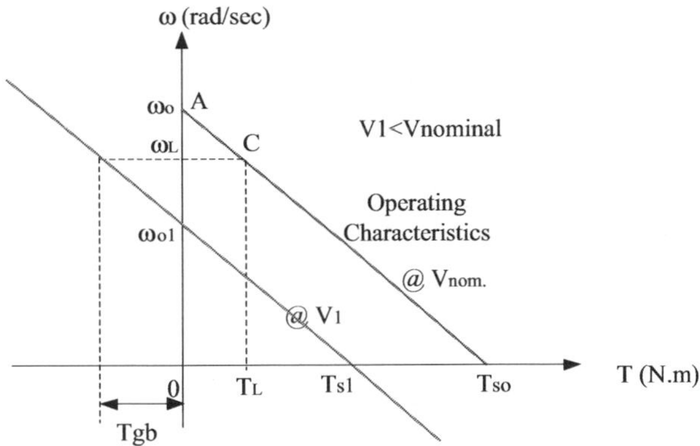

FIGURE 9.35
Torque-speed characteristics in case generator braking.

$$I_a = \frac{V - E}{R_a} = \frac{V - K\Phi\omega}{R_a}.$$

From the supply to the motor producing a torque:

$$T = K\Phi \cdot I_a = K\Phi\frac{V - E}{R_a} = K\Phi\frac{V - K\Phi\omega}{R_a}.$$

In the same direction as that of the speed (motor operation). When the voltage V is changed to V_1, where $V_1 < V$ such that:

$$\omega_{o\,new} = \frac{V_1}{K\Phi} < \omega_L$$

$$I_{a\,new} = \frac{V_1 - E}{R_a} = \frac{V_1 - K\Phi\omega}{R_a}$$

$$\omega_{o\,new} = \frac{V_1}{K\Phi} < \omega_L$$

$$\therefore I_{a\,new} = \frac{V_1 - E}{R_a} = \frac{V_1 - K\Phi\omega_L}{R_a} = -ve. \tag{9.49}$$

So new $(k\phi \cdot \omega_o) = V_1$.

For example, the current will return to the supply and produced torque will be $T_{gb} = k\phi$. $I_{a\,new} = k\phi(-I_{a\,new})$ = negative torque, i.e., oppose the motion, so the motor will brake.

Similar results when ϕ increase to give $\omega_{o\,new} < \omega_L$. Notice that the flux remains constant in value and direction. If the motors are shunt motor, similar operation can be activated. Series motor can never be operated in such mode of operation since its speed will be dangerously high.

9.6.2 Supply Reversing Braking

A separately excited DC motor operates at certain mechanical characteristics at point (T_L, ω_L) as shown in Figure 9.36. By changing the polarity of the armature supply voltage, the motor will produce a very large brake torque as shown in Figure 9.37.

Suppose point A is the operating point on the selected characteristics A(T_L, ω_L).

The armature current will be from the supply to the machine (motor) and equal to:

$$I_a = \frac{V - E}{R_a}$$

$$E = \text{brake EMF} = k\phi \cdot \omega_L,\ V > E.$$

When the supply polarity is reversed (ϕ kept constant in amplitude and direction), so $V \rightarrow -V$ (in the same direction as E which remain the same at $t = 0$).

$$I_a = \frac{-V - E}{R_a}$$

$$I_a = -\frac{V + E}{R_a} \tag{9.50}$$

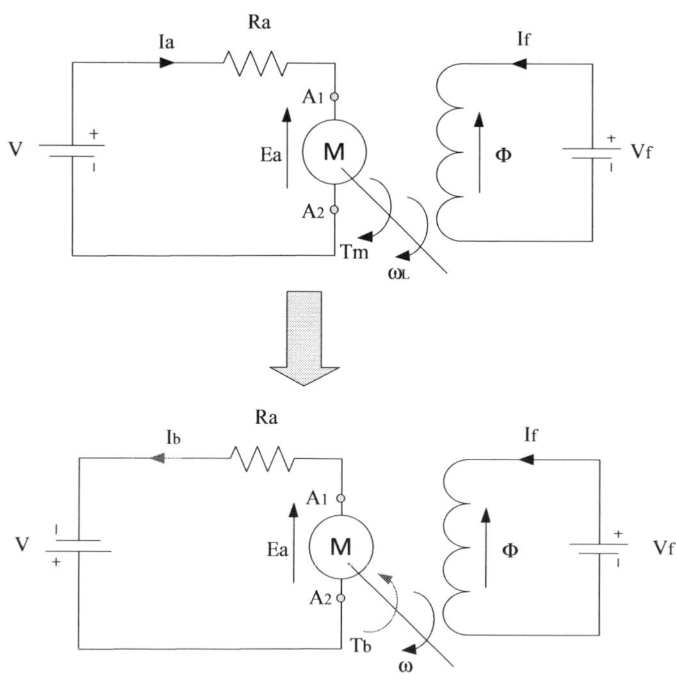

FIGURE 9.36
Supply reversing braking.

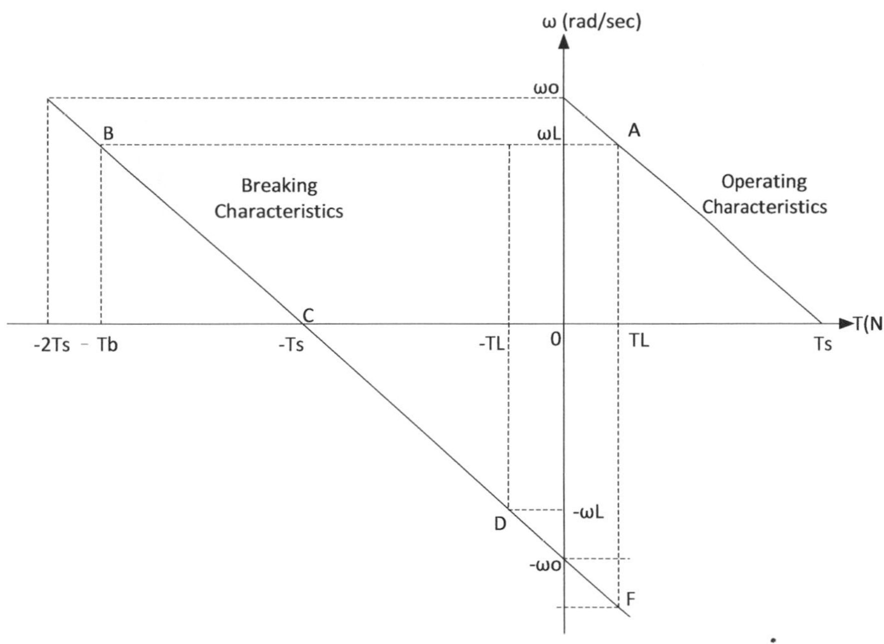

FIGURE 9.37
Torque-speed characteristics in case supply reversing braking.

Passing through the armature from the supply, but in opposite direction to that for the motor, (For motor from A1 to A2 while for supply reversing from A2 to A1). So, the resulting torque:

$$T = K\Phi I_a = -K\Phi\frac{V+E}{R_a} = -ve \tag{9.51}$$

which mean opposing the motion (brake).

So, the resultant brake torque:

$$T_b = K\Phi I_a = K\Phi\frac{V+E}{R_a} \tag{9.52}$$

At t = 0, and as this brake will reduce the speed quickly to zero, E = kϕ·ω will reduce and the braking torque through the braking process will reduce too.

The maximum possible brake torque in such method is the case of braking by supply reversing at no-load. For example, if the motor is at no-load, then V = E, $I_a = 0$. When V is reversed then:

$$I_a = \frac{-V-V}{R_a}$$

$$I_a = -2\frac{V}{R_a}$$

$$I_a = -2I_s \tag{9.53}$$

$$Tb = k\phi \cdot Ia = -2\ k\phi \cdot Is = -2\ Ts \text{ (negative means brake).} \tag{9.54}$$

So, if the operating point at (0, ω_o), the brake torque at t = 0 = $2T_s$ and the armature current = 2 I_s. So, $2T_s \geq T_b \geq T_s$ depending on the position of the operating point A.

At t = 0, the corresponding point on the braking characteristic is B and the corresponding brake torque is T_b. As ω_L reduced E ↓, T_b ↓ and the point will move on the brake characteristics from B to C.

At point C the load speed is zero, therefore if the armature supply V is switched OFF the system will stop and the time taken for stopping is very short (with respect to the previous method and which stopped due to frictional force only). If the supply is not switched OFF, then the motor will have a torque:

$$T = -T_s\left(K\Phi\frac{-V}{R_a}\right).$$

Therefore, the motor will be starting to operate in reverse direction. The final operating point such case will be point D($-T_L$, $-\omega_L$) if the load torque T_L is passive torque (frictional), otherwise, the operating point will be the point (F), F(T_L, $-\omega_L$), and o $\omega_L > \omega$ (generator braking).

An example of such case is braking a lift move upward by supply reversing. Supply voltage reversing is applied to shunt or series motor as shown in Figure 9.38.

The motor will not have braked because the supply polarity reversing cause the armature current and the field current both to change their direction then ϕ will change to $-\phi$ and I_a will change also to $-I_a$ which give:

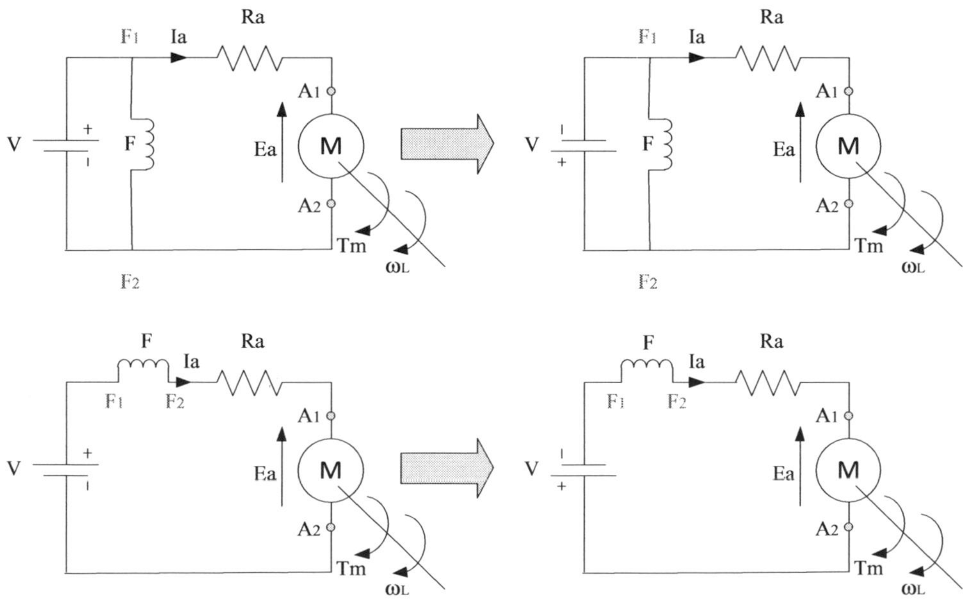

FIGURE 9.38
Supply voltage reversing is applied to shunt or series motor.

$$T = k\phi I_a \rightarrow k(-\phi)(-I_a) = k\phi I_a$$

For example, the shunt and series motor will not have affected by supply reversing. Therefore, for shunt and series motor, supply reversing braking can be done only if either the direction of armature current or field current is kept the same (unchanged). This can be done using two similar fields winding one wound opposite the other and used alternatively when the supply polarity reversing.

In such case, the current of the motor operation passes through the armature from A1 to A2 (Figure 9.39) and through the upper field winding from F1 to F2 causing a motor torque T_L in the same direction as ω_L. When the supply polarity is reversed, the upper field winding is disconnected and the lower field winding is connected, so the armature current direction will be reversed (from A2 to A1) which means $-I_a$ with respect to the first case, but the field current in the lower field winding is still from F1 to F2 which means ϕ is not changed. So,

$$T = k\phi I_a \Rightarrow T_b = k\phi(-I_a)$$

9.6.3 Dynamic Braking

For the separately excited motor, operate with a load T_L at speed ω_L at point A of the selective mechanical characteristics can be braking by switching OFF the supply armature voltage and closing the armature terminal through a short circuit or through R_d as shown with field kept constant (unchanged).

For motor operation and from Figure 9.40:

$$I_a = \frac{V - E}{R_a}$$

$$E = K\Phi \cdot \omega_L$$

and $T_L = k\phi \cdot I_a$.

When V = 0 and the terminal of the armature is short-circuited:

$$I_a = \frac{0 - E}{R_a} = -\frac{E}{R_a} \tag{9.55}$$

In the reverse direction, since ϕ is kept the same (constant)

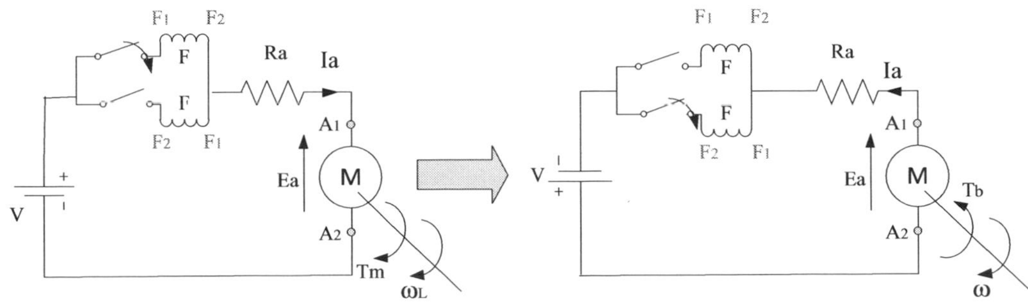

FIGURE 9.39
Supply voltage reversing is applied to the series motor.

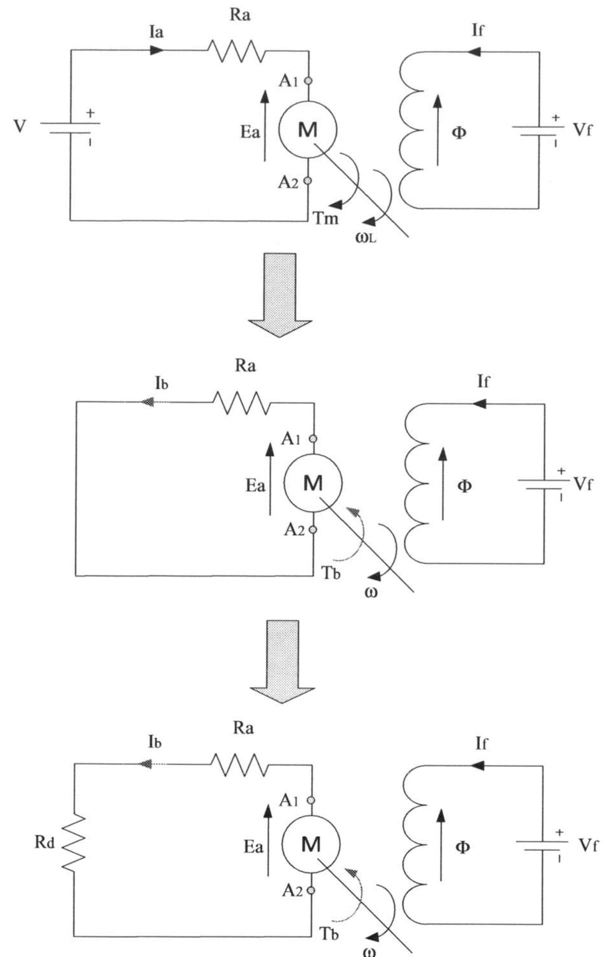

FIGURE 9.40
Motor with dynamic braking.

so,

$$T = k\phi\, I = k\phi\, E/Ra \tag{9.56}$$

As $\omega_L \downarrow$, E $\downarrow$ and $T_b \downarrow$.

For terminating the armature through R_d:

$$I_b = \frac{0 - E}{R_a + R_d} = -\frac{E}{R_a + R_d} \tag{9.57}$$

In the opposite direction, but less than that with short circuit:

$$T_b = K\Phi\, I_b = K\Phi\left(-\frac{E}{R_a + R_d}\right) \tag{9.58}$$

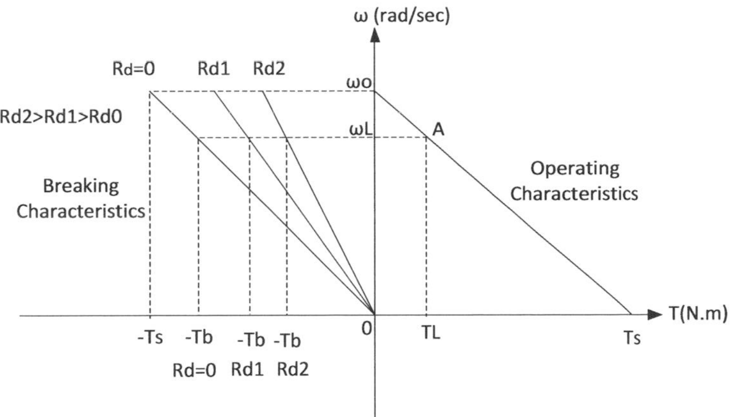

FIGURE 9.41
Torque-speed characteristics in case dynamic braking.

As $\omega\downarrow$, $I_a \downarrow$, $T_b \downarrow$.

Figure 9.41 shows torque-speed characteristics in case dynamic braking.

At $\omega = 0$, $I_a = 0$, and $T_b = 0$. The maximum possible:

$$I_b = \frac{V}{R_a} \tag{9.59}$$

At no-load, since E = V, and the brake torque is (T_s), as $R_d \uparrow$, $I_a \downarrow$, and $T_b \downarrow$.

If the load is passive load, the final position is the origin, but if the load is active load the final operating point will move to the 4th quadrant at (T_L,$-\omega_L$).

From Figure 9.42, operating in position (1) represented motor operation in a certain direction, while operating in position (2) represented also motor operation in the reverse direction.

Changing from position (1) to position (2) or vice versa through operation will represent supply reversing braking.

Changing from position (1) to position (4) or from (2) to (3) will represent dynamic braking through R_d which can be selected as required.

Example 9.10

A separate excited 150 Volt, 3 KW, 960 rpm DC motor has a total armature resistance of 1 Ω is used to lift a load of 0.8 Tr at speed of 10% of its no-load speed. Find:

1. The required additional resistance for such operation
2. If the supply is reverse braking is used, what will be the initial brake torque and what will be the final operation?
3. If the supply is switched OFF and the armature is terminated through resistance = 3 Ω, what will be the initial brake torque and what will be the final position?

Solution

$$\omega_r = \frac{2\pi\, n_r}{60} = \frac{2\pi \times 960}{60} = 100.53\, rad/sec$$

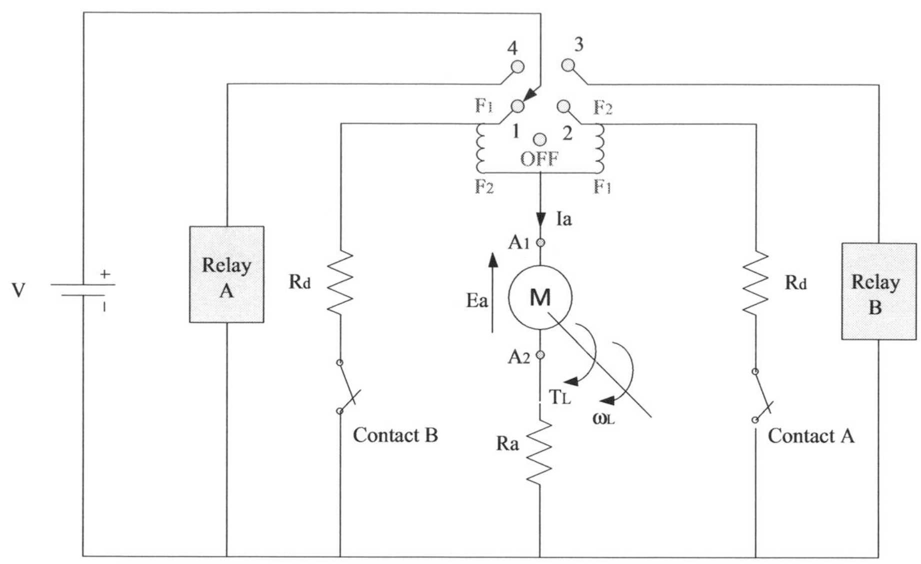

FIGURE 9.42
Dynamic braking modes.

$$T_r = \frac{P_r}{\omega_r} = \frac{3\times10^3}{100.53} = 29.84\,N\cdot m$$

$$\omega_L = 10\%\,\omega_o$$

$$\omega_L = 10\%\times100.53 = 10.053\,rad/sec$$

$$T_L = 0.8T_r$$

$$T_L = 0.8\times29.84 = 23.87\text{ N}\cdot\text{m}$$

$$\omega_r = \frac{V}{K\Phi} - \frac{R_a}{(K\Phi)^2}\cdot T_r$$

$$100.53 = \frac{150}{K\Phi} - \frac{1.2}{(K\Phi)^2}\cdot 29.84$$

$$100.53\,(K\Phi)^2 - 150\,K\Phi + 29.84 = 0$$

$K\Phi = 1.255$ or $K\Phi = 0.236$ (neglecting very high ω_o and $T_s \approx T_r$)

$$\omega_o = \frac{V}{K\Phi} = \frac{150}{1.255} = 119.52\,rad/sec$$

$$\omega_L = 12\,rad/sec.$$

To find R_d:

$$\omega_L = \frac{V}{K\Phi} - \frac{R_a + R_d}{(K\Phi)^2} \cdot T_L$$

$$12 = \frac{150}{1.255} - \frac{1 + R_d}{(1.255)^2} \cdot 23.87$$

$$R_d = 6\ \Omega.$$

For supply reversing the initial braking is represented by point B (T_b):

$$I_b = \frac{-V - E}{R_{at}}$$

$$I_b = \frac{-150 - 1.255 \times 12}{7} = 23.58\,A$$

(In the opposite direction to that for motor operation.).

So, $T_b = k\phi \cdot I_b = 1.25 \times 23.58 = 29.6$ N·m,

Since the load is the potential load, therefore, the final operating position will be represented by point C (general braking for reverse operation).

$$-\omega_L = \frac{V}{K\Phi} + \frac{R_{at}}{(K\Phi)^2} \cdot T_L$$

$$-\omega_L = 120 + \frac{7}{(1.255)^2}.23.87 = 226.93\,rad/sec.$$

For dynamic braking through 3 Ω:

$$I_b = \frac{0 - 1.255 \times 12}{7 + 3} = -1.506\,A.$$

In the opposite direction to motor operation.

So, $T_b = l\phi I_b = 1.255 \times 1.5 = 1.89$ N·m. (Point D) (Figure 9.43 torque-speed characteristics).

The final operating position will also be generator braking (pot. load).

$$-\omega_L = \frac{V}{K\Phi} + \frac{R_{at} + 3}{(K\Phi)^2} \cdot T_L$$

$-\omega_L = 120 + 7 + 3/(1.255)^2 \cdot 23.87 = 271.55\,rad/sec$ at point (E) (see Figure 9.44 torque-speed characteristics).

Example 9.11

The separately excited DC motor is shown in Figure 9.45 was operated at no-load, and the following data were recorded ω_m = 1000π/30 rad/sec, I_a = 0.95 A, V = 240 V, and V_β = 150 V. The field voltage V_s is unchanged, but the motor is loaded so that it supplies an output power P = 10 HP at 1000 rpm to a coupled mechanical load. At this load

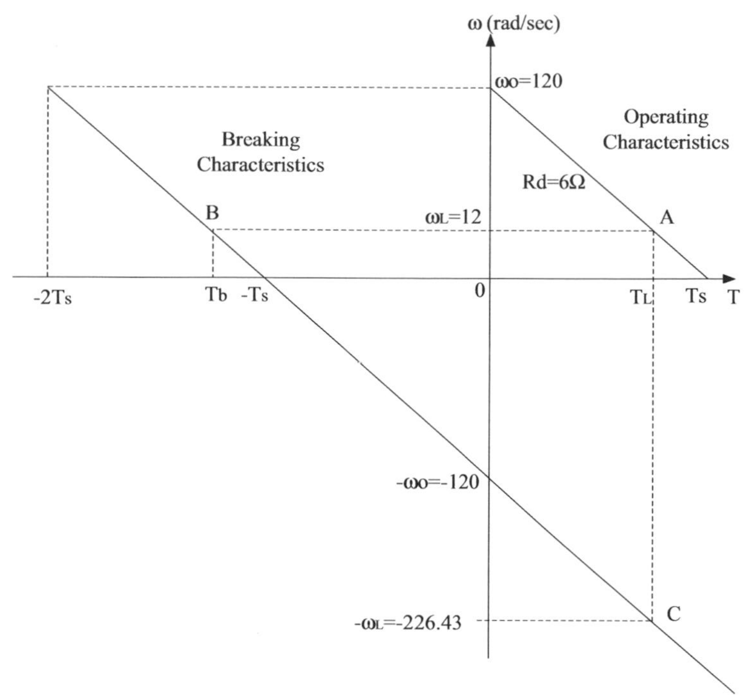

FIGURE 9.43
Torque-speed characteristics Example 9.10 i.

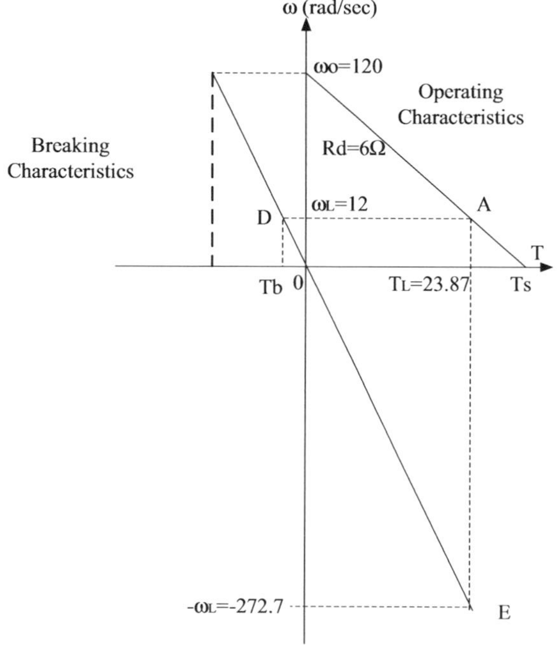

FIGURE 9.44
Torque-speed characteristics Example 9.10 ii.

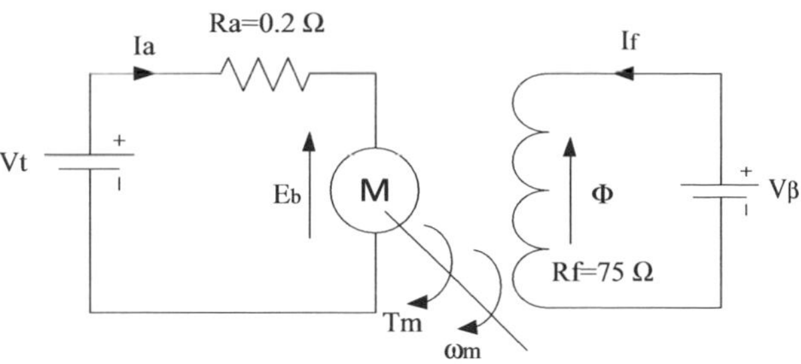

FIGURE 9.45
Motor circuit for Example 9.11.

point, determine the rotational losses, armature current, the no-load terminal voltage, efficiency. Also, calculate braking current when adding 3 Ω resistance in series with armature circuit. Neglect armature reaction.

Solution

Given data at no-load:

Speed = 1000 × Pi/30 rad/sec,

$$I_a = 0.95 \text{ A}, V_t = 240 \text{ V}, V_\beta = 150 \text{ V}.$$

Generated back EMF at no-load $E_b = V_t - I_a . R_a = 240 - 0.95 \times 0.2 = 239.81$ V.

No-load copper loss = $I_a{}^2 R_a = 0.952 \times 0.2 = 0.1805$ W.

Total input power at no-load = $V \times I_a = 240 \times 0.95 = 228$ W.

No-load losses (friction and windage losses) = 228−0.1805 = 227.8195 W.

At given load:

Speed and field are constants. It means back EMF is constant.

$$\therefore E_b = 239.81 \text{ V}.$$

Power output = $E_b \times I_a$ + friction and windage losses = 10 × 746 W

$$E_b \times I_a = 7460 - 227.8195 = 7687.81 \text{ W}.$$

$$\text{Therefore, } I_a = 7687.81/(239.81) = 32.05 \text{ A}$$

$$\text{copper losses} = 32.052 \times 0.2 = 205.4405 \text{ W}$$

$$\text{input power} = V_s \times I_a = 240 \times 32.05 = 7693.91 \text{ W}$$

$$\text{efficiency } \eta = P_{out}/P_{input}$$

$$\eta = 7460/7693.91$$

$$\eta = 96.95\%$$

$$I_b = \frac{-V_t - E_b}{R_{at}} = \frac{-240 - 239.81}{3.2} = -149.94 A.$$

Problems

9.1 A separate excited, 3 kW, 175 Volt, 1600 rpm, DC motor has an armature resistance = 1 Ω is to be used in the drive through a reducer of Nm/N_L=10. If the load torque is 140 N·m, find the required R_d added to the armature to operate the load at 1 rad/sec? What voltage reduction is used for such operation if no R_d is added?

9.2 A series 120 Volt, 2.2 KW, 850 rpm motor has R_a = 0.5 Ω and RF = 0.3 Ω. The field is linearly proportional to armature current. Find the required additional armature resistance to drive a load of 20 N·m at speed of 250 rpm.

9.3 A separately excited DC motor has KΦ = 1, R_a = 1.2 Ω. When it is operating at 150 rad/sec, the armature current is I_a = 45 A. The terminal voltage V_t is held constant under all conditions.

1. Compute V_t
2. Compute the no-load speed in rad/sec.

9.4 A separately excited DC motor is supplied from 220 V, 50 sources through a single-phase half-wave controlled converter. Its field is fed through i-phase semi converter with zero degrees firing-angle delay. Motor resistance R_a = 0.1 Ω, and motor constant = 0.6 V-sec/rad. For rated load torque of 15 N·m at 1200 rpm and for continuous ripple free currents, determine:

1. Firing-angle delay of the armature converter
2. The RMS value of thyristor and freewheeling diode currents
3. Input power factor of the armature converter. Solution: (a) motor constant = 0.5 V-sec/rad = 0.5 Nm/A =K_m, but motor torque, $T_e = K_m \cdot I_a$.

9.5 A separately excited DC motor drives in Figure 9.46, a rated load torque of 40 Nm at 1200 rpm. The field circuit resistance is 200 Ω and armature circuit resistance is 0.05 Ω. The field winding, connected to single-phase, 120 V source, is fed through 1-phase full converter with zero degrees firing angle. The armature circuit is also fed through another full converter from the same converter single-phase, 120 V source. With magnetic saturation neglected, the motor constant is 1.2 V-sec /A-rad. For ripple free armature and field currents, determine:

1. Rated armature current
2. Firing angle delay of armature converter at rated load

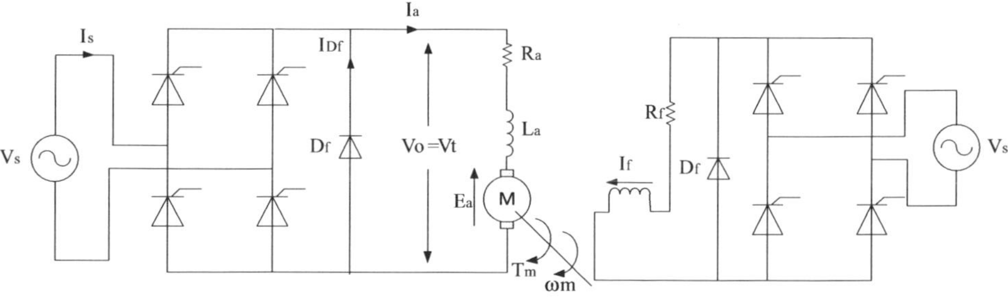

FIGURE 9.46
Separately excited DC motor drives of Problem 9.5.

3. Speed regulation at full load
4. Input PF of the armature converter and the drive at rated load.

9.6 A separate excited 150 Volt, 3 kW, 960 rpm DC motor has a total armature resistance of 1 Ω is used to lift a load of 0.8 Tr at speed of 10% of its no-load speed. Find:

1. The required additional resistance for such operation
2. If the supply is reverse braking is used, what will be the initial brake torque and what will be the final operation?
3. If the supply is switched OFF and the armature is terminated through resistance = 3 Ω, what will be the initial brake torque, and what will be the final position?

9.7 A step-down (buck) converter shown in Figure 9.47 is used to supply 80 V DC motor and an inductive load of $L = 5\text{mH}$, if $f_s = 6$ kHz, torque $= 40\text{N}\cdot\text{m}$, and input DC voltage is 200 V, calculate:

1. V_{Lmean} and speed of duty cycle range $0.5-1$.
2. Trace the output voltage and current of case D = 0.7.

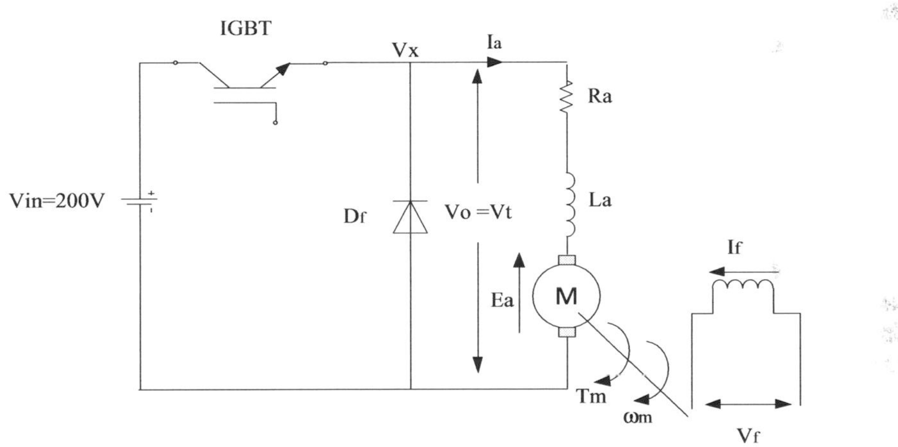

FIGURE 9.47
Step-down (buck) converter of Problem 9.7.

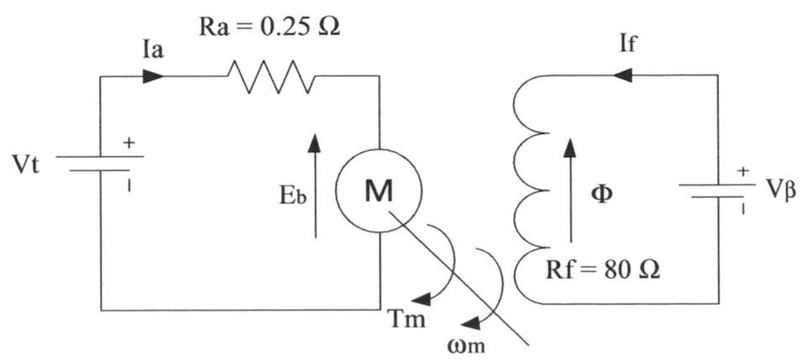

FIGURE 9.48
Motor circuit for Problem 9.8.

9.8 The separately excited DC motor is shown in Figure 9.48 was operated at no-load and the following data were recorded $\omega_m = 1000\pi/30$ rad/sec, $I_a = 1.1$ A, $V = 250$ V, and $V_\beta = 160$ V. The field voltage V_s is unchanged, but the motor is loaded so that it supplies an output power $P = 10$ HP at 1000 rpm to a coupled mechanical load. At this load point, determine the rotational losses, armature current, the no-load terminal voltage, efficiency. Also, calculate braking current when adding 2 Ω resistance in series with armature circuit. Neglect armature reaction.

10

AC Drives

Everyone enjoys doing the kind of work for which he is best suited.

Napoleon Hill

Three-phase induction motors are admirably suited to fulfill the demand of loads requiring substantially constant speed. Several industrial applications, however, need adjustable speeds for their efficient operation. The objective of the present chapter is to describe the basic principles of speed control techniques employed in three-phase induction motors using power electronics converters.

10.1 Advantages of AC Drives

1. AC motors are lighter in weight as compared to DC motors in the same rating.
2. AC motors are low maintenance as well as less expensive in comparison to equivalent DC motors.
3. AC motors are more effective in a hazardous environment like chemical, petrochemical conditions, but DC motors are unsuitable for such environments because of commutation sparking.

10.2 Disadvantages of AC Drives

1. AC drives detect some complication to control speed, voltage, as well as the whole controls, compared to DC motors.
2. Power converters for AC drives are more expensive then DC motors for more horsepower ratings.
3. Power converters for AC drives generate harmonics in the supply system of a load circuit.

10.3 Speed Control of Three-Phase Induction Motor

The induction motor (IM) is required to drive various kinds of loads, each of them having a different torque versus speed characteristics among a number of methods that exist for this purpose Figure 10.1, some commonly used ones are:

1. Stator voltage control
2. Stator frequency control
3. Stator current control
4. Controlling the stator voltage and the frequency Vs/f kept constant
5. Controlling the induced voltage and the frequency Es/f kept constant
6. Rotor resistance control
7. Slip energy recovery scheme.

In the modern industrial application and with the help of the development of semiconductor devices, changing the supply frequency (f) continuously to change the speed of the induction motor is now convincible. It is also manageable to change the supplied voltage of stator windings with the continuously changed supply frequency. Nowadays, the induction motor is progressively interchanging the variable speed DC drive—in the industry.

The magnitude of the alternator voltage can be controlled by controlling the alternator field excitation, but at constant excitation, this voltage is proportional to the frequency.

The induced back Electric Motive Force (EMF) equation of an induction motor is:

$$E = V_{ph} = 2.22\ f\Phi Z_{ph} \tag{10.1}$$

where:

f = supply frequency

ϕ = flux per pole

Z_{ph} = a number of series conductors per phase.

Hence, if the only f is varied keeping the applied voltage constant, the air-gap flux varies. If f is decreased from rated value, φ increase and the machine get saturated and the magnetizing current increase.

Hence, it is usual to keep the air-gap flux at its normal value. For this purpose, we must keep $E1/f$, i.e., $V1/f$ ratio constant, so that flux approximately remain constant. Let $K = (f/f_1) \leq 1$, where (f) is the operating frequency and (f_1) the normal rated frequency for which the machine is designed (i.e., 50). The synchronous speed is now K n_s, the applied voltage KV_1, and all the reactance KX. Substituting there in torque equation and simplifying, we get:

$$T = \frac{1}{2\pi \frac{n_s}{60}} \cdot \frac{3V_1^2}{\left(\frac{R_1}{K} + \frac{R_2'}{KS}\right)^2 + \left(X_1 + X_2'\right)^2} \cdot \frac{R_2'}{KS} \tag{10.2}$$

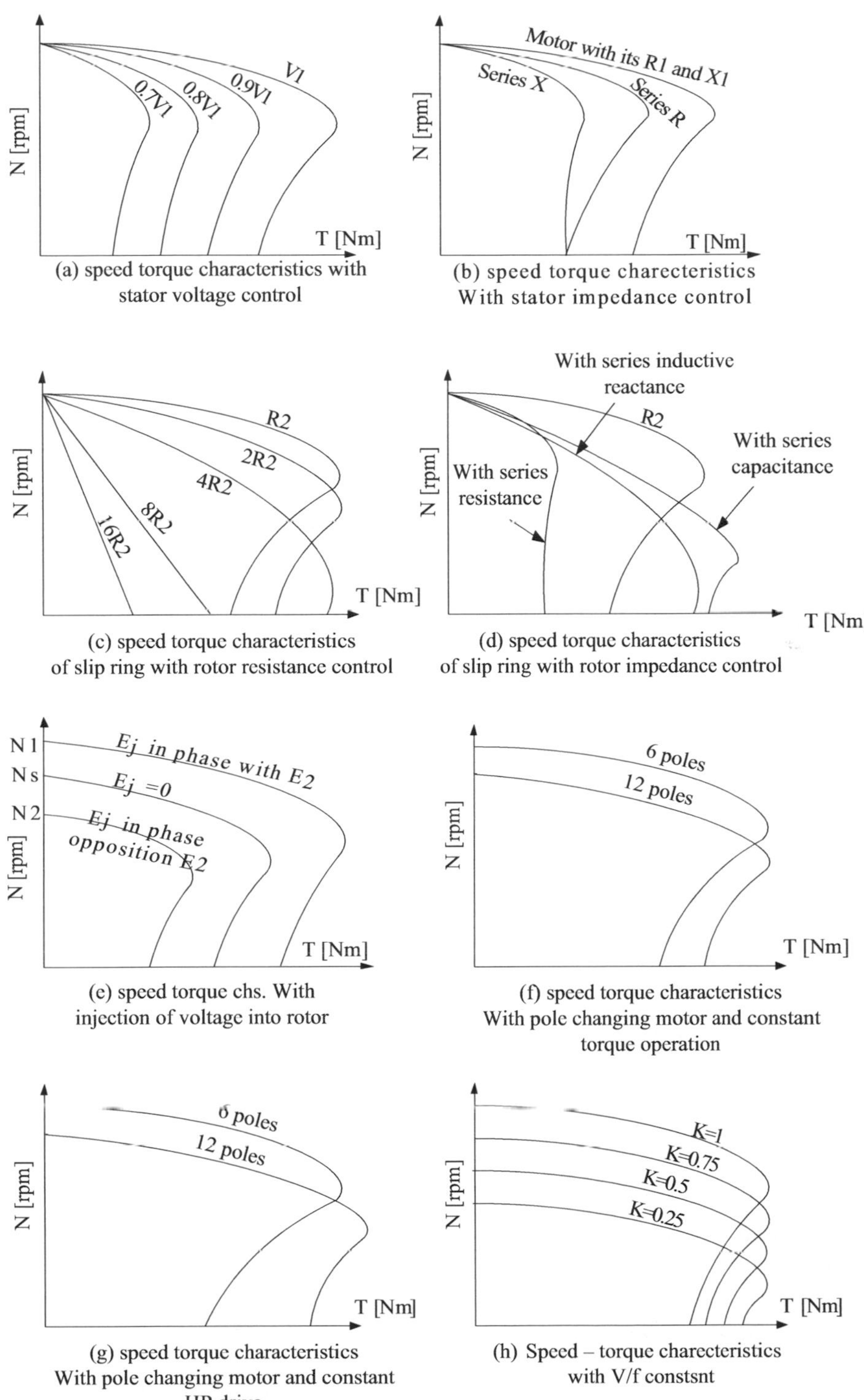

FIGURE 10.1
Torque-speed characteristics of an induction motor for a different method of speed control.

It may be seen that this expression is of the same as the original torque equation, but resistances have become larger by a factor K. Similar results can be obtained for starting torque, maximum torque, and maximum slip:

$$T_{st} = \frac{3}{2\pi \frac{n_s}{60}} \cdot \frac{V_1^2 \frac{R_2'}{K}}{\left(\frac{R_1}{K} + \frac{R_2'}{K}\right)^2 + (X_1 + X_2')^2} \tag{10.3}$$

$$T_{\max} = \frac{3}{2\pi \frac{n_s}{60}} \cdot \frac{V_1^2}{2\left[\frac{R_1}{K} + \sqrt{\left(\frac{R_1}{K}\right)^2 + (X_1 + X_2')^2}\right]} \tag{10.4}$$

The slip at which maximum torque occurs becomes:

$$S_{\max} = \frac{\frac{R_2'}{K}}{\sqrt{\left(\frac{R_1}{K}\right)^2 + (X_1 + X_2')^2}} \tag{10.5}$$

The slip at which maximum torque occurs when the frequency becomes larger. When the operating frequency decreases and the maximum torque decreases vaguely. In the beginning, torque starts to increase for a slight decrease in frequency, after reaching the maximum point, the torque starts to decrease even with the further decreasing frequency. This can be happened due to an increase of resistance of the machine and decrease of air-gap flux which is ignored in the torque equation above. And, the reduction in air gap resulted from the voltage drop in the stator impedance.

The speed-torque characteristics can be modified by:

1. By varying the supply voltage [Figure 10.1a]
2. Varying the stator parameters (R_1, X_1) [Figure 10.1b]
3. Varying the rotor parameters (R_2, X_2) [Figure 10.1c and d]

The last method of speed control in the case of slip ring induction motors.

Another method of speed control in the case of slip ring induction motor is by injection of voltage into rotor circuit by an external frequency converter or by special construction as in the case of Schrage motor [Figure 10.1e]. These methods require additional machine or special construction.

Another method of controlling the speed of squirrel cage induction motor is by changing the poles, i.e., the stator winding may be connected for different pole number and hence the synchronous speed of the motor can be changed (Figure 10.1f and g). This method is not possible in slip ring induction motor. This method has its own limitation:

1. The speed can be controlled only in steps.
2. It is possible to have only two different synchronous speed for one winding.
3. In cascade control, we can have speeds corresponding to P_1, P_2, and $P_1 \pm P_2$ pole numbers, but this method requires two machines mechanically coupled.

And the last method of controlling the speed of the induction motor is by keeping the V/f ratio constant as in Figure 10.1h.

10.4 Methods of Control Techniques

Three-phase induction motors are more commonly employed in adjustable-speed drives than three-phase synchronous motors. There are two types of three-phase inductor motors: squirrel-cage induction motors (SCIMs) and slip-ring (or wound-rotor) induction motors (SRIMs). The rotor of SCIM is built of copper or aluminum bars which are short-circuited by two end rings. The rotor of SRIM carries three-phase winding connected to three slip rings on the rotor shaft. The speed control of three-phase induction motors different according to the type of motor as follows:

10.4.1 Speed Control of Three-Phase Induction Motors

There are several methods of speed control through semiconductor devices are given below:

1. Stator voltage control
2. Stator frequency control
3. Stator voltage and frequency control
4. Stator current control
5. Static rotor-resistance control
6. Slip-energy recovery control.

SCIMs and SRIMs methods are valid to (1) to (4) both SCIMs and SRIMs, but methods (5) and (6) can only be valid for SRIMs. These methods are now described in what follows.

The angular speed in rad/s of an induction AC machine mechanical speed is given by:

$$\omega_m = (1 - S).\omega_s \tag{10.6}$$

From the equation, it can be seen that there are two possibilities for speed regulation of an induction motor. One can be achieved by modifying the slips (typical <5%) or the other can be achieved by changing synchronous speed $\omega_{s.}$ When the motor is connected to the main circuit with the constant frequency, the speed can be found at the intersection of the motor and load torque (Figure 10.2).

To get the valid estimation for motor drives, we may ignore the stator resistance and the magnetizing impedance in the equivalent circuit model of an induction motor. After an effective calculation of motors, simplification of an equation for the shaft torque in N_m.

$$T_e = {}^2T_{e\,max}\frac{S_m S}{S_m + S} \tag{10.7}$$

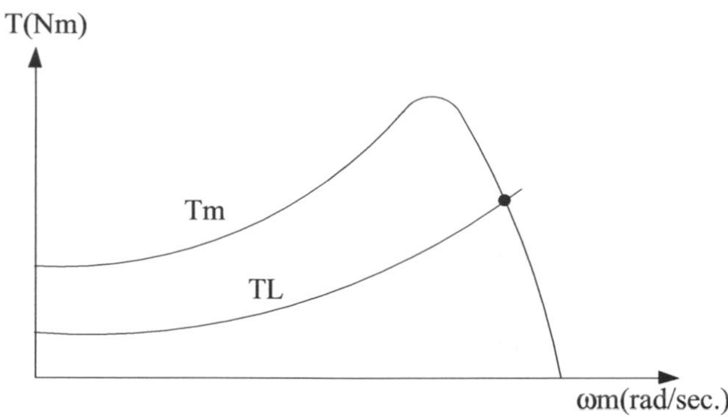

FIGURE 10.2
Typical motor torque-load characteristics.

where the maximal torque and corresponding slip are, respectively, given by:

$$T_{e\,max} = \frac{3|V_1|^2}{2\omega_s X_\sigma} \tag{10.8}$$

$$S_m = \frac{R_2'}{X_\sigma} \tag{10.9}$$

The variation of the slip S is usually small. Since the slip from no load to full load is very small which is about 0.05, the speed of an induction motor is almost constant specifically when frequency and number of poles are constant.

There is another method of speed control which is known as in the case of Schrage motor, in the case of slip ring induction motor. These schemes need some added machine configurations. In these methods, the rotor circuit is required to inject voltages by the external frequency converter.

- ***Voltage control***

  As the voltage decreases, the torque decreases (the torque is developed in an induction motor is proportional to the square of the terminal voltage). Practically, this is confined to 80%–100% control. Unfortunately, this is not an effective control.

- ***Frequency control***

  Frequency control or speed control is the most effective control among all the control. However, machine saturation is not acceptable. Since the flux is proportional to V/f, this control has to assure that the magnitude of the voltage is proportional to the speed. Power electronic circuits are best suited for this kind of control.

- ***Vector control***

  In vector control, the magnetizing current (I_d) and the current that produced by torque (I_q) are controlled in two different control loops. As I_d lags voltage by 90° and I_q are always in phase with voltage, the two vectors I_d and I_q which are always 90° apart. Afterward, then vector sum is sent to the modulator which modifies the vector information into a rotating Pulse Width Modulation (PWM) modulated 3-phase system with the correct frequency and voltage. Consequently, this phenomenon helps to get the fast-dynamic response for the induction motor by reducing torque pulsation and robust control.

10.4.1.1 Stator Voltage Control

Stator voltage control of an induction motor is used generally for three purposes: (a) to control the speed of the motor, (b) to control the starting and braking behavior of the motor, and (c) to maintain optimum efficiency in the motor when the motor load varies over a large range. Here, we are discussing speed control of the motor. At a given load, if the voltage applied to the motor decreases, keeping the frequency constant, air-gap flux decreases. This result in a reduction in torque or power developed. As the rotor speed decreases, increasing the value of slip at which the torque developed will balance the load torque. Thus, the variation of voltage results in a variation of slip frequency and speed control of the motor. For speed control in a reasonably wide range, the rotor should have a large resistance. The torque equation is given by:

$$T_e = \frac{3}{\omega_s} \cdot \frac{(K.V)^2}{\left(r_1 + \frac{r_2}{S}\right)^2 + (x_1 + x_2)^2} \cdot \frac{r_2}{S} \tag{10.10}$$

The equation of torque defines the value of torque is dependent on the square of the supplied voltage. When supply voltage decreases, the motor torque and the speed will be reduced accordingly. To control the speed, changed voltages are applied to a 3-phase IM and 3-phase voltage controller is mostly installed to implement the purpose. Figure 10.3 shows a 3-phase voltage controller feeding a 3-phase IM. From the figure, it can be seen that the thyristor is anti-parallelly connected with every phase. Therefore, the root mean square (RMS) value of the voltage can be regulated with the adjustment of the firing angle of the thyristor. Also, the motor torque and the speed of the machine is controlled in. The torque-speed characteristics of the three-phase induction motors for varying supply voltage and also for the fan load are shown in Figure 10.4.

This process holds few shortcomings in the field of speed control of the device. It controls speed only when a speed control below the normal rated speed as the operation of the voltages. Because the higher rated voltage is not allowable to execute the whole process. This method is suitable where the alternating operation of the drive is required and also for the fan and pump drives. In fan and pump drives, the load torque changes as the square of the speed. These types of drives required low torque at lower speeds. This condition can be attained by applying a lower voltage without exceeding the motor current. Three pairs of thyristor are needed in this induction where each pair consisting of two

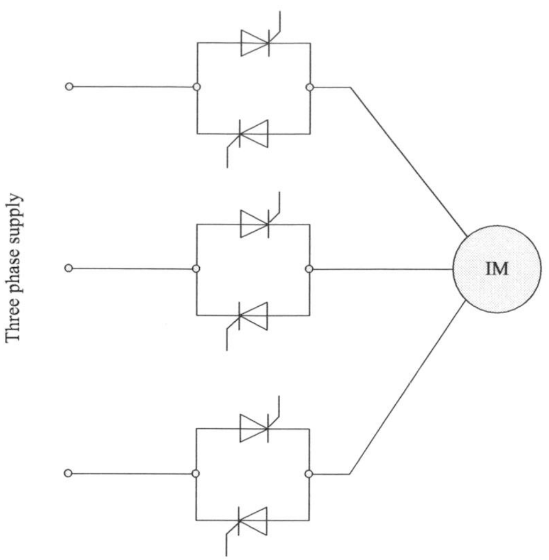

FIGURE 10.3
Thyristor voltage controller drive.

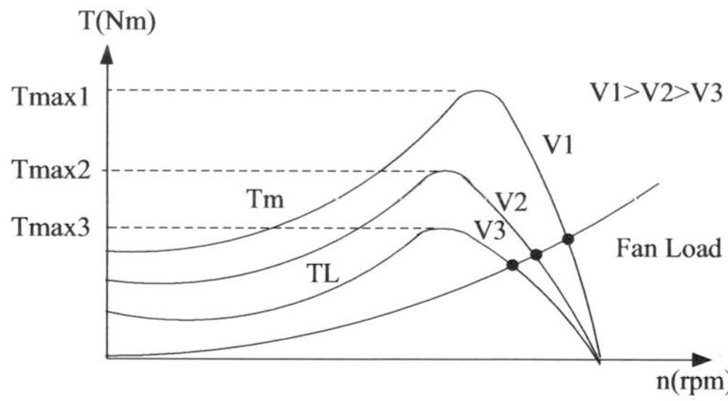

FIGURE 10.4
Torque-speed characteristics for different values of the supply voltage.

thyristors. The diagram below illustrates that the stator voltage is controlled by the thyristor voltage controller, speed control is acquired by varying the firing angle of the triac. These controllers are identified as *solid-state fan regulators.* As the solid-state regulators are more compact and efficient as compared to the conventional variable regulator. Thus, they are preferred over the normal regulator.

The variable voltage for speed control of small size motors mainly for single phase can be obtained by the following methods given below.

- By connecting an external resistance in the stator circuit of the motor
- By using an autotransformer
- By using a thyristor voltage controller
- By using a triac controller.

Nowadays, the thyristor voltage control method is selected for varying the firing angle of the thyristor. Triac is used to control speed variation, this method is also known as the phase angle control method. Additionally, energy consumption can also be controlled by controlling AC power.

10.4.1.2 Stator Frequency Control

By changing the supply frequency, motor synchronous speed can be altered and thus torque and speed of a 3-phase induction motor can be controlled. For a three-phase induction motor, the per-phase supply voltage is:

$$E_1 = \sqrt{2}\pi f_1.N_1\Phi k_{w1} \tag{10.11}$$

In this equation, if the frequency is reduced with constant V_1, then the air-gap flux increases. As a result, the induction motor magnetic circuit gets saturated. Furthermore, the motor parameters will change leading to inaccurate speed-torque characteristics. Thus, at low frequency, the reactance will flow high leading currents. Consequently, it increases loss and makes the system ineffective. In particular, speed control with constant supply voltage and reduced supply frequency are rarely suitable. But with the constant supply voltage and the increased supply frequency, the synchronous speed and therefore motor speed rises. But, with an increase in frequency and decreased flux and torque, IM performance at constant voltage and increase: frequency can be obtained by neglecting magnatizem reactance (Xm) and r_1 from the equivalent circuit induction motor. This assumption is not going to introduce any noticeable error as magnetizing current at high frequency is quite small. Thus:

$$I_2 = \frac{V_1}{\left[(r_2/S)^2 + (x_1 + x_2)^2\right]^{1/2}} \tag{10.12}$$

Synchronous speed,

$$\omega_s = \frac{4\pi.f_1}{P} = \frac{2\omega_1}{P} \tag{10.13}$$

Motor torque:

$$T_e = \frac{3}{\omega_s}.I_2^2.\frac{r_2}{S} \tag{10.14}$$

$$T_e = \frac{3P}{\omega_s}.\frac{(V_1)^2}{(r_2/S)^2 + (x_1 + x_2)^2}.\frac{r_2}{S} \tag{10.15}$$

$$S = \frac{f_2}{f_1} = \frac{\omega_2}{\omega_1} \tag{10.16}$$

or

$$\omega_2 = S.\omega_1$$

Here, f_2 and ω_2 are the rotor frequencies in and rad/sec, respectively. Substituting the value ω of slip $s = \omega_2/\omega_1$ in Equation 10.15, to get:

$$T_e = \frac{3P}{2\omega_1^2}.\frac{(V_1)^2.\omega_2}{r_2^2 + \omega_2^2.(l_1 + l_2)^2}.r_2 \tag{10.17}$$

Slip at which maximum torque occurs is given as:

$$S_m = \frac{r_2}{x_1 + x_2} \tag{10.18}$$

Rotor frequency in rad/sec at which maximum torque occurs is given by:

$$\omega_{2m} = S_m.\omega_1 = \frac{\omega_1.r_2}{\omega_1.(l_1 + l_2)} = \frac{r_2}{(l_1 + l_2)} \tag{10.19}$$

Note that ω_{2m} does not depend on the supply frequency ω_1. Substituting $r_2 = \omega_{2m}^2.(l_1 + l_2)$ in Equation 10.19:

$$T_{e.m} = \frac{3P}{2\omega_1^2}.\frac{(V_1)^2.\omega_{2m}^2.(l_1 + l_2)}{\omega_{2m}^2.(l_1 + l_2)^2 + \omega_{2m}^2.(l_1 + l_2)^2} \tag{10.20}$$

$$T_{e.m} = \frac{3P}{4\omega_1^2}.\frac{V_1^2}{l_1 + l_2} \tag{10.21}$$

Equation 10.21 indicates that $T_{e.m}$ is inversely proportional to supply-frequency squared. Also,

$$T_{e.m}.\omega_1^2 = \frac{3P}{4}.\frac{V_1^2}{l_1 + l_2} \tag{10.22}$$

At given source voltage V1, $\frac{3P}{4}.\frac{V_1^2}{l_1+l_2}$ is constant, therefore, (Tem. ω^2) is also constant. As the operating frequency ω_1 is increased, $T_{e.m}.\omega_1^2$ remains constant, but maximum torque at increased frequency ω_1 gets reduced as shown in Figure 10.5. Such type of IM behavior is

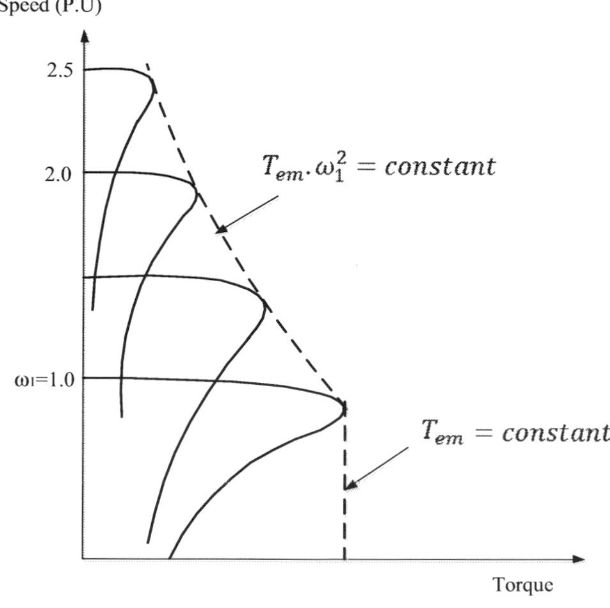

FIGURE 10.5
Speed torque characteristics of a three-phase IM with stator frequency control with constant supply voltage.

similar to the working of DC series motors. With constant voltage and increased-frequency operation, air-gap flux gets reduced, therefore, during this control, IM is said to be working in field-weakening mode. Constant voltage and variable frequency control of Figure 10.5 can be obtained by feeding 3-phase IM through three-phase.

Example 10.1

A 3-phase, 208 V, 20 kW, 1450 rpm, 60 Hz, star-connected induction motor has rotor leakage impedance of 0.4 + J1.6 Ω. Stator leakage impedance and rotational losses are assumed negligible.

If this motor is energized from 120 Hz, 400 V, 3-phase source, then calculate:

1. The motor speed at rated load stator frequency control with constant supply voltage
2. The slip at which maximum torque occurs, and
3. The maximum torque.

Solution

1.

$$T_e = \frac{P}{\omega_m} = \frac{60 \times 20000}{2\pi \times 1450} = 131.71 N.m$$

$$\omega_s = \frac{4\pi . f_1}{P} = \frac{2\omega_1}{P} = \frac{4\pi \times 120}{4} = 377\ rad/sec$$

$$Z_{120\ Hz} = 0.4 + j1.5 \times \frac{120}{60} = 0.\ 4 + j3$$

$$131.71 = \frac{3}{377} . \frac{(208/\sqrt{3})^2}{(3)^2} . \frac{0.4}{S}$$

$$S = 0.0387$$

2.

$$S_m = \frac{r_2}{x_2} = \frac{0.4}{3} = 0.1333$$

3.

$$T_{e.m} = \frac{3}{2\omega_s} . \frac{V_1^2}{x_2} = \frac{3}{2 \times 377} . \frac{\left(\frac{208}{\sqrt{3}}\right)^2}{3} = 19.12 N.m$$

Example 10.2 For the MATLAB/Simulink shown in Figure 10.6, draw the speed response.

Solution

Figure 10.7 shows a speed response of three-phase IM with stator frequency control with constant supply voltage using MATLAB/Simulink circuit shown in Figure 10.6.

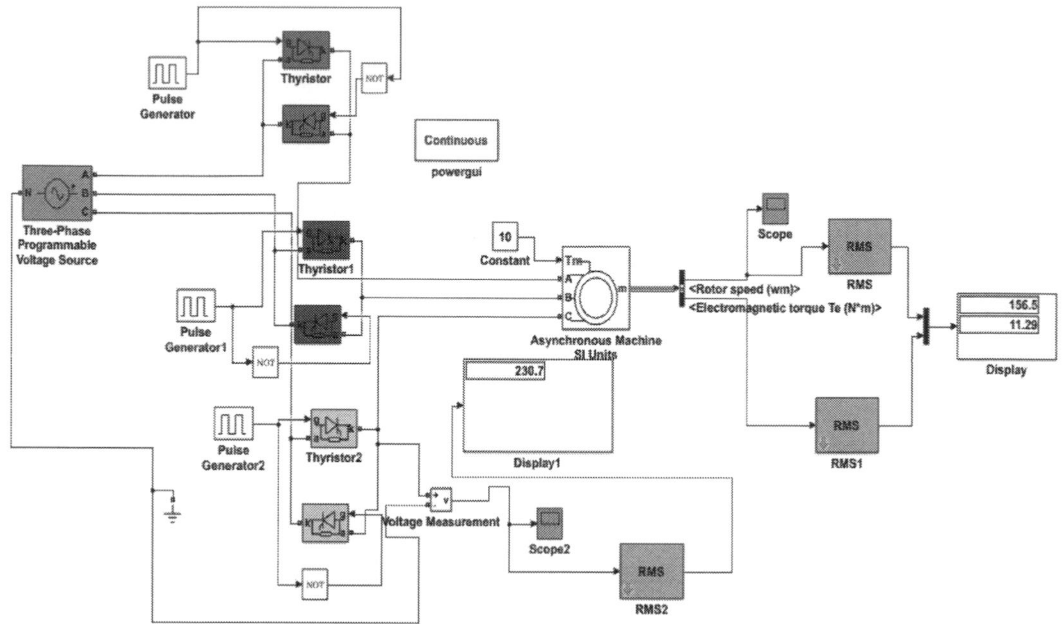

FIGURE 10.6
Thyristor voltage controller Simulink drive.

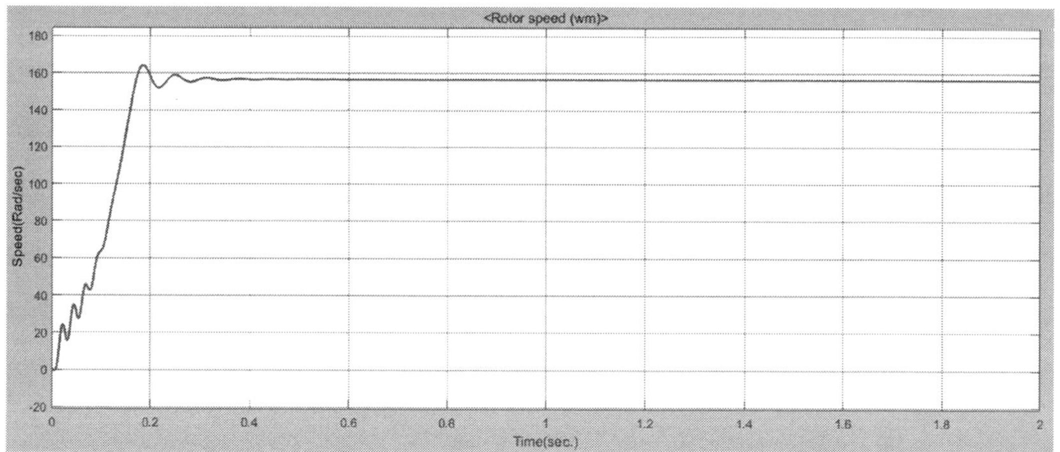

FIGURE 10.7
Speed response of Example 10.2.

10.4.1.3 Stator Voltage and Frequency Control

Synchronous speed can be controlled by varying the supply frequency. The voltage induced in the stator is $E_1 \propto \Phi.f$, where Φ is the air-gap flux and f is the supply frequency. As we can neglect the stator voltage drop, we obtain terminal voltage $V_1 \propto \Phi.f$. Thus, reducing the frequency without changing the supply voltage will lead to an increase in the air-gap flux which is undesirable. Hence, whenever frequency is varied in order to control speed, the terminal voltage is also varied so as to maintain the V/f ratio constant. Thus, by maintaining a constant V/f ratio, the maximum torque of the motor becomes constant for changing speed.

V/f control is the most popular and has found widespread use in industrial and domestic applications because of its ease-of-implementation. However, it has an inferior dynamic performance compared to vector control. Thus, in areas where precision is required, V/f control is not used. The various advantages of V/f control are as follows:

1. It provides a good range of speed.
2. It gives good running and transient performance.
3. It has a low starting current requirement.
4. It has a wider stable operating region.
5. Voltage and frequencies reach rated values at base speed.
6. The acceleration can be controlled by controlling the rate of change of supply frequency.
7. It is cheap and easy to implement.

Induction motors are the most widely used motors for appliances, industrial control, and automation. Hence, they are often called *the workhorse* of the motion industry.

They are robust, reliable, and durable. When power is supplied to an induction motor at the recommended specifications, it runs at its rated speed. However, many applications need variable speed operations. For example, a washing machine may use different speeds for each wash cycle. Historically, mechanical gear systems were used to obtain variable speed. Recently, electronic power and control systems have matured to allow these components to be used for motor control in place of mechanical gears. These electronics not only control the motor's speed, but can improve the motor's dynamic and steady-state characteristics. In addition, electronics can reduce the system's average power consumption and noise generation of the motor.

Induction motor control is complex due to its nonlinear characteristics. While there are different methods for control, variable voltage variable frequency (VVVF) or *V/f* is the most common method of speed control in open loop. This method is most suitable for applications without position control requirements or the need for high accuracy of speed control. Examples of these applications include heating, air conditioning, fans, and blowers. *V/f* control can be implemented by using low-cost peripheral interface controllers (PIC), rather than using costly digital signal processors (DSPs).

Many PIC micro microcontrollers have two hardware PWM, one less than the three required to control a 3-phase induction motor. In this application, we generate a third PWM in software, using a general purpose timer and an Input/Output (I/O) pin resource that are readily available on the PIC micro microcontroller.

10.4.1.4 V/f Control Theory

In the speed-torque characteristics, the induction motor draws the rated current and delivers the rated torque at the base speed. When the load is increased, the speed drops, and the slip increases. As we have discussed in the earlier section, the motor can take up to 2.5 times the rated torque with around 20% drop in the speed. Any further increase of load on the shaft can stall the motor.

The torque established by the motor is directly proportional to the magnetic field produced by the stator. So, the voltage applied to the stator is directly proportional to the product of stator flux and angular velocity. This makes the flux produced by the stator proportional to the ratio of applied voltage and frequency of supply.

By varying the frequency, the speed of the motor can be varied. Therefore, by varying the voltage and frequency by the same ratio, flux, and hence, the torque can be kept constant throughout the speed range.

$$\text{Stator voltage} = E_1 \approx V_1 \propto \Phi.2\pi.f \tag{10.23}$$

So:

$$\Phi \propto \frac{V}{f} \tag{10.24}$$

Above relation presents *V/f* as the most common speed control. Figure 10.8 indicates the torque curve between voltage and frequency. Figure 10.8 validates the voltage and frequency are reached up to the base speed. At base speed, the voltage and frequency reach the rated values as listed in the nameplate. It can be possible to run the motor beyond base speed by increasing the frequency further. However, there is a limit to reach the voltage, it cannot across the rated voltage. As a result, the only parameter can be increased that is frequency. When the motor runs above base speed, the torque becomes complex as the friction and windage losses increase drastically changes due to high speeds. Therefore, the torque curve gives a nonlinear pattern with respect to speed or frequency.

10.4.1.5 Static Rotor-Resistance Control

In a SRIM, a three-phase variable resistor R_2 can be inserted in the rotor circuit as shown in Figure 10.9a. By varying the rotor circuit resistance R_2. the motor torque can be controlled as shown in Figure 10.9b. The starting torque and starting current can also be varied by controlling the rotor circuit resistance, Figure 10.9b and c. This method of speed control is used when speed drop is required for short time, as for example, in overhead cranes, in load equalization.

The disadvantages of this method of speed control are:

1. Reduced efficiency at low speeds
2. Speed changes vary widely with load variation
3. Unbalances in voltages and currents if rotor circuit resistances are not equal

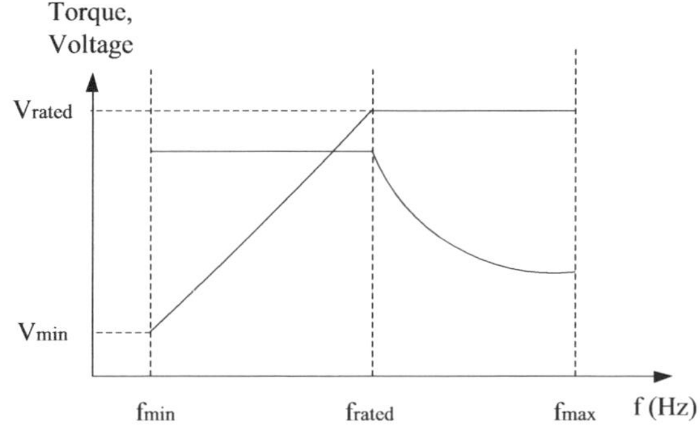

FIGURE 10.8
Relation between the voltage and torque versus frequency.

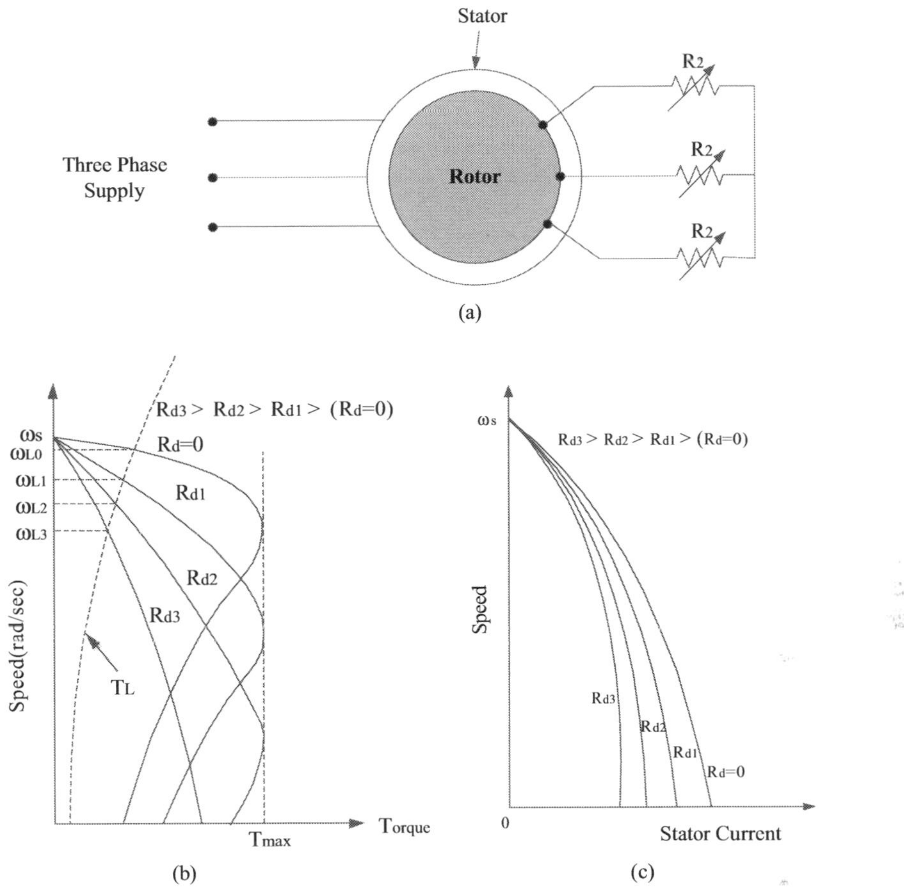

FIGURE 10.9
Three-phase IM speed control by rotor resistance. (a) Circuit arrangement. (b) Effect on developed torque. (c) Effect on stator current.

The three-phase resistor of Figure 10.9a may be replaced by a three-phase diode rectifier, chopper, and one resistor as shown in Figure 10.10. In this figure, the function of inductor L_d is to smoothen the current I_d. A Turn ON OFF (GTO) chopper allows the effective rotor circuit resistances to be varied for the speed control of SRIM. Diode rectifier converts slip frequency input power to DC at its output terminals. When chopper is on, $V_{dc} = V_d = 0$ and resistance R gets short-circuited. When chopper is off, $V_{dc} = V_d$ and resistance in the rotor circuit are R. This is shown in Figure 10.11.

c) From this figure, effective external resistance Re is:

$$Re = R.\ T_{off}/T$$

$$= R.\ (T\text{-}Ton)/T$$

$$= R.\ (1\text{–}k),$$

where k = Ton/T = duty cycle of chopper.

The equivalent circuit for 3-phase IM, diode rectifier, and chopper circuit is shown in Figure 10.12. If stator and rotor leakage impedances are neglected as compared to inductor

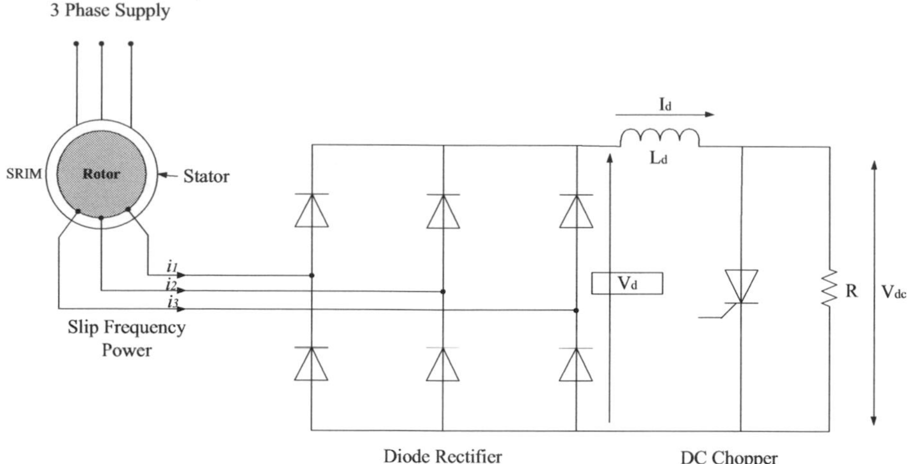

FIGURE 10.10
Static rotor-resistance control circuit drive.

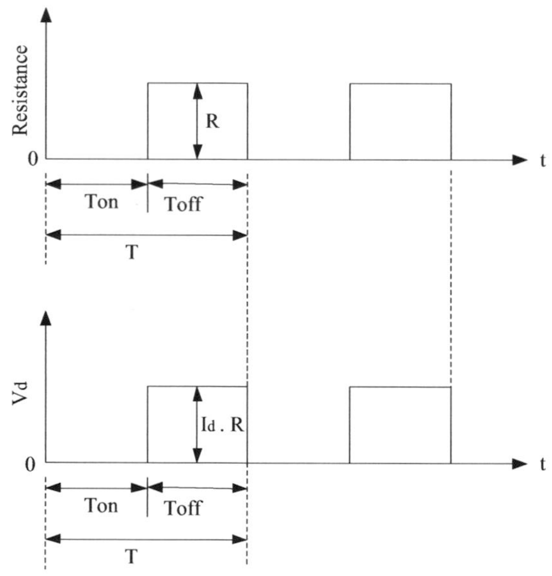

FIGURE 10.11
Resistance and rectifier voltage.

L_d, the equivalent circuit of Figure 10.12b is obtained. Stator voltage V_1, when referred to rotor circuit, gives slip-frequency voltage as:

$$s.\ V_1\ N_1/N_2 = s.\ \alpha.\ V_1 = s.\ E_2,$$

where E_2 = rotor induced EMF per phase at the standstill.

V_1 = stator voltage per phase.

α = (rotor effective turns, N_2)/(Stator effective turns, N_1) = per phase turns ratio from rotor to stator.

Voltage $S.E_2 = S.\alpha.V_1$, after rectification by a three-phase diode, appears as V_d (rectifier output voltage).

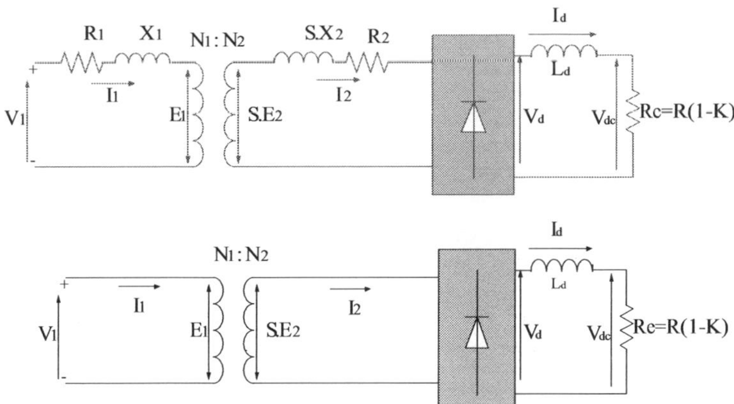

FIGURE 10.12
Equivalent circuit for 3-phase IM, diode rectifier, and chopper circuit.

$$V_d = \frac{3V_m}{\pi} = \frac{3.\sqrt{3}V_{m\,ph}}{\pi} = 2.339\,S\,a\,V_1 \tag{10.25}$$

For lossless rectifier:

$$\text{Total slip power} = 3S\;.P_g = V_d.I_d \tag{10.26}$$

$$P_m = (1-S).P_d \tag{10.27}$$

$$P_m = (1-S).\frac{V_d.I_d}{3S} \tag{10.28}$$

And:

$$P_m = T_e.\omega_r = T_e.\omega_s.(1-S) \tag{10.29}$$

So:

$$P_m = (1-S).\frac{V_d.I_d}{3S} = T_e.\omega_s.(1-S)$$

$$I_d = T_e.\omega_s.3S / V_d$$

Or

$$I_d = \frac{T_L.\omega_s}{2.339\,a\,V_1} \tag{10.30}$$

And:

$$V_d = I_d.R.(1-k)$$

So:

$$V_d = 2.339\ S\ a\ V_1 = I_d.R.(1-k)$$

$$\therefore S = \frac{I_d.R.(1-S)}{2.339a\ .V_1} \tag{10.31}$$

And $\omega_m = \omega_s(1 - s)\omega_m = \omega_s(1 - s)$.

So the motor speed:

$$\omega_m = \omega_s\left[1 - \left(\frac{T_L.\omega_s.R.(1-k)}{(2.339aV_1)^2}\right)\right] \tag{10.32}$$

The relationship of the torque-speed characteristic with a variable rotor resistance is given in Figure 10.13, the critical slip happens when motor produces maximum torque is proportional to the rotor resistance. As the rotor resistance increases, the critical slip value increase. As we can see from the diagram, for example: assume a constant load torque (T_L), the motor operating points shift from A to B to C with the change of resistance, respectively. According to the changes of resistances, the changed speeds are N_1, N_2, and N_3 with N_1>N_2>N3. In this way, the speed of the induction motor decreases with increases in the rotor resistance.

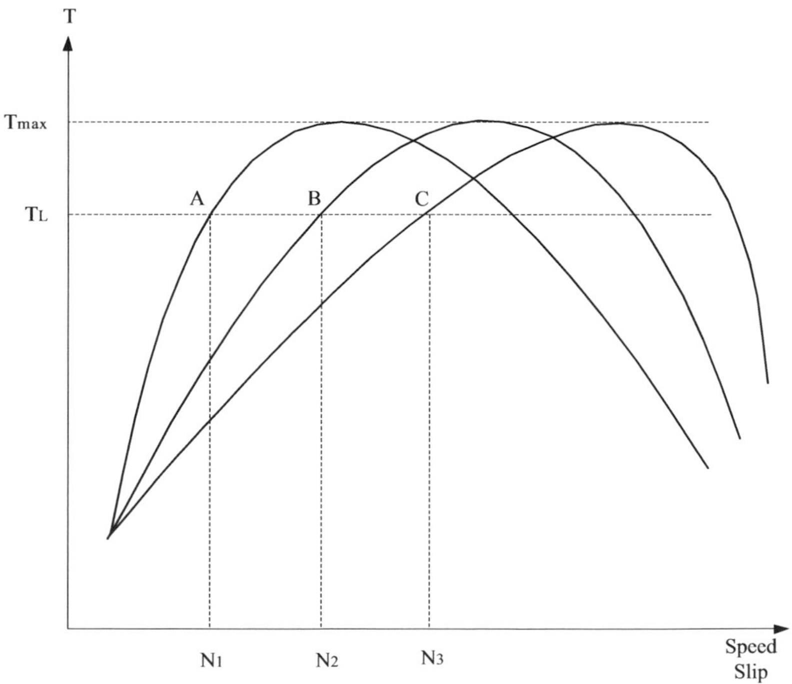

FIGURE 10.13
Torque-speed characteristics with a variable rotor resistance.

10.4.1.6 Slip-Energy Recovery Control

Due to output, the slip power losses are induced at the terminal of slip ring induction motor. Also, slip power losses are more often due to using variable resistances. To control these losses or to recovery the slip power, the power losses at the terminal are needed to be reduced.

In the recent days, there are two methods have introduced to control the speed of the slip ring. One of them is static rotor-resistance control and another one is slip-energy recovery control. In between two of them, the slip-energy recovery control method is developed to reduce the slip energy losses and to improve overall poor drive efficiency of the system. External resistance control by using three rheostats is a primitive method which is easy to control. In this method, the speed of the wound rotor of the induction motor is varied with the external resistance at the terminal. When the external resistance value is zero, the slip rings start to act as a short circuit. At the rated load, the internal torque slip curve of the machine starts to provide speed at this point. As the external resistance is getting started to increase, the curve would start to look flat. Eventually, at the high resistance, there will be no speed detected. When output slip terminals are connected to each other with any coil and the motor work as a full load, then we can use a variable rheostat at the terminal to control the speed. The speed of the motor can be controlled by varying the rheostat. As the rotor circuit needs energy and the power losses occur, the slip control is not an effective method to control slip (Figure 10.14).

Changing the resistance by using a diode like bridge rectifier and the chopper is more efficient rather than changing it mechanically. Normally, a bridge rectifier, the chopper is directly connected to power supply, whereas the slip voltage is rectified to DC by diode rectifier in the rotor circuit. With the connection of series inductor, the DC voltage is converted to the current source then it is used to feed to IGBT shunt chopper with shunt resistance shown. When the IGBT is off, the DC link current flows through it. On the contrary, when the device is ON, the resistance is short-circuited and the current is bypassed through it. Consequently, the established torque and the speed can be controlled by the variation of the duty cycle of the chopper. This can be assumed that electronic control of rotor resistance is definitely advantageous compared to rheostat controlled. Nevertheless, the problem of poor drive efficiency cannot be changed. This specific structure is effective in limited range like in intermittent speed-controlled application.

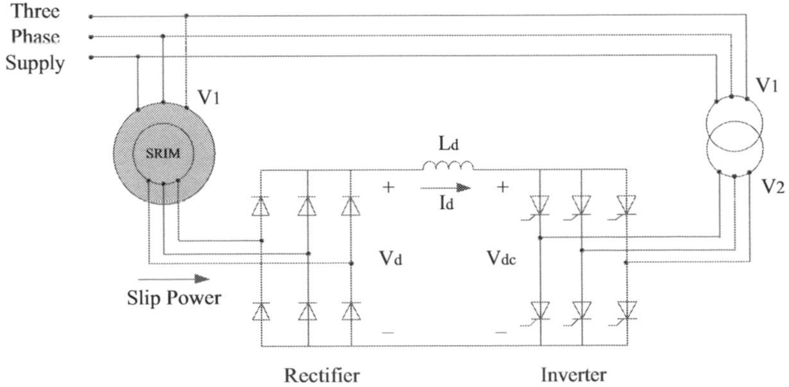

FIGURE 10.14
Slip power recovery drive.

To summarize, the speed control mechanism for high-power SRIM and low-power SRIM is analyzed as an extension and nonslip recovery mechanism to the available, respectively. The mechanism can be employed either at the rotor side or stator side. The mechanism that has engaged with a single thyristor with ON/OFF state at the rotor side of the slip ring induction motor. The basic projected outline here is to control the high-power machine by directly varying the load current or the rotor side current by the single thyristor after converting at the available load side current into the DC current with the help of three-phase rectifier bridge. As the three-phase rectifier is made of six diodes, respectively, all the AC power is converted into simple DC current and is controlled by the thyristor ON/OFF state time. The switch ON/OFF state time contribute the needed speed control as the rotor current is directly proportional to speed and torque in an increased power scenario as in this case. Although, the losses due to nonslip power recovery can be ignored because of the machines high-power status. In this scheme, we are able to simplify control of a high-power SRIM by just changing the amount of ON/OFF time of the thyristor at the rotor end that operates efficiently. Moreover, it is the best suited for very high inertia loads where a pull-out torque requires to be zero speed and accelerates to full speed with minimum current drawn in a very short time period.

Problems

10.1 List out some examples of prime movers.

10.2 List out some advantages of electric AC drives.

10.3 Give some examples of electric drives.

10.4 What are the advantages of static Kramer system, over static Scherbius system?

10.5 What is the function of a conventional Kramer system?

10.6 What is meant by slip power?

10.7 A 3-phase, 208 V, 25 kW, 1650 rpm, 60 Hz, star-connected induction motor has rotor leakage impedance of 0.4 + J1.6 Ω. Stator leakage impedance and rotational losses are assumed negligible. If this motor is energized from 120 Hz, 400 V, 3-phase source, then calculate:

1. The motor speed at rated load stator frequency control with constant supply voltage
2. The slip at which maximum torque occurs
3. The maximum torque.

10.8 A 3-phase, 208 V, 60 Hz, 6 poles, Y connected round-rotor synchronous motor has Zs = 0 + J2 Ω. Load torque, proportional to speed squared, is 300 Nm at rated synchronous speed. The speed of the motor is lowered by keeping *V/f* constant and maintaining unity Power Factor (pf) by field control of the motor. For the motor operation at 600 rpm, calculate:

1. The supply voltage
2. The armature current
3. The excitation voltage

4. The load angle
5. The pullout torque, neglect rotational losses.

10.9 A static Kramer drive is used for the speed control of a 4-pole SRIM fed from 3-phase, 300 V, 60 Hz supply. The inverter is connected directly to the supply. If the motor is required to operate at 1400 rpm, find the firing advance angle of the inverter. The voltage across the open-circuited slip rings at stand-still is 750 V. Allow a voltage drop of 0.7 V and 1.5 V across each of the diodes and thyristors, respectively. Inductor drop is neglected.

11

Special Machines

Live your life and forget your age.

Norman Vincent Peale

There are some special motors with their electrically special applications in the modern era. This chapter delivery a brief fundamental to electrical special machines which have special applications. Stepper motors, brushless *direct current* DC motor, and switched reluctance motor are the examples of some special machines that are mostly used. Additionally, this chapter has a short description of servomotors, synchro motors, and resolvers.

Among all the motors, electric motor plays a vital role in our daily life. The main mechanism of this kind of motor is basically to convert electrical energy into mechanical energy. There are several devices in our daily life whose movement is produced by the electric motor. Hair dryer, Video Cassette Recorder (VCR), and a disk drive in a computer are few examples that are using the mechanism of the electric motor to run the specific applications.

There are various types of an electric motor according to the nature of the supplied current and sizes, namely, *alternating current (AC)* electric motors and *DC* electric motors. According to the name, DC electric motor will not run when AC current supplied. On the other hand, AC electric motor will not operate with the DC supplied current. Furthermore, AC electric motors are also subdivided into *single phase* and *three-phase* motors. Single phase AC electrical supply is what is typically supplied in a home and three-phase electrical power is commonly only available in a factory setting.

11.1 Stepper Motors

In the early 1960s, stepper motors were developed and introduced as a low-cost alternative to position servo systems in the emerging computer peripheral industry. To achieve the accurate position control without knowing the position feedback, the stepper motor is effective to use. However, it reduces the cost of a position control system by running "open-loop." Also, stepper motors apply a doubly salient topology, which means they have "teeth" on both the rotor and stator. Magnetizing stator teeth are used to generate electrical torque, and the permanent-magnet rotor teeth are utilized to line up with the stator teeth. There are many different configurations of stepper motors and even more different ways to drive them. There is a most common stator configuration which consists of two coils which are arranged around the circumference of the stator in such a way that if they are driven with square waves which have a quadrature phase relationship between them, the motor will rotate. To rotate the rotor motor in the opposite direction, simply reverse the phase relationship between the two coils signals. This alteration of either square wave causes the rotor to move by a small amount or a "step." Therefore, the motor is itself known as a *stepper motor.* The size of this step depends on the teeth arrangement of the motor, but a common value is 1.8° or 200 steps per revolution. Speed control is achieved by simply varying the frequency of the square-waves. As stepper motors can be driven with square waves, they are easily controlled by inexpensive digital circuitry and do not even require PWM (pulse width modulation). For this reason, stepper motors have often been inappropriately referred to as "digital motors." However, the quadrature changes into sine and cosine waveforms by utilizing power modulation techniques which may provide more resolution. This is termed as "micro-stepping," where each discrete change in the sine and cosine levels constitutes one micro-step.

A stepper motor, also known as a stepping motor, a pulse motor, or digital motor, is an electromechanical device which rotates a discrete step angle when energized electrically. Stepper motors are synchronous motors in which rotor's positions depend directly on driving signal.

The main difference between the stepping motor and a general motor is that the stepping motor only powered by a fixed driving voltage that does not rotate. Also, stepping motor displays excellent functions such as accurate driving, rapid stopping, and rapid starting. The stepper motor provides controllable speed or position in response to input step pulses commonly applied from an appropriate control circuit. Stepping motors are driven by a pulse signal. When a digital pulse signal is used as an input into the stepping motor, the rotor of the stepping motor is rotated by a fixed angle, that is, a well-known stepping angle. Since the stepper motor increments in a precise amount with each step pulse, it converts digital information, as represented by the input step pulses, to the corresponding incremental rotation. By increasing the rate of the step pulses, it is possible to increase the speed of the motor.

11.1.1 Step Angle

The angle through which the motor shaft rotates for each command pulse is called *the step angle β*. The parameter step angle can determine the number of steps per revolution and the accuracy of the position. It is observed, the smaller the step angle, the greater the number of steps per revolution and higher the resolution or accuracy of positioning obtained. The step angles can be as small as 0.72° or as large as 90°. But the most common step sizes are 1.8°, 2.5°, 7.5°, and 15°.

The value of step angle can be expressed either in terms of the rotor and stator poles (teeth) Nr and Ns, respectively, or in terms of the number of stator phases (m) and the number of rotor teeth.

$$\beta = \frac{(N_s - N_r)}{N_s \, . \, N_r} \times 360° \tag{11.1}$$

$$\beta = \frac{360°}{\text{m.N}_\text{r}} = \frac{360°}{\text{No. of stator phases} \times \text{No. of rotor teeth}} \tag{11.2}$$

For example, if Ns = 10 and Nr = 6, β = (10−6) × 360°/8 × 6=30°.

The resolution is given by the number of steps needed to complete one revolution of the rotor shaft. The higher the resolution, the greater the accuracy of positioning of objects by the motor.

$$\text{Resolution} = \text{No. of steps/revolution} = 360°/\beta \tag{11.3}$$

A stepping motor also possess extraordinary ability to operate at very high stepping rates up to 20,000 steps per second in some motors and yet to remain fully in synchronism with the command pulses. When the pulse rate is high the shift rotation seems continuous and operation at a high rate is called *slewing*. When in the slewing range, the motor generally emits an audible whine having a fundamental frequency equal to the stepping rate. If f is the stepping frequency (or pulse rate) in pulses per second (pps) and β is the step angle, when motor shaft speed is given by:

$$n = \frac{\beta . f}{360^\circ} = \text{pulse frequncy revolution} \tag{11.4}$$

If the stepping rate is increased too quickly, the motor loses synchronism and stops. The same thing happens if when the motor is slewing, command pulses are suddenly stopped instead of being progressively slowed.

Stepping motors are designed to operate for long periods with the rotor held in a fixed position and with rated current flowing in the stator windings. It means that stalling is no problem for such motors, whereas for most of the other motors, stalling results in the collapse of back Electric Motive Force (EMF) (E_b) and a very high current which can lead to a quick burn-out.

Example 11.1

A hybrid variable reluctance (VR) stepping motor has eight main poles which have been castellated to have five teeth each. If the rotor has 60 teeth, calculate the stepping angle.

Solution

$$Ns = 8 \times 5 = 40; \text{ and } Nr = 60$$

$$\beta = \frac{(N_s - N_r)}{N_s . N_r} \times 360^\circ$$

$$\beta = \frac{(N_r - N_s)}{N_s . N_r} \times 360^\circ = \frac{(60 - 40)}{40.60} \times 360^\circ = 3^\circ$$

Example 11.2

A stepper motor has a step angle of 2.5°. Determine (a) resolution, (b) number of steps required for the shaft to make 25 revolutions, and (c) shaft speed, if the stepping frequency is 3600 pps.

Solution

a. resolution = 360°/β = 360°/2.5° = 144 steps/revolution
b. steps/revolution = 144. Hence, steps required for making 25 revolutions = 144 × 25 = 3600
c. n = β × f/360° = 2.5 × (3600/360°) = 25 rps.

11.1.2 How Stepper Motors Work

Stepper motors consist of a permanent-magnet rotating shaft, called *the rotor* and electromagnets on the stationary portion that surrounds the motor called *the stator*. Figure 11.1 illustrates one complete rotation of a stepper motor. At position 1, we can see that the rotor

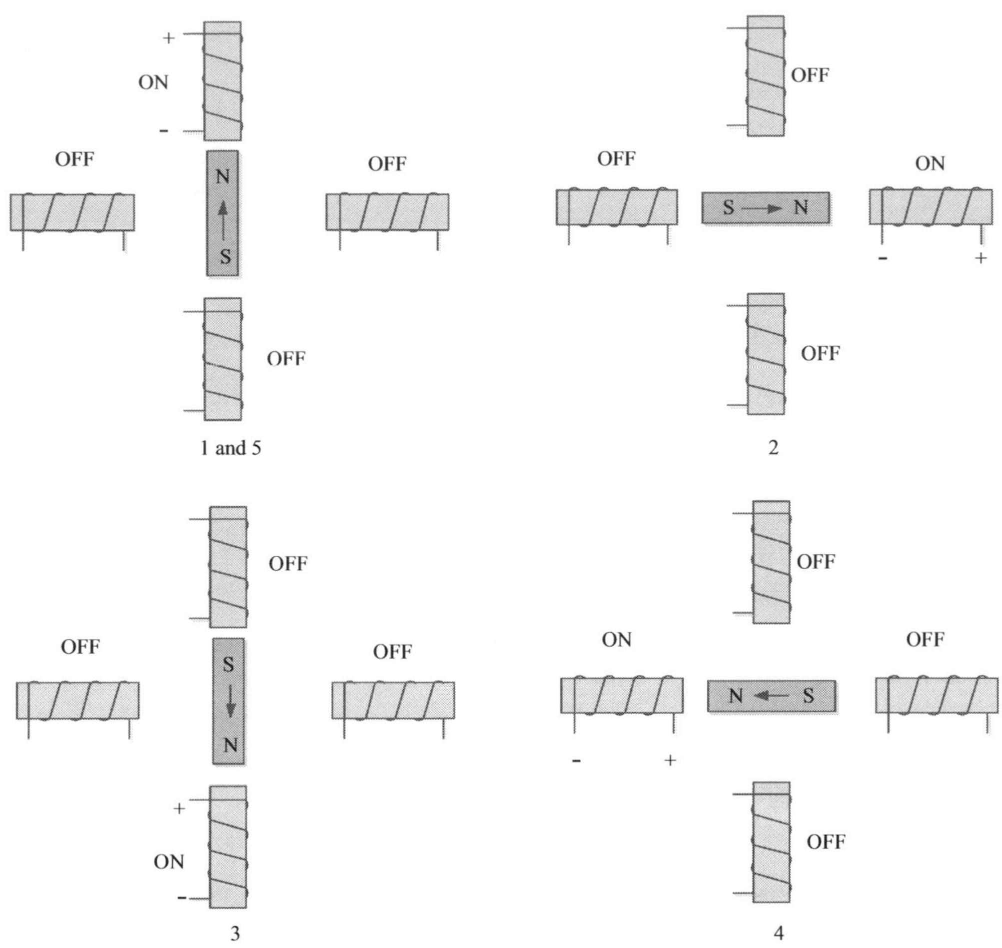

FIGURE 11.1
One complete rotation of a stepper motor. Rotor position Case 1&5, Rotor position Case 2, Rotor position Case 3, Rotor position Case 4.

is beginning at the upper electromagnet, which is currently active (has voltage applied to it). To move the rotor clockwise (CW), the upper electromagnet is deactivated and the right electromagnet is activated, causing the rotor to move 90°CW, aligning itself with the active magnet. This process is repeated in the same manner at the south and west electromagnets until we once get the starting position again.

In the above example, we used a motor with a resolution of 90° for demonstration purposes. In reality, this would not be a very practical motor for most applications. The average stepper motor's resolution—a number of degrees rotated per pulse—is much higher than this. For example, a motor with a resolution of 5° would move its rotor 5° per step, thereby requiring 72° pulses (steps) to complete a full 360° rotation.

You may perhaps double the resolution of some motors by a method known as "half-stepping." Instead of switching the next electromagnet in the rotation on one at a time, you can turn on both electromagnets by the help of half stepping phenomena that results in an equal attraction between thereby doubling the resolution. As you can see in Figure 11.2, in the first position only the upper electromagnet is active, and the rotor

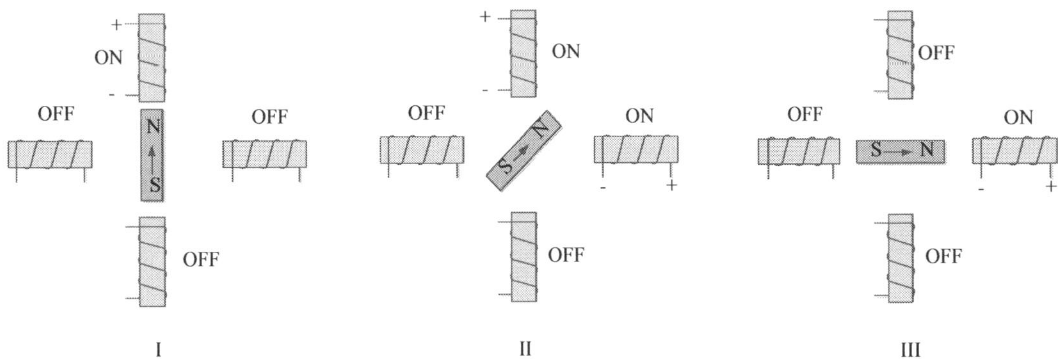

FIGURE 11.2
Rotation of stepper motor.

is drawn completely to it. In position 2, both the top and right electromagnets are active, causing the rotor to position itself between the two active poles. Finally, in position 3, the top magnet is deactivated and the rotor is drawn all the way right. This process can then be repeated for the entire rotation.

Among all types of stepper motors—4-wire stepper motors basically contain only two electromagnets. In fact, the operation is more complicated than the motors which contain three or four magnets because the driving circuit must be able to reverse the current after each step. For our purposes, we will be using a 6-wire motor. For example, motors which rotated 90° per step, real-world motors employ a series of mini-poles on the stator and rotor to increase resolution. Although this may seem to add more complexity to the process of driving the motors, the operation is identical to the simple 90° motor we used in our example. An example of a multipole motor can be seen in Figure 11.3. In position 1, the north pole of the rotor's permanent magnet is aligned with the south pole of the stator's

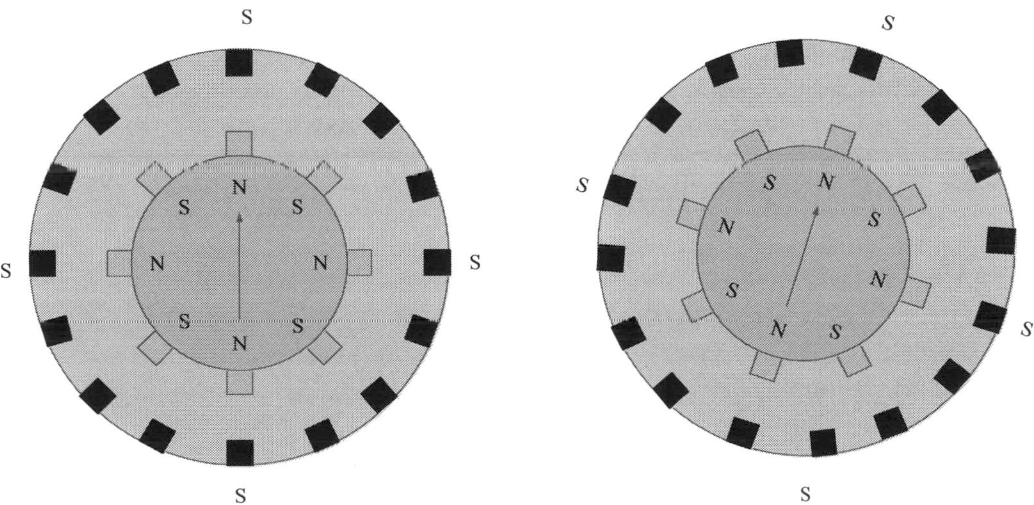

FIGURE 11.3
An example of a multipole motor.

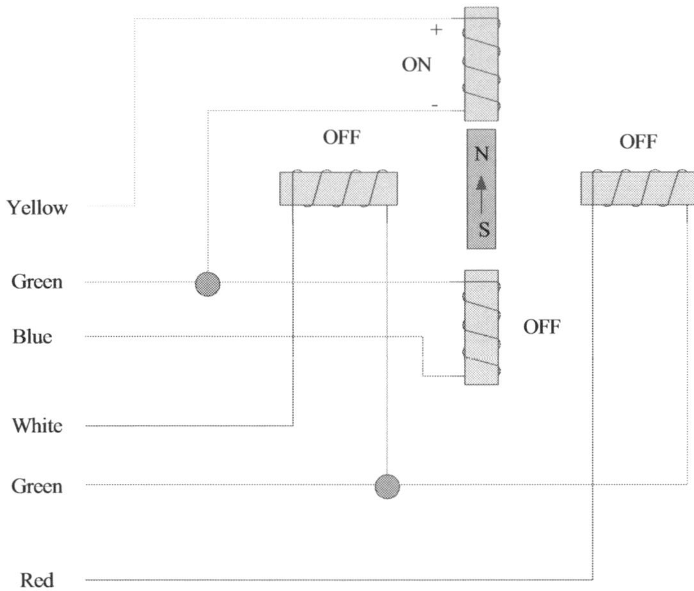

FIGURE 11.4
The electrical equivalent of the stepper motor.

electromagnet. Also, it is noted that multiple positions are aligned at once. In position 2, the upper electromagnet is deactivated and the next one to its immediate left is activated, causing the rotor to rotate a precise amount of degrees. In this example, after eight steps the sequence repeats.

The specific stepper motor we are using for our experiments (5° per step) has six wires coming out of the casing. If we follow Figure 11.4, the electrical equivalent of the stepper motor, we can see that three wires go to each half of the coils and that the coil windings are connected in pairs. This is true for all four-phase stepper motors.

11.1.3 DC Motors versus Stepper Motors

- Stepper motors are operated open-loop, where most of the DC motors are operated closed loop.
- Stepper motors are easy to control with microprocessors, however, logic and drive electronics are more complex.
- Stepper motors are brushless and brushes contribute several problems, e.g., wear, sparks, electrical transients.
- DC motors have a continuous displacement and can be accurately positioned, whereas stepper motor motion is incremental, and its resolution is limited to the step size.
- Stepper motors can slip if became overloaded and the error can remain undetected (A few stepper motors use closed-loop control.).
- Feedback control with DC motors gives a much faster response time compared to stepper motors.

11.1.4 Advantages of Stepper Motors

- The resulted position error is noncumulative, high accuracy of motion is possible at even under open-loop control
- Huge savings in the sensor (measurement system) and a controller which cost are affordable when the open-loop mode is used
- Because of the incremental nature of command and motion, stepper motors are easily adaptable to digital control applications
- No serious stability problems exist, even under open-loop control
- Torque capacity and power requirements can be optimized and the response can be controlled by electronic switching
- Brushless construction has obvious advantages.

11.1.5 Disadvantages of Stepper Motors

- They have low torque capacity (typically less than 2,000 oz-in) compared to DC motors
- They have limited speed (limited by torque capacity and by pulse-missing problems due to faulty switching systems and drive circuits)
- Large errors and oscillations can result when a pulse is missed under open-loop control
- They have high vibration levels due to the stepwise motion.

11.1.6 Specification of Stepping Motor Characteristics

In this section, some technical terms are used for specifying the characteristics of a stepping motor are discussed.

11.1.6.1 Static Characteristics

The characteristics relating to stationary motors are called *static characteristics.*

11.1.6.1.1 T/Θ Characteristics

Firstly, the stepping motor is kept stationary at a rest (equilibrium) position by supplying a current in a specified mode of excitation with the assumption with single-phase or two-phase excitation. If an external torque is applied to the shaft, an angular displacement will occur. The relation between the external torque and the displacement may be plotted as in Figure 11.5. This curve is conventionally called *the* T/θ *characteristic curve,* and the maximum of static torque is termed the "holding torque," which occurs at $\theta = \theta_M$ Figure 11.6. At displacements larger than θ_M, the static torque does not act in a direction toward the original equilibrium position, but in the opposing direction toward the next equilibrium position. The holding torque is rigorously defined as "the maximum static torque that can be applied to the shaft of an excited motor without causing continuous motion." The angle at which the holding torque is produced is not always separated from the equilibrium point by one step angle.

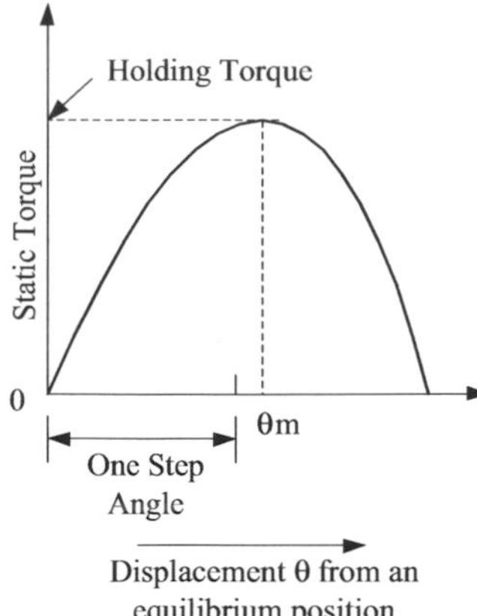

FIGURE 11.5
T/θ characteristics.

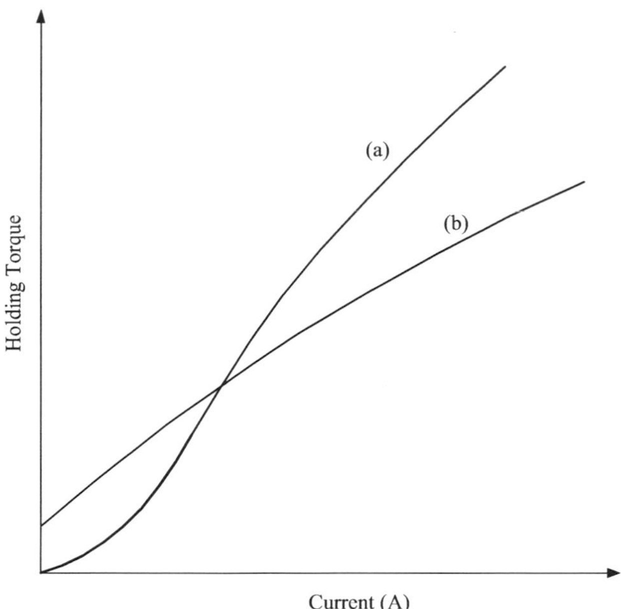

FIGURE 11.6
Examples of T/I characteristics. (a) a 1.8° four phase variable reluctance motor; and (b) a 1.8° four phase hybrid motor.

11.1.6.1.2 T/I Characteristics

The holding torque increases with current, and this relation is conventionally referred to as Torque/Amper (T/I) characteristics. Figure 11.6 compares the T/I characteristics of a typical hybrid motor with those of a variable-reluctance motor, the step angle of both being 18°. The maximum static torque appearing in the hybrid motor with no current is the detent torque, which is defined as the maximum static torque that can be applied to the shaft of an unexcited motor without causing continuous rotation.

11.1.6.2 Dynamic Characteristics

The characteristics relating to motors which are in motion or about to start are called *dynamic characteristics.*

11.1.6.2.1 Pull-In Torque Characteristics

These are alternatively called *the starting characteristics* and refer to the range of frictional load torque at which the motor can start and stop without losing steps for various frequencies in a pulse train. The number of pulses in the pulse train used for the test is 100 or so. The reason why the word "range" is used here, instead of "maximum," is that the motor is not capable of starting or maintaining a normal rotation at small frictional loads in certain frequency ranges as indicated in Figure 11.8. When the pull-in torque is measured or discussed, it is also necessary to identify the driving circuit clearly, the method of measuring, the method of coupling, and the inertia to be coupled to the shaft. Generally, the self-starting range decreases with the increasing inertia.

11.1.6.2.2 Pull-Out Torque Characteristics

This characteristic is also alternatively called *the slewing characteristic.* After the test motor is started by a specified driver in the specified excitation mode in self-starting range, the pulse frequency is gradually increased, the motor will eventually run out of synchronism. The relation between the frictional load torque and the maximum pulse frequency with which the motor can synchronize is called *the pull-out characteristic* (see Figure 11.7). The pull-out curve is greatly affected by the driver circuit, coupling, measuring instruments, and other conditions.

11.1.6.2.3 The Maximum Starting Frequency

This is defined as the maximum control frequency at which the unloaded motor can start and stop without losing steps.

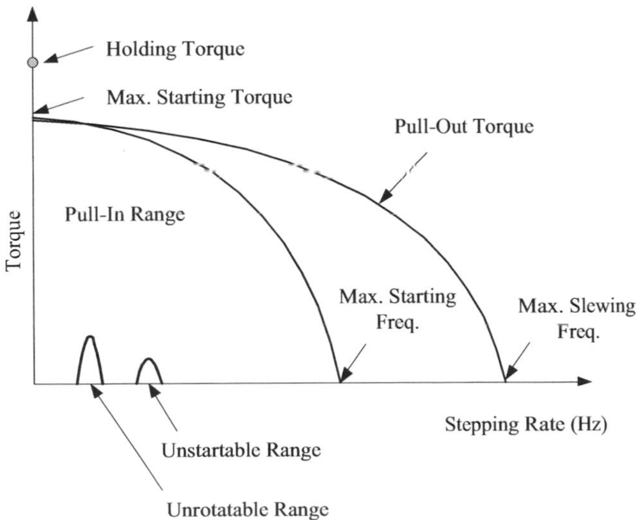

FIGURE 11.7
Dynamic characteristics.

11.1.6.2.4 Maximum Pull-Out Rate

This is defined as the maximum frequency (stepping rate) at which the unloaded motor can run without losing steps and is alternatively called *the maximum slewing frequency.*

11.1.6.2.5 Maximum Starting Torque

This is alternatively called "maximum pull-in torque" and is defined as the maximum frictional load torque with which the motor can start and synchronize with the pulse train of a frequency as low as 10.

11.1.7 Steady State Phasor Analysis

Figure 11.8 illustrates the equivalent circuit in the time domain of a phase winding. The circuit equation for one phase excitation can be written as:

$$V = R.i + L\frac{di}{dt} + \frac{dL(\theta)}{d\theta}.\omega_r i + \frac{d\lambda_m}{dt} \quad (11.5)$$

where L is the stator winding inductance, λ_m the stator winding flux linkage due to the permanent magnet, and:

$$e = \frac{dL(\theta)}{d\theta}.\omega_r i + \frac{d\lambda_m}{dt} \quad (11.6)$$

11.1.7.1 Phasor Expression of Variable Reluctance Stepping Motor

For a variable reluctance stepping motor, we have $\lambda_m=0$,
and:

$$L(\theta) = L_o + L_1 sin(N_r\theta) \quad (11.7)$$

Since the unipolar drive is employed, we may express the fundamental components of the voltage and current in the stator phase winding as:

$$v(t) = V_o + V_1\cos(\omega t) \quad (11.8)$$

and

$$i(t) = I_o + I_1\cos(\omega t - \delta - \alpha) \quad (11.9)$$

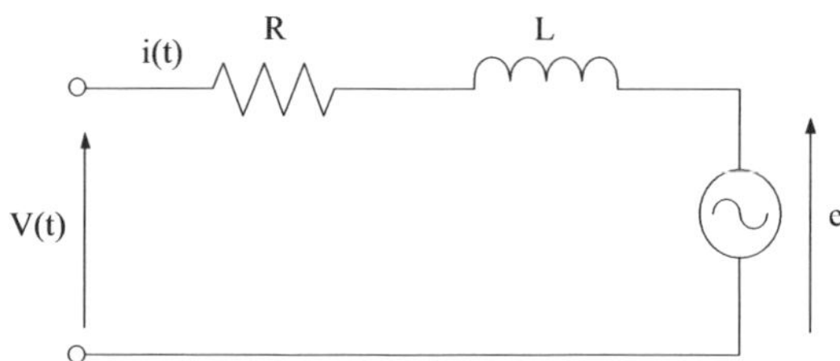

FIGURE 11.8
Per phase equivalent circuit in the time domain.

Substituting and neglecting the high-frequency terms, we obtain the voltage and current relations as:

$$V_o = R.I_o \tag{11.10}$$

$$V_1 \cos(\omega t) = RI_1 \cos(\omega t - \delta - \alpha) - \omega L_o I_1 \sin(\omega t - \delta - \alpha) + \omega L_1 I_o \cos(\omega t - \delta) \tag{11.11}$$

In phasor expression, the above voltage-current relationship becomes:

$$V = RI + j\omega L_o I + E \tag{11.12}$$

$$\mathrm{E} = \omega\ \mathrm{L}_1\ \mathrm{I}_o \angle -\delta \tag{11.13}$$

11.1.7.2 Phasor Expression of PM and Hybrid Stepping Motors

For permanent magnet (PM) and hybrid motors, L can be considered as independent of the rotor position.

The fundamental component of the voltage and current can be expressed as:

$$v(t) = V \cos(\omega t) \tag{11.14}$$

and

$$i(t) = I \cos(\omega t - \delta - \alpha) \tag{11.15}$$

Assuming the flux linkage of the stator winding due to the permanent magnet is:

$$\lambda_m = \lambda_m^{'} sin(\omega t - \delta) \tag{11.16}$$

$$V_1 \cos(\omega t) = RI \cos(\omega t - \delta - \alpha) - \omega LI \sin(\omega t - \delta - \alpha) + \omega \lambda_m \cos(\omega t - \delta) \tag{11.17}$$

In phasor expression, the above voltage-current relationship becomes:

$$V = RI + j\omega LI + E \tag{11.18}$$

where

$$E = \omega \lambda_m^{'} \angle -\delta \tag{11.19}$$

11.1.7.3 Equivalent Circuit in Frequency Domain

A common phasor expression for all stepping motors is:

$$V = R\ I + j\ \omega\ L\ I + E \tag{11.20}$$

where

$$E = \omega\ L_1\ I_o\ \angle -\delta \tag{11.21}$$

For a VR stepping motor, and:

$$E = \omega \lambda_m^{'} \angle -\delta \tag{11.22}$$

and

$$\beta = tan^{-1} \frac{X_L}{R}$$

For a PM or hybrid stepping motor, Figure 11.9 shows the corresponding phasor diagram.

11.1.7.4 Pull-Out Torque Expression

From the phasor diagram, it can be derived that the electromagnetic torque of a stepping motor can be expressed as:

$$T = \frac{pmEI\,cos(\beta - \delta)}{\omega\sqrt{R^2 + \omega^2 L^2}} - \frac{pmE^2 R}{\omega(R^2 + \omega^2 L^2)} \tag{11.23}$$

m is the number of phases, and $p = N_r/2$ the pole pairs of the motor.

The pull-out torque is the maximum torque for a certain speed and can be determined by letting $\delta = \beta$. Therefore:

$$T_{max} = \frac{pmEI}{\omega\sqrt{R^2 + \omega^2 L^2}} - \frac{pmE^2 R}{\omega(R^2 + \omega^2 L^2)} \tag{11.24}$$

Figure 11.10 plots the predicted pull-out torque against the rotor speed by equation (11.24), where:

$$K = \frac{L_1 V_o}{L_o V_1} \tag{11.25}$$

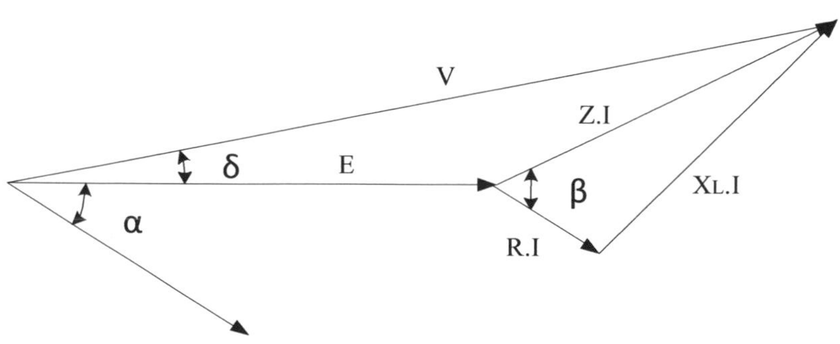

FIGURE 11.9
Phasor diagram of stepping motors.

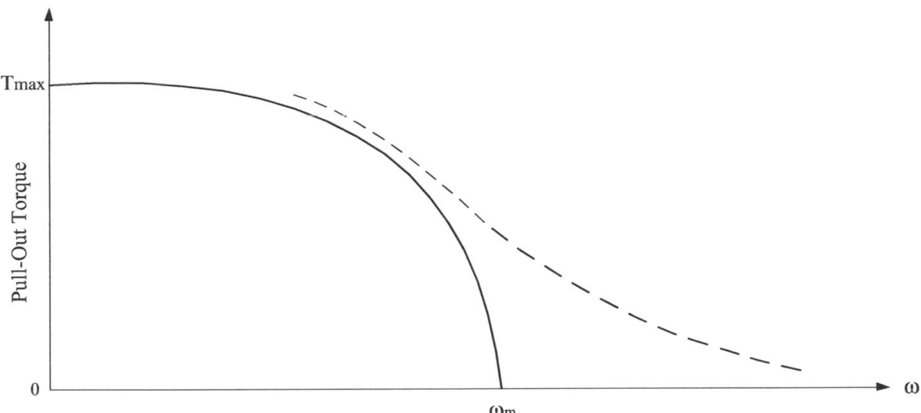

FIGURE 11.10
Predicted pull-out torque against rotor speed.

For a VR stepping motor, and:

$$K = \frac{\lambda_m'}{VL / R} \tag{11.26}$$

For a PM or hybrid stepping motor.

11.1.8 Applications

Stepper motors can be found almost anywhere. Most of us use every day without even realizing it. As for instance, steppers power "analog" wristwatches, disc drives, printers, robots, cash points, machine tools, CD players, profile cutters, plotters, and much more. Unlike other electric motors, they do not simply rotate smoothly when switched on. Every revolution is divided into a number of steps (typically 200), and the motor must be sent a separate signal for each step. It can only take one step at a time, and the size of each step is the same. Therefore, step motors may be considered a digital device.

11.2 Permanent-Magnet DC Motor

A PMDC motor is similar to an ordinary DC shunt motor except that its field is provided by permanent magnets instead of a salient-pole wound-field structure. Figure 11.11a shows 2-pole PMDC motor ,whereas Figure 11.11b shows a 4-pole wound-field DC motor for comparison purposes.

11.2.1 Construction

As shown in Figure 11.11a, the permanent magnets of the PMDC motor are supported by a cylindrical steel stator which also serves as a return path for the magnetic flux. The rotor (i.e., armature) has winding slots, commutator segments, and brushes as in conventional machines.

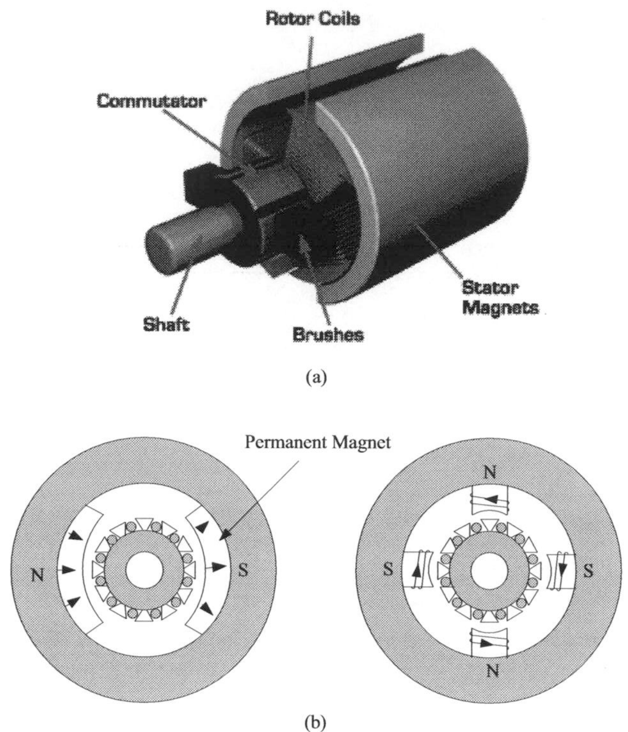

FIGURE 11.11
Permanent-magnet DC motor (a) Motor construction (b) Direction of field and current. (https://www.electrical4u.com/permanent-magnet-dc-motor-or-pmdc-motor/)

There are three types of permanent magnets used for such motors. The materials used have residual flux density and high coercively.

1. Alnico magnets: These magnets which are in having the ratings in the range of 1 KW to 150 KW used in motors
2. Ceramic (ferrite) magnets: This material is much more economical in fractional kilowatt motors
3. Rare-earth magnets: These magnets are made of samarium cobalt and neodymium iron cobalt which have the highest energy product. Such magnetic materials are costly, but are the best economic choice for small as well as large motors

Another form of the stator construction is the one in which permanent-magnet material is cast in the form of a continuous ring instead of in two pieces as shown in Figure 11.11b.

11.2.2 Working

Most of these motors usually run on 6 V, 12 V, or 24 V DC supply obtained either from batteries or rectified alternating current. In such motors, torque is produced by the interaction between the axial current-carrying rotor conductors and the magnetic flux produced by the permanent magnets.

11.2.3 Performance

The speed-torque curve is a straight line which makes this motor ideal for a servomotor. Moreover, input current increases linearly with load torque. The efficiency of such motors is higher as compared to wound-field DC motors because, in their case, there is no field copper loss.

11.2.4 Speed Control

Since flux remains constant, the speed of a PMDC motor cannot be controlled by using flux control method. The only way to control its speed is to vary the armature voltage with the help of an armature rheostat or electronically by using choppers. Consequently, such motors are found in systems where speed control below base speed only is required.

11.2.5 Advantages

1. In very small ratings, use of permanent-magnet excitation results in lower manufacturing cost
2. In many cases, a PMDC motor is smaller in size than a wound-field DC motor of equal power rating
3. Since field excitation current is not required, the efficiency of these motors is generally higher than that of the wound-field motors
4. Low-voltage PMDC motors produce less air noise
5. When designed for low-voltage (12 V or less), these motors produce very little radio and TV interference.

11.2.6 Disadvantages

1. Since their magnetic field is active at all times even when the motor is not being used, these motors are made totally enclosed to prevent their magnets from collecting magnetic junk from the neighborhood. Hence, as compared to wound-field motors, their temperature tends to be higher. However, it may not be much of a disadvantage in situations where the motor is used for short intervals
2. A more serious disadvantage is that the permanent magnets can be demagnetized by armature reaction Magnetic Motive Force (MMF) causing the motor to become inoperative. Demagnetization can result from (a) improper design, (b) excessive armature current caused by a fault or transient or improper connection in the armature circuit, (c) improper brush shift and, (d) temperature effects.

11.2.7 Applications

1. Small, 12-V PMDC motors are used for driving automobile heater and air conditioner blowers, windshield wipers, windows, fans, radio antennas, etc. They are also used for electric fuel pumps, marine motor starters, wheelchairs, and cordless power tools
2. Toy industry uses millions of such motors which are also used in other appliances such as the toothbrush, food mixer, ice crusher, portable vacuum cleaner, and shoe polisher and also in portable electric tools such as drills, saber saws, hedge trimmers, etc.

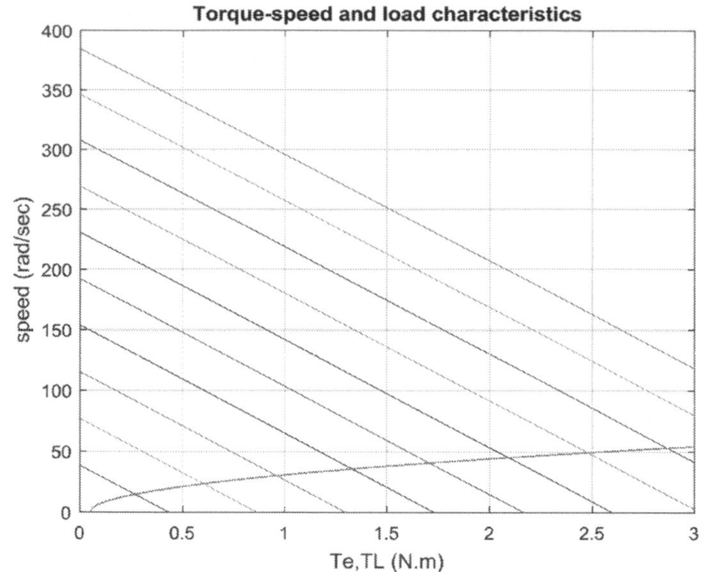

FIGURE 11.12
Permanent-magnet DC motor torque-speed and load characteristics.

MATLAB program for permanent-magnet DC motor.

```
Ra=1.5; ka=0.13;
Te=0:0.01:3;
for Va=5:5:50;
wr=Va/ka-(Ra/ka^2)*Te;
wrl=0:1:100;
Tl=0.05 + 0.001*wrl.^2;
Plot(Te, wr,'-',Tl, wrl,'-');
hold on;
axis([0, 3, 0, 400]);
end;
grid
```

Results see Figure 11.12.

11.3 Low-Inertia DC Motors

These motors are so designed as to make their armature mass very low. This permits them to start, stop and change direction, and speed very quickly making them suitable for instrumentation applications. The two common types of low-inertia motors are:

1. shell-type motor and
2. printed circuit (PC) motor.

11.3.1 Shell-Type Low-Inertia DC Motor

The armature of this kind of DC motor is made up of flat aluminum or copper coils bonded together to form a hollow cylinder. This hollow cylinder is not basically attached physically to its iron core which is stationary and is located inside the shell-type rotor. Since iron does not form part of the rotor, the rotor inertia is very small.

11.3.2 Printed-Circuit (Disc) DC Motor

It is a low-voltage DC motor which has its armature (rotor) winding and commutator printed on a thin disk of nonmagnetic insulating material. This disk-shaped armature contains no iron and etched copper conductors are printed on its both sides. It uses permanent magnets to produce the necessary magnetic field. The magnetic circuit is completed through the flux-return plate which also supports the brushes. Brushes mounted in an axial direction bear directly on the inner parts of the armature conductors which thus serve as a commutator. Since the number of armature conductors is very large, the torque produced is uniform even at low speeds. Typical sizes of these motors are in the fractional and sub-fractional horsepower ranges.

The speed can be controlled by varying either the applied armature voltage or current. Because of their high efficiency, fan cooling is not required in many applications. But the motor brushes need periodic inspection and replacement. The rotor disk which carries the conductors and commutator are very thin has a limited life. Hence, it requires replacing after some time.

The main features of this motor are:

1. Very low-inertia
2. High overload current capability
3. Linear speed-torque characteristic
4. Smooth torque down to near-zero speed
5. Very suitable for direct drive control applications
6. High torque/inertia ratio.

Advantages

1. High efficiency
2. Simplified armature construction
3. Being of low-voltage design produces a minimum radio and TV interference.

Disadvantages

1. Restricted to low-voltages only
2. Short armature life
3. Suited for intermittent duty cycle only because motor overheats in a very short time since there is no iron to absorb excess heat
4. Liable to burn out if stalled or operated with the wrong supply voltage.

Applications

These low-inertia motors have been developed specifically to provide high performance characteristics when used in direct-drive control applications. Examples are:

1. High-speed paper tape readers
2. Oscillographs
3. *X-Y* recorders
4. Layer winders
5. Point-to-point tool positioners, i.e., as positioning servomotors
6. With in-built optical position encoder, it competes with stepping motor
7. In high rating is being manufactured for heavy-duty drives such as lawn mowers, battery-driven vehicles, etc.

11.4 Servo Motors

The servo motors are used to turn the electrical signal to mechanical displacement of the rotor. The electrical signal is basically the voltage applied to the primary coil. These motors are extensively used in control systems.

These motors are specially established in control systems to measure devices, and in all types of controls, and their response is quicker in terms of change in speed for any change in voltage where the motor is mechanically connected to the load axis directly or through the gears which it is moving. This motor helps some electric devices that are unable to generate sufficient torque help move the load and do the job instead. Figure 11.13 illustrates different forms of service motors.

FIGURE 11.13
Different forms of servo motors. (https://www.efxkits.us/different-types-servo-motor-applications/).

It is a two-phase inductive drive and it contains two windings in the fixed between the 90° angle, and each coil is independent of the each other where the source of the voltage is subjected to fixed value and frequency on one of the coils called (excitation windings), and variable voltage can be controlled on the other coil, called (control windings).

The rotor has the windings of the squirrel cage, usually a small diameter, and a drop much less than its length to reduce the torque of the inertia and to be responsive.

There are several types of servo motors, but in this section just deal with a simple DC type here. If you take a normal DC motor that can be bought at Radio Shack, it has one coil (two wires). If you attach a battery to those wires the motor will spin. This is very different from a stepper already. Reversing the polarity will reverse the direction. Attach that motor to the wheel of a robot and watch the robot move to note the speed. When adding a heavier payload to the robot, the robot will slow down due to the increased load. The computer inside of the robot would not know this happened unless there was an encoder on the motor keeping track of its position.

So, in a DC motor, the speed and current draw are affected by the load. For applications that the exact position of the motor must be known, a feedback device like an encoder must be used. The control circuitry to perform good servo control of a DC motor is much more complex than the circuitry that controls a stepper motor.

One of the main differences between servo motors and stepper motors is that servo motors, by definition, run using a control loop and require feedback of some kind. A control loop uses feedback from the motor to help the motor get to the desired state (position, velocity, and so on). There are many different types of control loops. Generally, the PID (Proportional, integral, and derivative) control loop is used for servo motors.

When using a control loop such as PID, you may need to tune the servo motor. Tuning is the process of making a motor response in a desirable way. Tuning a motor can be a very difficult and tedious process, but is also an advantage in that it lets the user have more control over the behavior of the motor.

Since servo motors have a control loop to check what state they are in, they are generally more reliable than stepper motors. When a stepper motor misses a step for any reason, there is no control loop to compensate in the move. The control loop in a servo motor is constantly checking to see if the motor is on the right path and, if it is not, it makes the necessary adjustments.

In general, servo motors run more smoothly than stepper motors except when microstepping is used. Also, as speed increases, the torque of the servo remains constant, making it better than the stepper at high speeds (usually above 1000 rpm).

Some of the advantages of servo motors over stepper motors are as follows:

1. High intermittent torque
2. High torque to inertia ratio
3. High speeds
4. Work well for speed control
5. Available in all sizes
6. Quiet.

Some of the disadvantages of servo motors compared with stepper motors are as follows:

1. More expensive than stepper motors
2. Cannot work open-loop—feedback is required
3. Require tuning of control loop parameters
4. More maintenance due to brushes on brushed DC motors.

11.4.1 Mathematical Model of Servo Motor

The differential equations and transfer functions for DC servo motors are more complicated than those of AC servo motors. Since DC servo motors have time lags because of both the armature inductance and the winding, while AC motors have only a single time constant. DC servo motors are described by three differential equations: The developed torque [T(t)] is described by:

$$T(t)=K_2.i_{f(t)} \tag{11.27}$$

where $i_f(t)$ is the current through the field and K_2 is constant. The field voltage [$V_f(t)$] is described by:

$$V_f(t)=R_f.i_f(t)+L_F\frac{d}{dt}i_{f(t)} \tag{11.28}$$

where R_f is the field resistance and L_f is the field inductance. Lastly, the mechanical torque [T(t)] is described by:

$$T(t) = J\frac{d^2}{dt^2}\theta(t) + B.\frac{d}{dt}\theta(t) \tag{11.29}$$

where J is the motor's moment of inertia, B is the motor's viscous damping, and θ(t) is the motor's angular position. By assuming zero initial conditions and then Laplace transforming each of these equations, s-domain equations are reached. The developed torque is now:

$$T(s)=K_2.I_f(s) \tag{11.30}$$

the field voltage is now:

$$V_f(s)=(L_f.s+R_f).I_f(s) \tag{11.31}$$

and the mechanical torque is now:

$$T(s)=(J.s^2+R_f.s).\Theta(s) \tag{11.32}$$

where all of the constants have the same meaning as in the time-domain differential equations, I_f is used in place of i_f and Θ is used in place of θ. By substituting and solving, the transfer function of the motor is found to be:

$$\frac{\Theta(s)}{V_f(s)}=\frac{K_m}{s.(\tau_m.s+1).(\tau_e.s+1)} \tag{11.33}$$

where:

$$\tau_m = \frac{J}{B} \tag{11.34}$$

is the mechanical time constant of the motor:

$$\tau_e = \frac{L_f}{R_f} \tag{11.35}$$

is the electrical time constant of the motor, and:

$$K_m = \frac{K_2}{B.R_f} \tag{11.36}$$

is another constant. This is the transfer function that will be used for the control analysis of the DC servo motor in the next section.

The differential equations and transfer functions for AC servo motors are considerably less complicated than those for DC servo motors because AC servo motors only have a single time constant while DC servo motors have two. AC motors are described by two differential equations: The torque [T(t)] is described by:

$$T(t) = K.v(t) - m.\frac{d}{dt}\theta(t) \tag{11.37}$$

where K is a constant, v(t) is the voltage provided to the motor, θ(t) is the angular position of the motor, and m is described by:

$$\text{m} = \frac{\text{stall torque (at rated voltage)}}{\text{no-load speed (at rated voltage)}}$$

where stall torque (at rated voltage) and no-load speed (at rated voltage) are characteristics of any specific AC motor. The torque is also described by:

$$T(t) = J\frac{d^2}{dt^2}\theta(t) + B.\frac{d}{dt}\theta(t) \tag{11.38}$$

which is identical to the third differential equation that describes DC motors and has the same meaning. By equating the two AC motor equations, assuming zero initial conditions, and then taking the Laplace transform of the resultant equation, the transfer function of an AC motor is found to be:

$$\frac{\Theta(s)}{V(s)} = \frac{K_m}{s.(\tau.s+1)} \tag{11.39}$$

where:

$$K_m = \frac{K}{m+B} \tag{11.40}$$

is a constant, and:

$$\tau = \frac{J}{m+B} \tag{11.41}$$

is the time constant of the motor. This is the transfer function that will be used for the control analysis of the AC motor in the next section.

11.4.2 The Difference between Stepper Motors and Servos Motor

Stepper motors are less expensive and typically easy to use rather than a servo motor of a similar size. Because of moving into discrete steps, this kind of motor is called *stepper motor.* Therefore, a stepper motor needs a stepper drive and control to control the motor. Controlling a stepper motor requires a stepper drive and a controller.

The drive then interprets these signals and drives the motor. Stepper motors can be run in an open-loop configuration (no feedback) and are good for low-cost applications. In general, a stepper motor will have high torque at low speeds, but low torque at high speeds.

Movement at low speeds is also choppy unless the drive has the micro-stepping capability. At higher speeds, the stepper motor is not as choppy, but it does not have as much torque. When idle, a stepper motor has a higher holding torque than a servo motor of similar size, since current is continuously flowing in the stepper motor windings.

A stepper motor's shaft has permanent magnets attached to it. Around the body of the motor is a series of coils that create a magnetic field that interacts with the permanent magnets. When these coils are turned on and off the magnetic field cause the rotor to move. As the coils are turned on and off in sequence the motor will rotate forward or reverse. This sequence is called *the phase pattern* and there are several types of patterns that will cause the motor to turn. Common types are a full-double phase, full-single phase, and half step.

To make a stepper motor rotate, you must constantly turn on and off the coils. If you simply energize one coil the motor will just jump to that position and stay there resisting change. This energized coil pulls full current even though the motor is not turning. The stepper motor will generate a lot of heat at standstill. The ability to stay put at one position rigidly is often an advantage of stepper motors. The torque at standstill is called *the holding torque.*

Because steppers can be controlled by turning coils on and off, they are easy to control using digital circuitry and micro-controller chips. The controller simply energizes the coils in a certain pattern and the motor will move accordingly. At any given time the computer will know the position of the motor since the number of steps given can be tracked. This is true only if some outside force of greater strength than the motor has not interfered with the motion.

An optical encoder could be attached to the motor to verify its position, but steppers are usually used open-loop (without feedback). Most stepper motor control systems will have a home switch associated with each motor that will allow the software to determine the starting or reference "home" position.

Some of the advantages of stepper motors over servo motors are as follows:

1. Low cost
2. Low maintenance (brushless)
3. Excellent holding torque (eliminated brakes/clutches)
4. Can work in an open-loop (no feedback required)
5. Excellent torque at low speeds

6. Very rugged—any environment
7. Excellent for precise positioning control
8. No tuning required.

Some of the disadvantages of stepper motors in comparison with servo motors are as follows:

1. Rough performance at low speeds unless you use micro-stepping. Consume current regardless of load
2. Limited sizes available
3. High noisy
4. Torque decreases with speed (you need an oversized motor for higher torque at higher speeds)
5. Stepper motors can stall or lose position running without a control loop.

11.5 Brushed DC Motors

Brushed DC motors are widely used in applications ranging from toys to push-button adjustable car seats. Brushed DC (BDC) motors are inexpensive, easy to drive, and are readily available in all sizes and shapes.

This section will discuss how a BDC motor works, how to drive a BDC motor, and how a drive circuit can be interfaced to a peripheral interface controller (PIC) microcontroller.

The last decade or two, servomotors have evolved from largely brush types to brushless. This has been driven by lower maintenance and higher reliability of brushless motors. As brushless motors have become more prevalent during this period, the circuit and system techniques used to drive them have evolved as well. The variety of control schemes has led to a similar variety of buzzwords that describe them.

Most high-performance servo systems employ an inner control loop that regulates torque. This inner torque loop will then be enclosed in outer velocity and position loops to attain the desired type of control. While the designs of the outer loops are largely independent of motor type, the design of the torque loop is inherently specific to the motor being controlled.

The torque produced by a brush motor is fairly easy to control because the motor commutates itself. Torque is proportional to the DC current into the two terminals of the motor, irrespective of speed. Torque control can, therefore, be implemented by a Proportional-Integration (P-I) feedback loop which adjusts the voltage applied to the motor in order to minimize the error between requested and measured motor currents.

11.5.1 Stator

The stator generates a stationary magnetic field that surrounds the rotor. This field is generated by either permanent magnets or electromagnetic windings. The different types of BDC motors are distinguished by the construction of the stator or the way the electromagnetic windings are connected to the power source.

11.5.2 Rotor

The rotor, also called *the armature*, is made up of one or more windings. When these windings are energized they produce a magnetic field. The magnetic poles of this rotor field will be attracted to the opposite poles generated by the stator, causing the rotor to turn. As the motor turns, the windings are constantly being energized in a different sequence so that the magnetic poles generated by the rotor do not overrun the poles generated in the stator. This switching of the field in the rotor windings is called *commutation*.

11.5.3 Brushless Motor Basics

A brushless DC motor consists of a permanent magnet, which rotates (the rotor) and surrounded by three equally spaced windings which are fixed (the stator). The flowing current in each winding produces a magnetic field vector, which sums up the fields from the other windings. By controlling currents in the three windings, a magnetic field of an arbitrary direction and magnitude can be produced by the stator. Torque is then produced by the attraction or repulsion between this net stator field and the magnetic field of the rotor.

For any position of the rotor, there is an optimal direction of the net stator field, which maximizes torque, there is also a direction, which will produce no torque. If the permanent magnet rotor in the same direction as the field produces the net stator field, no torque is produced. The fields interact to produce a force, but because the force is in line with the axis of rotation of the rotor, it only serves to compress the motor bearings, not to cause rotation. On the other hand, if the stator field is orthogonal to the field produced by the rotor, the magnetic forces work to turn the rotor and torque is maximized as shown in Figure 11.14.

11.5.4 Advantage and Disadvantage of the Brushless DC Motor

The most noticeable advantage of the brushless configuration is being without brushes, as they reducing the brush maintenance and emits any other problems regarding the brushes. As for example, brushes tend to yield radiofrequency interference and the sparking associated with them which are a source of ignition in inflammable environments.

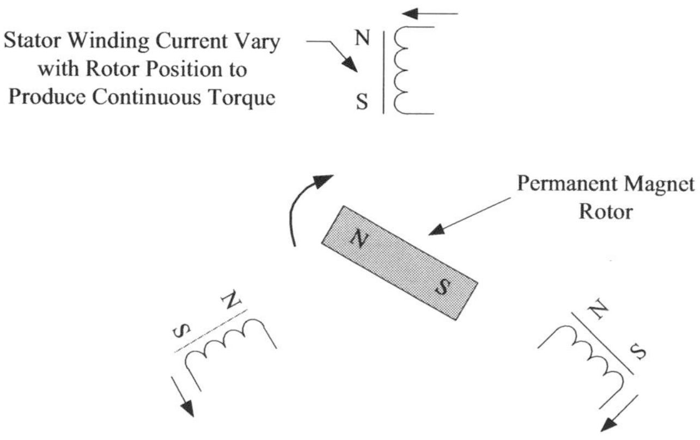

FIGURE 11.14
Brushless DC motor.

Additionally, brushless motors are potentially cleaner, faster, make less noise, and more reliable than induction motors. The rotor losses are very low, and the stator easily cooled because of the fine slot structure and the proximity of the outside air.

Their main disadvantages are: (i) the need for shaft position sensing and (ii) increased complexity in the electronic controller. There are few more disadvantages of brush DC motors which include inadequate heat dissipation, high rotor inertia, low-speed range due to limitations imposed by the brushes, and electromagnetic interference (EMI) generated by brush arcing. Brushless DC motors (BLDC) motors have a number of advantages over their brush brothers.

Problems

11.1 A hybrid VR stepping motor has ten main poles which have been castellated to have five teeth each. If the rotor has 80 teeth, calculate the stepping angle.

11.2 A hybrid VR stepping motor has 12 main poles which have been castellated to have six teeth each. If the rotor has 80 teeth, calculate the stepping angle.

11.3 A stepper motor has a step angle of 3°. Determine (a) resolution, (b) number of steps required for the shaft to make 25 revolutions, and (c) shaft speed, if the stepping frequency is 3600 pps.

11.4 A stepper motor has a step angle of 2.8°. Determine (a) resolution, (b) number of steps required for the shaft to make 30 revolutions, and (c) shaft speed, if the stepping frequency is 3600 pps.

11.5 What are the parts of brushed DC motors.

11.6 What are the advantage and disadvantage of the brushless DC motor.

11.7 Compare between stepper motors, servos motor, and brushless DC motor.

11.8 Give the mathematical model of the servo motor.

Appendix A: Mathematical Formula

This appendix, by no means exhaustive, serves as a handy reference. It does contain all the formulas needed to solve problems in this book.

A.1 Quadratic Formulas

The roots of the quadratic equation $ax^2 + bx + c = 0$:

$$x_1, x_2 = \frac{-b \pm \sqrt{b^2 - 4ac}}{2a}$$

A.2 Trigonometric Identities

$$\sin(-x) = -\sin x$$

$$\cos(-x) = \cos x$$

$$\sec x = \frac{1}{\cos x}, \qquad \csc x = \frac{1}{\sin x}$$

$$\tan x = \frac{\sin x}{\cos x}, \qquad \cot x = \frac{1}{\tan x}$$

$$\sin(x \pm 90^\circ) = \pm \cos x$$

$$\cos(x \pm 90^\circ) = \mp \sin x$$

$$\sin(x \pm 180^\circ) = -\sin x$$

$$\cos(x \pm 180^\circ) = -\cos x$$

$$\cos^2 x + \sin^2 x = 1$$

$$\frac{a}{\sin A} = \frac{b}{\sin B} = \frac{c}{\sin C} \qquad \text{(law of sines)}$$

$$a^2 = b^2 + c^2 - 2bc\cos A \qquad \text{(law of cosines)}$$

$$\frac{\tan\frac{1}{2}(A-B)}{\tan\frac{1}{2}(A+B)} = \frac{a-b}{a+b} \qquad \text{(law of tangents)}$$

$$\sin(x \pm y) = \sin x \cos y \pm \cos x \sin y$$

$$\cos(x \pm y) = \cos x \cos y \mp \sin x \sin y$$

$$\tan(x \pm y) = \frac{\tan x \pm \tan y}{1 \mp \tan x \tan y}$$

$$2\sin x \sin y = \cos(x-y) - \cos(x+y)$$

$$2\sin x \cos y = \sin(x+y) - \sin(x-y)$$

$$2\cos x \cos y = \cos(x+y) - \cos(x-y)$$

$$\sin 2x = 2\sin x \cos x$$

$$\cos 2x = \cos^2 x - \sin^2 x = 2\cos^2 x - 1 = 1 - 2\sin^2 x$$

$$\tan 2x = \frac{2\tan x}{1-\tan^2 x}$$

$$\sin^2 x = \frac{1}{2}(1-\cos 2x)$$

$$\cos^2 x = \frac{1}{2}(1+\cos 2x)$$

$$a\cos x + b\sin x = K\cos(x+\theta), \text{ where } K = \sqrt{a^2+b^2} \text{ and } \theta = \tan^{-1}\left(\frac{-b}{a}\right)$$

$$e^{\pm jx} = \cos x \pm j\sin x \qquad \text{(Euler's formula)}$$

$$\cos x = \frac{e^{jx} + e^{-jx}}{2}$$

$$\sin x = \frac{e^{jx} - e^{-jx}}{2j}$$

$$1\text{rad} = 57.296^\circ$$

A.3 Hyperbolic Functions

$$\sinh x = \frac{1}{2}\left(e^{x} - e^{-x}\right)$$

$$\cosh x = \frac{1}{2}\left(e^{x} + e^{-x}\right)$$

$$\tanh x = \frac{\sinh x}{\cosh x}$$

$$\coth x = \frac{1}{\tanh x}$$

$$\operatorname{csc} hx = \frac{1}{\sinh x}$$

$$\operatorname{sec} hx = \frac{1}{\cosh x}$$

$$\sinh\left(x \pm y\right) = \sinh x \ \cosh y \pm \cosh x \ \sinh y$$

$$\cosh\left(x \pm y\right) = \cosh x \ \cosh y \pm \sinh x \ \sinh y$$

$$\tan\left(x \pm y\right) = \frac{\tan x \pm \tan y}{1 \mp \tan x \ \tan y}$$

A.4 Derivatives

If $U = U(x)$, $V = V(x)$, and a = constant:

$$\frac{d}{dx}\left(aU\right) = a\frac{dU}{dx}$$

$$\frac{d}{dx}\left(UV\right) = U\frac{dV}{dx} + V\frac{dU}{dx}$$

$$\frac{d}{dx}\left(\frac{U}{V}\right) = \frac{V\frac{dU}{dx} - U\frac{dV}{dx}}{V^2}$$

$$\frac{d}{dx}\left(aU^{n}\right)=naU^{n-1}$$

$$\frac{d}{dx}\left(a^{U}\right)=a^{U}\,1na\frac{dU}{dx}$$

$$\frac{d}{dx}\left(e^{U}\right)=e^{U}\frac{dU}{dx}$$

$$\frac{d}{dx}(\sin U)=\cos U\frac{dU}{dx}$$

$$\frac{d}{dx}(\cos U)=-\sin U\frac{dU}{dx}$$

$$\frac{d}{dx}\tan U=\frac{1}{\cos^{2}U}\frac{dU}{dx}$$

A.5 Indefinite Integrals

If $U = U(x)$, $V = V(x)$, and a = constant:

$$\int a\,dx = ax + C$$

$$\int UdV = UV - \int VdU \qquad \text{(integration by parts)}$$

$$\int U^{n}dU=\frac{U^{n+1}}{n+1}+C, \qquad n\neq 1$$

$$\int \frac{dU}{U}=1nU+C$$

$$\int a^{U}dU=\frac{a^{U}}{1na}+C, \qquad a>0, a\neq 1$$

$$\int e^{ax}dx=\frac{1}{a}e^{ax}+C$$

$$\int xe^{ax}dx=\frac{e^{ax}}{a^{2}}(ax-1)+C$$

$$\int x^2 e^{ax} dx = \frac{e^{ax}}{a^3}\left(a^2x^2 - 2ax + 2\right) + C$$

$$\int 1nxdx = x1nx - x + C$$

$$\int \sin axdx = -\frac{1}{a}\cos ax + C$$

$$\int \cos axdx = \frac{1}{a}\sin ax + C$$

$$\int \sin^2 axdx = \frac{x}{2} - \frac{\sin 2ax}{4a} + C$$

$$\int \cos^2 axdx = \frac{x}{2} + \frac{\sin 2ax}{4a} + C$$

$$\int x \sin axdx = \frac{1}{a^2}\left(\sin ax - ax\cos ax\right) + C$$

$$\int x \cos axdx = \frac{1}{a^2}\left(\cos ax + ax\sin ax\right) + C$$

$$\int x^2 \sin axdx = \frac{1}{a^3}\left(2ax\sin ax + 2\cos ax - a^2x^2\cos ax\right) + C$$

$$\int x^2 \cos axdx = \frac{1}{a^3}\left(2ax\cos ax - 2\sin ax + a^2x^2\sin ax\right) + C$$

$$\int e^{ax} \sin bxdx = \frac{e^{ax}}{a^2 + b^2}\left(a\sin bx - b\cos bx\right) + C$$

$$\int e^{ax} \cos bxdx = \frac{e^{ax}}{a^2 + b^2}\left(a\cos bx + b\sin bx\right) + C$$

$$\int \sin ax \sin bxdx = \frac{\sin(a-b)x}{2(a-b)} - \frac{\sin(a+b)x}{2(a+b)} + C, \qquad a^2 \neq b^2$$

$$\int \sin ax \cos bxdx = -\frac{\cos(a-b)x}{2(a-b)} - \frac{\cos(a+b)x}{2(a+b)} + C, \qquad a^2 \neq b^2$$

$$\int \cos ax \cos bxdx = \frac{\sin(a-b)x}{2(a-b)} + \frac{\sin(a+b)x}{2(a+b)} + C, \qquad a^2 \neq b^2$$

$$\int \frac{dx}{a^2+x^2} = \frac{1}{a}\tan^{-1}\frac{x}{a} + C$$

$$\int \frac{x^2dx}{a^2+x^2} = x - a\tan^{-1}\frac{x}{a} + C$$

$$\int \frac{dx}{\left(a^2+x^2\right)^2} = \frac{1}{2a^2}\left(\frac{x}{x^2+a^2} + \frac{1}{a}\tan^{-1}\frac{x}{a}\right) + C$$

A.6 Definite Integrals

If *m* and *n* are integers:

$$\int_0^{2\pi} \sin ax \; dx = 0$$

$$\int_0^{2\pi} \cos ax \; dx = 0$$

$$\int_0^{\pi} \sin^2 ax \; dx = \int_0^{\pi} \cos^2 axdx = \frac{\pi}{2}$$

$$\int_0^{\pi} \sin mx \;\; \sin nx \; dx = \int_0^{\pi} \cos mx \;\; \cos nx \; dx = 0, \qquad m \neq n$$

$$\int_0^{\pi} \sin mx \;\; \cos nx \; dx = \begin{cases} 0, & m+n = even \\ \dfrac{2m}{m^2-n^2}, & m+n = odd \end{cases}$$

$$\int_0^{2\pi} \sin mx \sin nx dx = \int_{-\pi}^{\pi} \sin mx \;\; \sin nx \; dx = \begin{cases} 0, & m \neq n \\ \pi, & m \neq n \end{cases}$$

$$\int_0^{\infty} \frac{\sin ax}{x} dx = \begin{cases} \dfrac{\pi}{2}, & a > 0 \\ 0, & a = 0 \\ -\dfrac{\pi}{2}, & a < 0 \end{cases}$$

$$\int_0^{\infty} \frac{\sin^2 x}{x} dx = \frac{\pi}{2}$$

$$\int_0^\infty \frac{\cos bx}{x^2+a^2}dx = \frac{\pi}{2a}e^{-ab}, \qquad a>0, b>0$$

$$\int_0^\infty \frac{x \sin bx}{x^2+a^2}dx = \frac{\pi}{2}e^{-ab}, \qquad a>0, b>0$$

$$\int_0^\infty \operatorname{sinc} x dx = \int_0^\infty \operatorname{sinc}^2 x dx = \frac{1}{2}$$

$$\int_0^\pi \sin^2 nx dx = \int_0^\pi \sin^2 x dx = \int_0^\pi \cos^2 nx dx = \int_0^\pi \cos^2 x dx = \frac{\pi}{2}, \qquad n = \text{an integer}$$

$$\int_0^\pi \sin mx \ \sin nx \ dx = \int_0^\pi \cos mx \ \cos nx \ dx = 0, \qquad m \neq n, m, n \text{ integers}$$

$$\int_0^\pi \sin mx \ \cos nx \ dx = \begin{cases} \frac{2m}{m^2-n^2}, & m+n = \text{odd} \\ 0, & m+n = \text{even} \end{cases}$$

$$\int_{-\infty}^\infty e^{\pm j2\pi tx} dx = \delta(t)$$

$$\int_0^\infty x^n e^{-ax} dx = \frac{n!}{a^{n+1}}$$

$$\int_0^\infty e^{-a^2x^2} dx = \frac{\sqrt{\pi}}{2a}, \qquad a>0$$

$$\int_0^\infty x^{2n} e^{-ax^2} dx = \frac{1 \bullet 3 \bullet 5 \bullet\bullet\bullet (2n-1)}{2^{n+1}a^n}\sqrt{\frac{\pi}{a}}$$

$$\int_0^\infty x^{2n+1} e^{-ax^2} dx = \frac{n!}{2a^{n+1}}, \qquad a>0$$

A.7 L'Hopital's Rule

If *f(0)* = *0* = *h(0)*, then:

$$\lim_{x\to 0}\frac{f(x)}{h(x)} = \lim_{x\to 0}\frac{f'(x)}{h'(x)},$$

where the prime indicates differentiation.

A.8 Taylor and Maclaurin Series

$$f(x)=f(a)+\frac{(x-a)}{1!}f'(a)+\frac{(x-a)^2}{2!}f''(a)+\ldots$$

$$f(x)=f(0)+\frac{x}{1!}f'(0)+\frac{x^2}{2!}f''(0)+\ldots$$

where the prime indicates differentiation.

A.9 Power Series

$$e^x=1+x+\frac{x^2}{2!}+\frac{x^3}{3!}+\ldots+\frac{x^n}{n!}+\ldots$$

$$\sin x=x-\frac{x^3}{3!}+\frac{x^5}{5!}-\frac{x^7}{7!}+\ldots$$

$$\cos x=1-\frac{x^2}{2!}+\frac{x^4}{4!}-\frac{x^6}{6!}+\frac{x^8}{8!}-\ldots$$

$$\tan x=x+\frac{x^3}{3}+\frac{2x^5}{15}+\frac{17x^7}{315}+\ldots$$

$$(1+x)^n=1+nx+\frac{n(n+1)}{2!}x^2+\frac{n(n-1)(n-2)}{3!}x^3+\ldots+\binom{n}{k}x^k+\ldots+x^n$$

$$\approx 1+nx,\qquad |x|=1$$

$$\frac{1}{1-x}=1+x+x^2+x^3+\ldots,\qquad |x|<1$$

$$Q(x)=\frac{e^{-x^2/2}}{x\sqrt{2\pi}}\left(1-\frac{1}{x^2}+\frac{1\cdot 3}{x^4}-\frac{1\cdot 3\cdot 5}{x^6}+\ldots\right)$$

$$J_n(x)=\frac{1}{n!}\left(\frac{x}{2}\right)^n-\frac{1}{(n+1)!}\left(\frac{x}{2}\right)^{n+2}+\frac{1}{2!(n+2)!}\left(\frac{x}{2}\right)^{n+4}-\ldots$$

$$J_n(x)\approx\sqrt{\frac{2}{\pi x}}\cos\left(x-\frac{\pi}{4}-\frac{n\pi}{2}\right),\qquad x<1$$

A.10 Sums

$$\sum_{k=1}^{N} k = \frac{1}{2}N(N+1)$$

$$\sum_{k=1}^{N} k^2 = \frac{1}{6}N(N+1)(2N+1)$$

$$\sum_{k=1}^{N} k^3 = \frac{1}{4}N^2(N+1)^2$$

$$\sum_{k=0}^{N} a^k = \frac{a^{N+1}-1}{a-1} \qquad a \neq 1$$

$$\sum_{k=M}^{N} a^k = \frac{a^{N+1}-a^M}{a-1} \qquad a \neq 1$$

$$\sum_{k=0}^{N} \binom{N}{k} a^{N-k} b^k = (a+b)^N, \text{ where } \binom{N}{k} = \frac{N!}{(N-k)!k!}$$

A.11 Logarithmic Identities

$$\log xy = \log x + \log y$$

$$\log \frac{x}{y} = \log x - \log y$$

$$\log x^n = n \log x$$

$$\log_{10} x = \log x \ \text{(common logarithm)}$$

$$\log_e x = \ln x \ \text{(natural logarithm)}$$

A.12 Exponential Identities

$$e^x = 1 + x + \frac{x^2}{2!} + \frac{x^3}{3!} + \frac{x^4}{4!} + \cdots$$

where e = 2.7182

$$e^x e^y = e^{x+y}$$

$$(e^x)^n = e^{nx}$$

$$\ln e^x = x$$

A.13 Approximations

$$\sin x = x \qquad or \qquad \lim_{x \to 0} \frac{\sin x}{x} = 1$$

Appendix B: Complex Numbers

The ability to handle complex numbers is important in signals and systems. Although calculators and computer software packages such as MATLAB are now available to manipulate complex numbers, it is advisable that students be familiar with how to handle them by hand.

B.1 Representation of Complex Numbers

A complex number z may be written in *rectangular form* as:

$$z = x + jy \tag{B.1}$$

where $j = \sqrt{-1}$; x is the real part of z, while y is the imaginary part, that is:

$$x = \text{Re}(z), \qquad y = \text{Im}(z) \tag{B.2}$$

The complex number z is shown plotted in the complex plane in Figure B.1. Since $j = \sqrt{-1}$:

$$\begin{aligned}
\frac{1}{j} &= -j \\
j^2 &= -1 \\
j^3 &= j \cdot j^2 = -j \\
j^4 &= j^2 \cdot j^2 = 1 \\
j^5 &= j \cdot j^4 = j \\
&\vdots
\end{aligned} \tag{B.3}$$

A second way of representing the complex number z is by specifying it magnitude r and angle θ it makes with the real axis, as shown in Figure B.1. This is known as the *polar form*. It is given by:

$$z = |z| \angle \theta = r \angle \theta \tag{B.4}$$

where:

$$r = \sqrt{x^2 + y^2}, \qquad \theta = \tan^{-1} \frac{y}{x} \tag{B.5a}$$

or

$$x = r \cos \theta, \qquad y = r \sin \theta \tag{B.5b}$$

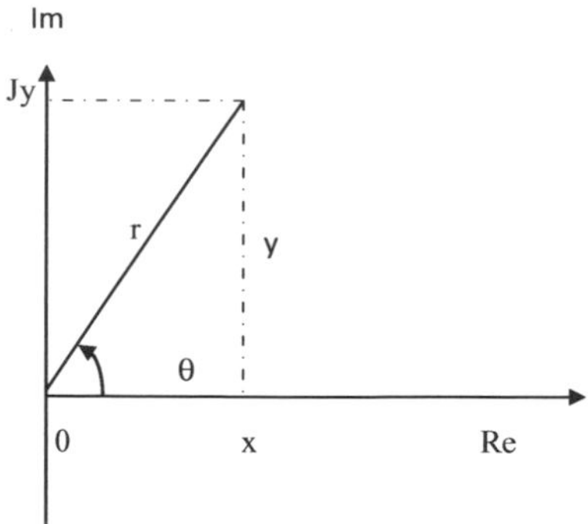

FIGURE B.1
Graphical representation of a complex number.

that is,

$$z = x + jy = r\angle\theta = r\cos\theta + jr\sin\theta \tag{B.6}$$

In converting from rectangular to polar form using Eq. (B.5), we must exercise care in determining the correct value of θ. These are the four possibilities:

$$\begin{aligned} z &= x + jy, && \theta = \tan^{-1}\frac{y}{x} && \text{(1st quadrant)} \\ z &= -x + jy, && \theta = 180^o - \tan^{-1}\frac{y}{x} && \text{(2nd quadrant)} \\ z &= -x - jy, && \theta = 180^o + \tan^{-1}\frac{y}{x} && \text{(3rd quadrant)} \\ z &= x - jy, && \theta = 360^o - \tan^{-1}\frac{y}{x} && \text{(4th quadrant)} \end{aligned} \tag{B.7}$$

assuming that x and y are positive.

The third way of representing the complex number x is the *exponential form*:

$$z = re^{j\theta} \tag{B.8}$$

This is almost the same as the polar form because we use the same magnitude r and the angle θ.

The three forms of representing a complex number are summarized as follows.

$$\begin{aligned} z &= x + jy, && (x = r\cos\theta,\ y = r\sin\theta) && \text{Rectangular form} \\ z &= r\angle\theta, && \left(r = \sqrt{x^2 + y^2}, \theta = \tan^{-1}\frac{y}{x}\right) && \text{Polar form} \\ z &= re^{j\theta}, && \left(r = \sqrt{x^2 + y^2}, \theta = \tan^{-1}\frac{y}{x}\right) && \text{Exponential form} \end{aligned} \tag{B.9}$$

B.2 Mathematical Operations

Two complex numbers $z_1 = x_1 + jy_1$ and $z_2 = x_2 + jy_2$ are equal if and only their real parts are equal, and their imaginary parts are equal, that is:

$$x_1 = x_2, \qquad y_1 = y_2 \tag{B.10}$$

The complex conjugate of the complex number z = x + jy is:

$$z^* = x - jy = r \angle -\theta = re^{-j\theta} \tag{B.11}$$

Thus, the complex conjugate of a complex number is found by replacing every j by –j.

Given two complex numbers $z_1 = x_1 + jy_1 = r_1 \angle \theta_1$ *and* $z_2 = x_2 + jy_2 = r_2 \angle \theta_2$, their sum is:

$$z_1 + z_2 = (x_1 + x_2) + j(y_1 + y_2) \tag{B.12}$$

and their difference is

$$z_1 - z_2 - (x_1 - x_2) + j(y_1 - y_2) \tag{B.13}$$

While it is more convenient to perform addition and subtraction of complex numbers in rectangular form, the product and quotient of two complex numbers are best done in polar or exponential form. For their product:

$$z_1 z_2 = r_1 r_2 \angle \theta_1 + \theta_2 \tag{B.14}$$

Alternatively, using the rectangular form:

$$\begin{aligned} z_1 z_2 &= (x_1 + jy_1)(x_2 + jy_2) \\ &= (x_1 x_2 - y_1 y_2) + j(x_1 y_2 + x_2 y_1) \end{aligned} \tag{B.15}$$

For their quotient,

$$\frac{z_1}{z_2} - \frac{r_1}{r_2} \angle \theta_1 - \theta_2 \tag{B.16}$$

Alternatively, using the rectangular form:

$$\frac{z_1}{z_2} = \frac{x_1 + jy_1}{x_2 + jy_2} \tag{B.17}$$

We rationalize the denominator by multiplying both the numerator and denominator by z_2^*:

$$\frac{z_1}{z_2} = \frac{(x_1 + jy_1)(x_2 - jy_2)}{(x_2 + jy_2)(x_2 - jy_2)} = \frac{x_1 x_2 + y_1 y_2}{x_2^2 + y_2^3} + j\frac{x_2 y_1 - x_1 y_2}{x_2^2 + y_2^3} \tag{B.18}$$

B.3 Euler's Formula

Euler's formula is an important result in complex variables. We derive it from the series expansion of e^x, $\cos\theta$, and $\sin\theta$. We know that:

$$e^x = 1+x+\frac{x^2}{2!}+\frac{x^3}{3!}+\frac{x^4}{4!}+\dots \tag{B.19}$$

Replacing x by jθ gives:

$$e^{j\theta} = 1+j\theta-\frac{\theta^2}{2!}-j\frac{\theta^3}{3!}+\frac{\theta^4}{4!}+\dots \tag{B.20}$$

Also:

$$\cos\theta = 1-\frac{\theta^2}{2!}+\frac{\theta^4}{4!}-\frac{\theta^6}{6!}+\dots \tag{B.21a}$$

$$\sin\theta = \theta-\frac{\theta^3}{3!}+\frac{\theta^5}{5!}-\frac{\theta^7}{7!}+\dots \tag{B.21b}$$

so that:

$$\cos\theta+j\sin\theta = 1+j\theta-\frac{\theta^2}{2!}-j\frac{\theta^3}{3!}+\frac{\theta^4}{4!}+j\frac{\theta^5}{5!}-\dots \tag{B.22}$$

Comparing eqs. (B.20) and (B.22), we conclude that:

$$e^{j\theta} = \cos\theta + j\sin\theta \tag{B.23}$$

This is known as Euler's formula. The exponential form of representing a complex number as in Eq. (B.8) is based on Euler's formula. From Eq. (B.23), notice that:

$$\cos\theta = \text{Re }(e^{j\theta}), \qquad \sin\theta = \text{Im }(e^{j\theta}), \tag{B.24}$$

and that

$$|e^{j\theta}| = \sqrt{\cos^2\theta+\sin^2\theta} = 1 \tag{B.25}$$

Replacing θ by –θ in eq. (B.23) gives:

$$e^{-j\theta} = \cos\theta - j\sin\theta \tag{B.26}$$

Adding eqs. (B.23) and (B.26) yields:

$$\cos\theta = \frac{1}{2}\left(e^{j\theta}+e^{-j\theta}\right) \tag{B.27}$$

Subtracting eq. (B.26) from eq. (B.23) yields:

$$\sin\theta = \frac{1}{2j}\left(e^{j\theta} - e^{-j\theta}\right) \tag{B.28}$$

The following identities are useful in dealing with complex numbers. If $z = x + jy = r\angle\theta$, then:

$$zz^* = |z|^2 = x^2 + y^2 = r^2 \tag{B.29}$$

$$\sqrt{z} = \sqrt{x + jy} = \sqrt{r}e^{j\theta/2} = \sqrt{r}\ \angle\theta/2 \tag{B.30}$$

$$z^n = (x + jy)^n = r^n\angle\, n\theta = r^n(\cos n\theta + j\sin n\theta) \tag{B.31}$$

$$z^{1/n} = (x + jy)^{1/n} = r^{1/n}\angle\ \theta/n + 2\pi k/n, \qquad k = 0,1,2,\ldots,n-1 \tag{B.32}$$

$$\ln\ (re^{j\theta}) = \ln r + \ln e^{j\theta} = \ln r + j\theta + j2\pi k\ \ (k\text{=integer}) \tag{B.33}$$

$$\begin{aligned} e^{\pm j\pi} &= -1 \\ e^{\pm j2\pi} &= 1 \\ e^{j\pi/2} &= j \\ e^{-j\pi/2} &= -j \end{aligned} \tag{B.34}$$

$$\begin{aligned} &\text{Re}\left(e^{(\alpha+j\omega)t}\right) - \text{Re}\left(e^{\alpha t}e^{j\omega t}\right) = e^{\alpha t}\cos\omega t \\ &\text{Im}\left(e^{(\alpha+j\omega)t}\right) = \text{Im}\left(e^{\alpha t}e^{j\omega t}\right) = e^{\alpha t}\sin\omega t \end{aligned} \tag{B.35}$$

MATLAB handles complex numbers quite easily as real numbers.

Example B.1. **Evaluate the complex numbers:**

a. $z_1 = \dfrac{j(3-j4)^*}{(-1+j6)(2+j)^2}$

b. $z_2 = \left[4\ \angle 30^\circ + 2 - j + 6e^{j\pi/4}\right]^{1/2}$.

Solution

a. This can be solved in two ways: working with z in rectangular form or polar form.
Method 1 (working in rectangular form):

$$\text{Let } z_1 = \frac{z_3 z_4}{z_5 z_6}$$

where

$$z_3 = j$$

$$z_4 = (3 - j4)^* = \text{the complex conjugate of } (3 - j4)$$

$$= (3 + j4)$$

$$z_5 = -1 + j6$$

$$z_6 = (2 + j)^2 = 4 + j4 - 1 = 3 + j4$$

Hence,

$$z_1 = \frac{j(3 + j4)}{(-1 + j6)(3 + j4)} = \frac{j3 - 4}{-3 - j4 + j18 - 24}$$

$$= \frac{-4 + j3}{-27 + j14}$$

Multiplying and dividing by $-27 - j14$ (rationalization), we have:

$$z_1 = \frac{(-4 + j3)(-27 - j14)}{(-27 + j14)(-27 - j14)} = \frac{150 - j25}{27^2 + 14^2}$$

$$= 0.1622 - j0.027 = 0.1644\angle -9.46^\circ$$

Method 2 (working in polar form)

$$z_3 = j = 1\angle 90^\circ$$

$$z_4 = (3 - j4)^* = (5\angle -53.13^\circ)^* = 5\angle 53.13^\circ$$

$$z_5 = (-1 + j6) = \sqrt{37}\angle 99.46^\circ$$

$$z_6 = (2 + j)^2 = (\sqrt{5}\angle 26.56^\circ)^2 = 5\angle 53.13^\circ$$

Hence,

$$z_1 = \frac{z_3 z_4}{z_5 z_6} = \frac{(1\angle 90^\circ)(5\angle 53.13^\circ)}{(\sqrt{37}\angle 99.46^\circ)(5\angle 53.13^\circ)} = \frac{1}{\sqrt{37}}\angle(90^\circ - 99.46^\circ)$$

$$= 0.1644\angle -9.46^\circ = 0.1622 - j0.027$$

b. Let

$$z_7 = 4\angle 30^\circ = 4\cos 30^\circ + j4\sin 30^\circ = 3.464 + j2$$

$$z_8 = 2 - j$$

$$z_9 = 6e^{j\pi/4} = 6\cos 45^\circ + j6\sin 45^\circ = 4.243 + j4.243$$

Then,

$$z_2 = \left[z_7 + z_8 + z_9\right]^{1/2}$$
$$= \left[3.464 + j2 + 2 - j + 4.243 + j4.243\right]^{1/2}$$
$$= (9.707 + j5.243)^{1/2} = (11.03\angle 28.374°)^{1/2}$$
$$= 3.32\angle 14.19°$$

Practice Problem B.1 Evaluate the following complex numbers:

(a) $j^3\left[\dfrac{1+j}{2-j}\right]^2$

(b) $6\angle 30° + j5 - 3 + e^{j45°}$

Answer: (a) 0.24 + j0.32, (b) 2.03 + j8.707

Appendix C: Introduction to MATLAB®

MATLAB has become a powerful tool for technical professionals worldwide. The term MATLAB is an abbreviation for MATrix LABoratory implying that MATLAB is a computational tool that employs matrices and vectors/arrays to carry out numerical analysis, signal processing, and scientific visualization tasks. Because MATLAB uses matrices as its fundamental building blocks, one can write mathematical expressions involving matrices just as easily as one would on paper. MATLAB is available for Macintosh, Unix, and Windows operating systems. A student version of MATLAB is available for Personal Computers (PCs). A copy of MATLAB can be obtained from:

The Mathworks, Inc.
3 Apple Hill Drive
Natick, MA 01760-2098
Phone:(508) 647-7000
Website: http://www.mathworks.com

A brief introduction to MATLAB is presented in this appendix. What is presented is sufficient for solving problems in this book. Other information on MATLAB required in this book is provided on the chapter-to-chapter basis as needed. Additional information about MATLAB can be found in MATLAB books and from online help. The best way to learn MATLAB is to work with it after one has learned the basics.

C.1 MATLAB Fundamentals

The Command window is the primary area where you interact with MATLAB. A little later, we will learn how to use the text editor to create M-files, which allow executing sequences of commands. For now, we focus on how to work in the Command window. We will first learn how to use MATLAB as a calculator. We do so by using the algebraic operators in Table C.1.

To begin to use MATLAB, we use these operators. Type commands to MATLAB prompt ">>" in the Command window (correct any mistakes by backspacing) and press the <Enter> key. For example,

```
» a=2; b=4; c=-6;
» dat = b^2 - 4*a*c
dat =
  64
» e=sqrt(dat)/10
e =
  0.8000
```

The first command assigns the values 2, 4, and −6 to the variables *a*, *b*, and *c*, respectively. MATLAB does not respond because this line ends with a colon. The second command sets *dat* to $b^2 - 4ac$ and MATLAB return the answer as 64. Finally, the third line sets *e* equal to

TABLE C.1

Basic Operations

Operation	MATLAB Formula	
Addition	a + b	
Division (right)	a/b	(means $a \div b$)
Division (left)	a\b	(means $b \div a$)
Multiplication	a × b	
Power	a^b	
Subtraction	a – b	

the square root of *dat* and divides by 10. MATLAB prints the answer as 0.8. As function *sqrt* is used here, other mathematical functions listed in Table C.2 can be used. Table C.2 provides just a small sample of MATLAB functions. Others can be obtained from the online help. To get help, type:

```
>> help
```

[a long list of topics come up]
and for a specific topic, type the command name. For example, to get help on a *log to base 2*, type:

TABLE C.2

Typical Elementary Math Functions

Function	Remark
abs(x)	Absolute value or complex magnitude of x
acos, acosh(x)	Inverse cosine and inverse hyperbolic cosine of x in radians
acot, acoth(x)	Inverse cotangent and inverse hyperbolic cotangent of x in radians
angle(x)	Phase angle (in radian) of a complex number x
asin, asinh(x)	Inverse sine and inverse hyperbolic sine of x in radians
atan, atanh(x)	Inverse tangent and inverse hyperbolic tangent of x in radians
conj(x)	Complex conjugate of x
cos, cosh(x)	Cosine and hyperbolic cosine of x in radian
cot, coth(x)	Cotangent and hyperbolic cotangent of x in radian
exp(x)	Exponential of x
fix	Round toward zero
imag(x)	Imaginary part of a complex number x
log(x)	Natural logarithm of x
log2(x)	Logarithm of x to base 2
log10(x)	Common logarithms (base 10) of x
real(x)	Real part of a complex number x
sin, sinh(x)	Sine and hyperbolic sine of x in radian
sqrt(x)	Square root of x
tan, tanh	Tangent and hyperbolic tangent of x in radian

```
>> help log2
```

[a help message on the log function follows]

Note that MATLAB is case sensitive so that sin(a) is not the same as sin(A).

Try the following examples:

```
>> 3^(log10(25.6))
>> y=2* sin(pi/3)
>>exp(y+4-1)
```

In addition to operating on mathematical functions, MATLAB easily allows one to work with vectors and matrices. A vector (or array) is a special matrix with one row or one column. For example:

```
>> a = [1 -3 6 10 -8 11 14];
```

is a row vector. Defining a matrix is similar to defining a vector. For example, a 3×3 matrix can be entered as:

```
>> A = [1 2 3; 4 5 6; 7 8 9]
```

or as:

```
>> A = [1 2 3
4 5 6
7 8 9]
```

In addition to the arithmetic operations that can be performed on a matrix, the operations in Table C.3 can be implemented.

Using the operations in Table C.3, we can manipulate matrices as follows.

```
» B = A'
B =
  1 4 7
  2 5 8
  3 6 9
» C = A + B
```

TABLE C.3

Matrix Operations

Operation	Remark
A′	Finds the transpose of matrix A
det(A)	Evaluates the determinant of matrix A
inv(A)	Calculates the inverse of matrix A
eig(A)	Determines the eigenvalues of matrix A
diag(A)	Finds the diagonal elements of matrix A
expm(A)	Exponential of matrix A

```
C =
  2 6 10
  6 10 14
  10 14 18
» D = A^3 - B*C

D =
  372 432 492
  948 1131 1314
  1524 1830 2136
» e= [1 2; 3 4]

e =
  1 2
  3 4
» f=det(e)

f =
  -2
» g = inv(e)

g =
  -2.0000 1.0000
  1.5000 -0.5000
» H = eig(g)

H =
  -2.6861
  0.18611
```

Note that not all matrices can be inverted. A matrix can be inverted if and only if its determinant is nonzero. Special matrices, variables, and constants are listed in Table C.4. For example, type:

```
>> eye(3)
```

TABLE C.4

Special Matrices, Variables, and Constants

Matrix/Variable/Constant	Remark
eye	Identity matrix
ones	An array of ones
zeros	An array of zeros
i or j	Imaginary unit or sqrt(–1)
pi	3.142
NaN	Not a number
inf	Infinity
eps	A very small number, 2.2e–16
rand	Random element

```
ans=
  1 0 0
  0 1 0
  0 0 1
```

to get a 3 × 3 identity matrix.

C.2 Using MATLAB to Plot

To plot using MATLAB is easy. For two-dimensional plot, use the plot command with two arguments as:

```
>> plot(xdata, ydata),
```

where *xdata* and *ydata* are vectors of the same length containing the data to be plotted.

For example, suppose we want to plot y = 10 × sin(2 × pi × x) from 0 to 5 × pi, we will proceed with the following commands:

```
>> x = 0:pi/100:5*pi;    % x is a vector, 0 <= x <= 5*pi, increments of pi/100
>> y = 10*sin(2*pi*x); % create a vector y
>> plot(x, y);          % create the plot.
```

With this, MATLAB responds with the plot in Figure C.1.

MATLAB will let you graph multiple plots together and distinguish with different colors.

This is obtained with the command plot (xdata, ydata, "color"), where the color is indicated by using a character string from the options listed in Table C.5.

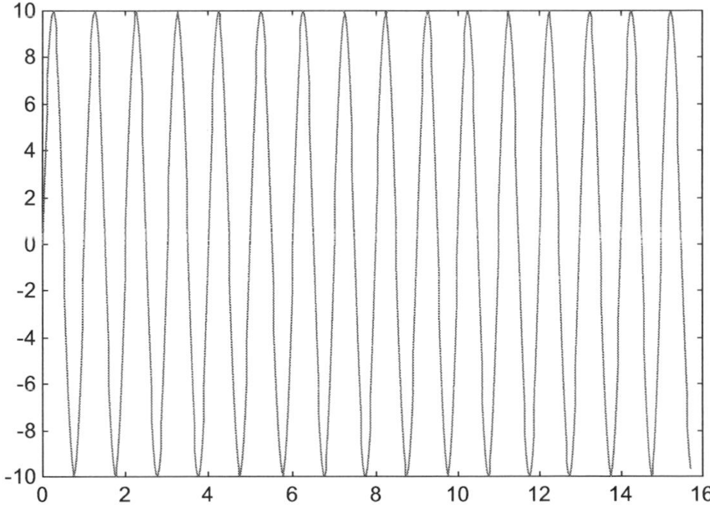

FIGURE C.1
MATLAB plot of y = 10 × sin(2 × pi × x).

TABLE C.5

Various Color and Line Types

y	yellow	.	point
m	magenta	o	circle
c	cyan	x	x-mark
r	red	+	plus
g	green	-	solid
b	blue	*	star
w	white	:	dotted
k	black	-.	dashdot
		--	dashed

For example,

```
>> plot(x1, y1, 'r', x2,y2, 'b', x3,y3, '--');
```

will graph data (x1,y1) in red, data (x2,y2) in blue, and data (x3,y3) in dashed line all on the same plot.

MATLAB also allows for logarithm scaling. Rather that the **plot** command, we use:

```
loglog      log(y) versus log(x)
semilogx    y versus log(x)
semilogy    log(y) versus x.
```

Three-dimensional plots are drawn using the functions *mesh* and *meshdom* (mesh domain). For example, draw the graph of z = x × exp(−x^2 −y^2) over the domain −1 <x, y<1, we type the following commands:

```
>> xx = -1:.1:1;
» yy = xx;
» [x, y] = meshgrid(xx, yy);
» z=x.*exp(-x.^2 -y.^2);
» mesh(z);.
```

(The dot symbol used in x. and y. allows element-by-element multiplication.) The result is shown in Figure C.2.

Other plotting commands in MATLAB are listed in Table C.6. The **help** command can be used to find out how each of these is used.

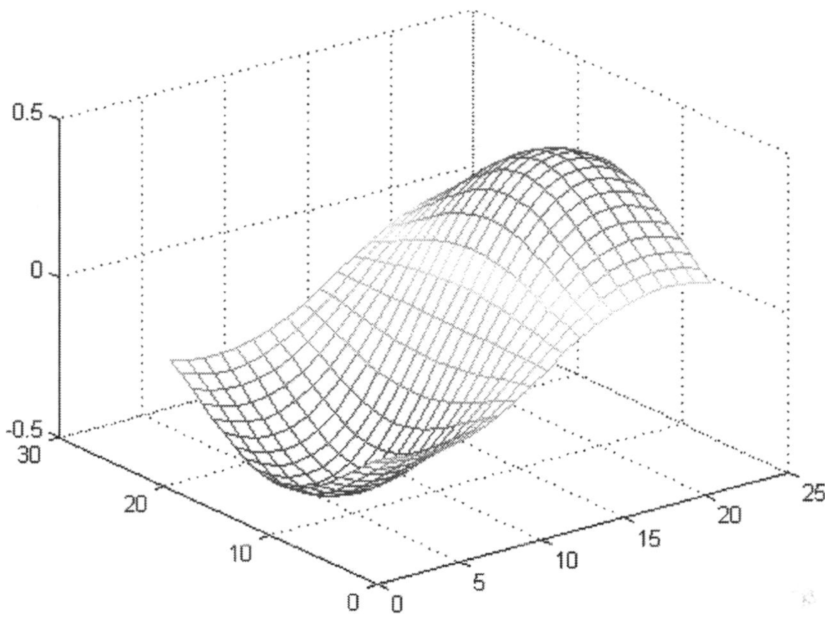

FIGURE C.2
A three-dimensional plot.

TABLE C.6
Other Plotting Commands

Command	Comments
Bar(x, y)	A bar graph
Contour(z)	A contour plot
Errorbar(x, y, l, u)	A plot with error bars
Hist(x)	A histogram of the data
Plot3(x, y, z)	A three-dimensional version of plot()
Polar(r, angle)	A polar coordinate plot
Stairs(x, y)	A stairstep plot
Stem(n, x)	Plots the data sequence as stems
Subplot(m, n, p)	Multiple (m-by-n) plots per window
Surf(x, y, x, c)	A plot of 3-D colored surface

C.3 Programming with MATLAB

So far MATLAB has been used as a calculator, you can also use MATLAB to create your own program. The command line editing in MATLAB can be inconvenient if one has several lines to execute. To avoid this problem, one creates a program which is a sequence of statements to be executed. If you are in Command window, click **File/New/M-files** to open a new file in the MATLAB Editor/Debugger or simple text editor. Type the program and save the program in a file with an extension.m, say filename.m; it is, for this reason, called an M-file. Once the program is saved as an M-file, exit the Debugger window. You are now back in Command window. Type the file without the extension.m to get results. For example, the plot that was made above can be improved by adding title and labels and typed as an M-files called example1.m.

```
x = 0:pi/100:5*pi;          % x is a vector, 0 <= x <= 5*pi, increments of pi/100
y = 10*sin(2*pi*x);         % create a vector y
plot(x, y);                 % create the plot
xlabel('x (in radians)'); % label the x-axis
ylabel('10*sin(2*pi*x)'); % label the y-axis
title('A sine functions'); % title the plot
grid                        % add grid.
```

Once it is saved as *example1.m* and we exit text editor, type:

```
>> example1,
```

in the Command window, and hit <Enter> to obtain the result shown in Figure C.3.

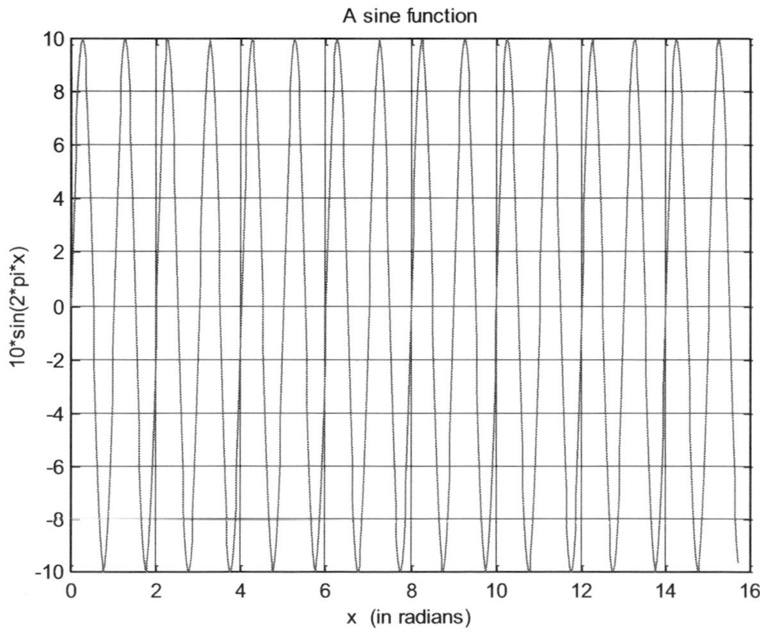

FIGURE C.3
MATLAB plot of y = 10 × sin(2 × pi × x) with title and labels.

TABLE C.7

Relational and Logical Operators

Operator	Remark
<	Less than
< =	Less than or equal
>	Greater than
> =	Greater than or equal
= =	Equal
~ =	Not equal
&	And
|	Or
~	Not

To allow flow control in a program, certain relational and logical operators are necessary. They are shown in Table C.7. Perhaps the most commonly used flow control statements are *for* and *if*. The *for* statement is used to create a loop or a repetitive procedure and has the general form:

```
for x = array
```

[commands]

```
End.
```

The *if* statement is used when certain conditions need be met before an expression is executed. It has the general form:

if expression
[commands if the expression is True]

```
else,
```

[commands if the expression is False]

```
end.
```

For example, suppose we have an array y(x) and we want to determine the minimum value of y and its corresponding index x. This can be done by creating an M-file as shown below.

```
% example2.m
% This program finds the minimum y value and its corresponding x index
x = [1 2 3 4 5 6 7 8 9 10]; %the nth term in y
y = [3 9 15 8 1 0 -2 4 12 5];
min1 = y(1);
for k=1:10
  min2=y(k);
  if(min2 < min1)
    min1 = min2;
    xo = x(k);
```

```
    else
      min1 = min1;
    end
  end
  diary
  min1, xo
  diary off.
```

Note the use of *for* and *if* statements. When this program is saved as example2.m, we execute it in the Command window and obtain the minimum value of y as –2 and the corresponding value of x as 7, as expected.

```
» example2
min1 =
      -2
xo =
   7
```

If we are not interested in the corresponding index, we could do the same thing using the command:

```
>> min(y).
```

The following tips are helpful in working effectively with MATLAB:

- Comment your M-file by adding lines beginning with a % character
- To suppress output, end each command with a semi-colon (;), you may remove the semi-colon when debugging the file
- Press up and down arrow keys to retrieve previously executed commands
- If your expression does not fit on one line, use an ellipse (…) at the end of the line and continue on the next line. For example, MATLAB considers:

$$y = \sin(x + \log10(2x + 3)) + \cos(x + \ldots$$

$$\log10(2x + 3));$$

as one line of expression

- Keep in mind that variable and function names are case sensitive.

C.4 Solving Equations

Consider the general system of n simultaneous equations as:

$$a_{11}x_1 + a_{12}x_2 + \bullet\bullet\bullet + a_{1n}x_n = b_1$$

$$a_{21}x_1 + a_{22}x_2 + \bullet\bullet\bullet + a_{2n}x_n = b_2$$

$$\cdots \qquad \cdots \qquad \cdots$$

$$a_{n1}x_1 + a_{n2}x_2 + \bullet\bullet\bullet + a_{nn}x_n = b_n$$

or in matrix form:

$$AX = B$$

where:

$$A=\begin{bmatrix} a_{11} & a_{12} & \bullet\bullet\bullet & a_{1n} \\ a_{21} & a_{22} & \bullet\bullet\bullet & a_{2n} \\ \bullet\bullet\bullet & \bullet\bullet\bullet & \bullet\bullet\bullet & \bullet\bullet\bullet \\ a_{n1} & a_{n2} & a_{n3} & a_{nn} \end{bmatrix},\ X=\begin{bmatrix} x_1 \\ x_2 \\ \bullet\bullet\bullet \\ x_n \end{bmatrix},\ B=\begin{bmatrix} b_1 \\ b_2 \\ \bullet\bullet\bullet \\ b_n \end{bmatrix}$$

A is a square matrix and is known as the coefficient matrix, while X and B are vectors. X is the solution vector we are seeking to get. There are two ways to solve for X in MATLAB. First, we can use the backslash operator (\) so that:

$$X = A\backslash B.$$

Second, we can solve for X as:

$$X = A^{-1}B,$$

which in MATLAB is the same as:

$$X = inv(A) \times B.$$

We can also solve equations using the command **solve**. For example, given the quadratic equation $x^2 + 2x - 3 = 0$, we obtain the solution using the following MATLAB command:

```
>> [x]=solve('x^2 + 2*x - 3 = 0')
x =
[-3]
[1].
```

Indicating that the solutions are x = –3 and x = 1. Of course, we can use the command **solve** for a case involving two or more variables. We will see that in the following example.

Example C.1 Use MATLAB to solve the following simultaneous equations

$$25x_1 - 5x_2 - 20x_3 = 50$$
$$-5x_1 + 10x_2 - 4x_3 = 0$$
$$-5x_1 - 4x_2 + 9x_3 = 0.$$

Solution

We can use MATLAB to solve this in two ways:

Method 1:

The given set of simultaneous equations could be written as:

$$\begin{bmatrix} 25 & -5 & -20 \\ -5 & 10 & -4 \\ -5 & -4 & 9 \end{bmatrix} \begin{bmatrix} x_1 \\ x_2 \\ x_3 \end{bmatrix} = \begin{bmatrix} 50 \\ 0 \\ 0 \end{bmatrix} \quad or \quad \mathrm{AX} = \mathrm{B}$$

We obtain matrix A and vector B and enter them in MATLAB as follows.

```
» A = [25 -5 -20; -5 10 -4; -5 -4 9]
A =
  25 -5 -20
  -5 10 -4
  -5 -4 9
» B = [50 0 0]'
B =
  50
  0
  0
» X = inv(A)*B
X =
  29.6000
  26.0000
  28.0000
» X = A\B
X =
  29.6000
  26.0000
  28.0000.
```

Thus, $x_1 = 29.6$, $x_2 = 26$, and $x_3 = 28$.

Method 2:

Since the equations are not many in this case, we can use the command **solve** to obtain the solution of the simultaneous equations as follows:

$[x_1, x_2, x_3] = \text{solve}(\text{'}25 \times x_1 - 5 \times x_2 - 20 \times x_3 = 50\text{'},$
$\text{'}-5 \times x_1 + 10 \times x_2 - 4 \times x_3 = 0\text{'}, \text{'}-5 \times x_1 - 4 \times x_2 + 9 \times x_3 = 0\text{'})$

$x_1 =$

148/5

$x_2 =$

26

$x_3 =$

28,

which is the same as before.

Practice Problem C.1 Solve the problem the following simultaneous equations using MATLAB:

$$3x_1 - x_2 - 2x_3 = 1$$
$$-x_1 + 6x_2 - 3x_3 = 0$$
$$-2x_1 - 3x_2 + 6x_3 = 6.$$

Answer: $x_1 = 3 = x_3$, $x_2 = 2$.

C.5 Programming Hints

A good program should be well documented, of reasonable size, and capable of performing some computation with reasonable accuracy within a reasonable amount of time. The following are some helpful hints that may make writing and running MATLAB programs easier.

- Use the minimum commands possible and avoid execution of extra commands. This is particularly true of loops
- Use matrix operations directly as much as possible and avoid *for, do,* and/or *while* loops if possible
- Make effective use of functions for executing a series of commands over several times in a program
- When unsure about a command, take advantage of the help capabilities of the software
- It takes much less time running a program using files on the hard disk than on a floppy disk
- Start each file with comments to help you remember what it is all about later
- When writing a long program, save frequently. If possible, avoid a long program, break it down into smaller subroutines.

C.6 Other Useful MATLAB Commands

Some common useful MATLAB commands which may be used in this book are provided in Table C.8.

TABLE C.8

Other Useful MATLAB Commands

Command	Explanation
Diary	Save screen display output in text format
Mean	Mean value of a vector
Min(max)	Minimum (maximum) of a vector
Grid	Add a grid mark to the graphic window
Poly	Converts a collection of roots into a polynomial
Roots	Finds the roots of a polynomial
Sort	Sort the elements of a vector
Sound	Play vector as sound
Std	Standard deviation of a data collection
Sum	Sum of elements of a vector

Appendix D: Answer to Odd-Numbered Problems

Chapter 1

1.5 (a) 7.21 V, (b) 7.987 V

1.7 65.89 m/sec

1.9 1.92 Wb

Chapter 2

2.3 $6.67 \times 10 - 3$ N

2.5 6.655 A

2.7 $\phi = 0.48$ Wb

2.9 $E = -0.2 \times 10^{-5}$ V

2.11 I = 1.5 mA

Chapter 3

3.7 f = 60 Hz

3.9 600 rpm

3.11 100 Ω

3.13 (i) 5.32 A, (ii) 10.6.4 V, (iii) 167.5 V, (iv) 37.6 Ω, and (v) 0.53 leading

3.15 0.6 lagging

3.17 0.318 Henry

Chapter 4

4.5 8.33 A, 104.13 A, 0.027 wb

Chapter 5

5.1 (i) Core area $A_i = 0.0536 * 10^6\,mm^2$
(ii) Window area $A_w = 0.3735 * 10^6\,mm^2$

5.3 (i) $Ai = 0.0246\,m^2$, $d = 0.1812$ m, $Aw = 0.0872\,m^2$, $Ww = 0.1867$ m
(ii) Tp = 134 turns, Ts = 15 turns
(iii) $Ap = 90\,mm^2$, $As = 833.33\,mm^2$

Chapter 6

6.17 Ra = 0.8 Ω

6.19 Ra = 1.8 Ω

6.21 N = 300 rpm

6.23 Rse = 0.3 Ω

Chapter 7

7.3 2p = 2

Chapter 8

8.3 6.366 V, 0.31 A

8.5 10.8 A

8.7 i.$V_{lmean} = 54.9\,V$, $I_{lmean} = 27.45\,A$, $v_{s\,rms} = 141.42\,V$

8.8 ii.$V_{lmean} = 59.59\,V$, $I_{lmean} = 29.8\,A$, $v_{o\,rms} = 98.61\,V, RF = 1.318$

8.9 $I_{min} = 92.28$ A, $I_{max} = 101.79$ A, the average load current $I_{av} = 96$ A, the average value of the diode current $I_D = 57.9$, the effective input resistance $R_i = 1.25$

Chapter 9

9.1 $R_d = 10\,\Omega$, V = 24.35 Volts

9.3 Vt = 279 V, $\omega_o = 186$ rad/sec

9.5 $I_a = 61.72\,$ A, $\alpha = 38.53°$
$N_R\% = 3.78\%$, pf = 0.704 lagging
Input power factor $= \frac{P}{S} = 0.712$ lagging

9.7

i.

Duty Cycle (D)	I_{Lmean} (A)	V_{Lmean} (V)	Speed (Rad/sec)
0.5	54.96	83.25	87.82
0.6	55.17	99.94	111.7

ii.

Duty Cycle (D)	I_{Lmean} (A)	V_{Lmean} (V)	Speed (Rad/sec)
100%	56	200	206.9

iii.

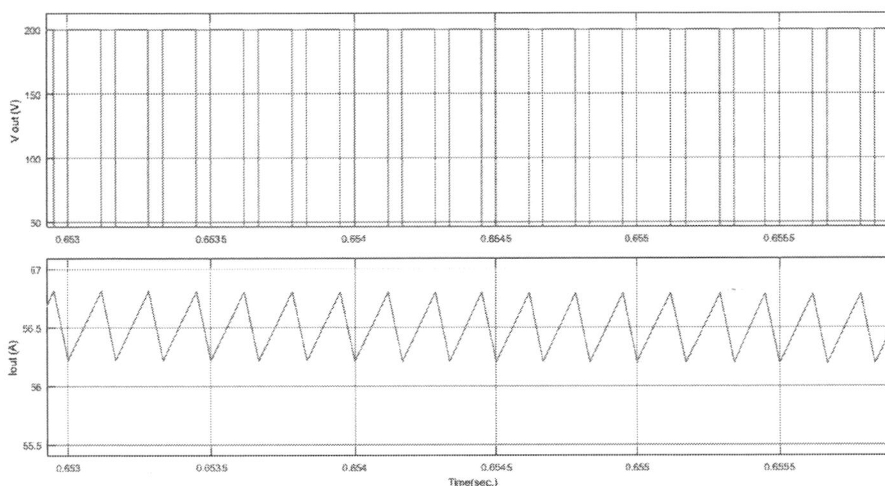

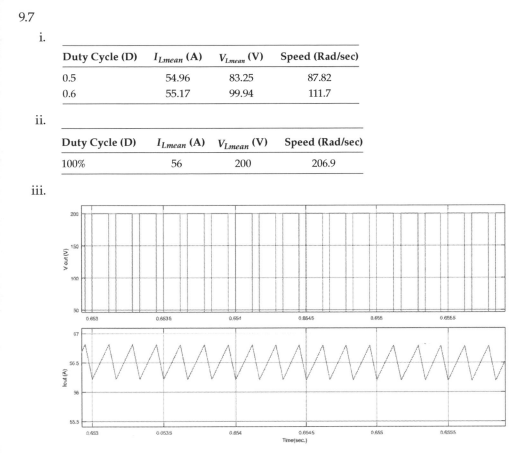

Chapter 10

10.7 (i) Te = 144.68 Nm, $\omega_s = 377\ rad / sec$

(ii) $S_m = 0.08$

(iii) $T_{e,m} = 11.47$ N.m

10.9 Firing advance angle of inverter = 17.95°

Chapter 11

11.1 $\beta = 2.7°$

11.3 (a) Resolution = 120 steps/revolution

(b) The steps required for making 25 revolutions = 3000

(c) $n = 20$ rps

Selected Bibliography

1. *"Stepper Motor Controller,"* Texas Instruments Incorporated.
2. Z. Qi, *"Design of a Driver of Two-Phase Hybrid Stepper Motor Based on THB6064H,"* 2017 2nd Asia Conference on Power and Electrical Engineering IOP.
3. X. H. WAN, *Electrical Machinery*. Beijing, China: China Machine PRESS, 2009.
4. N. Q. Le, and J. W. Jeon, "An open-loop stepper motor driver-based on FPGA," *International Conference on Control, Automation and Systems*, 2007 October 17–20, 2007 In COEX, Seoul, Korea.
5. AC Motors, (Http://Www. Allaboutcircuits.Com/Vol_2/Chpt_13/5.Html).
6. Hard Disk Drives, (Http://Www.Storagereview.Com/Guide2000/Ref/Hdd/Op/Actactuator.Html).
7. OSR Journal of Electrical And Electronics Engineering (IOSR-JEEE) Www.Iosrjournals.Org, International Conference On Advances In Engineering & Technology, 2014.
8. P. C. Krause, O. Wasynczuk, and S. D. Sudhoff, *"Analysis of Electric Machinery and Drive Systems,"* 2nd ed., Hoboken, NJ: Wiley Interscience, A John Wiley & Sons, INC. Publication.
9. A. Hughes, *"Electric Motors and Drives, Fundamentals, Types, and Applications,"* 3rd ed., New York: Elsevier, 2002.
10. P. S. Bimibhra, *"Power Electronics,"* 4th ed., 2007, New Delhi, India: Khanna Publishers.
11. "Introduction to power electronics," A Tutorial Burak Ozpineci Power Electronics and Electrical Power Systems Research Center, The U.S. Department of Energy.
12. R. W. Erickson, *"Fundamentals of Power Electronics "* University of Colorado, Boulder, 1997, New York: Chapman & Hall.
13. C. W. Lander, *"Power Electronics,"* 3rd ed., New York: McGraw-Hill, 1993.
14. J. Zhao, and Y. Yu. "Brushless DC Motor Fundamentals Application Note," 2011.
15. A. Nafees, Http://Eed.Dit.Googlepages.Com.
16. V. Larin, and D. Matveev "Analysis of transformer frequency response deviations using white-box modeling," *Conference: CIGRE Study Committee A2 COLLOQUIUM*, Cracow, Poland, October 2017.
17. J. C. Stephen, *"Electric Machinery Fundamentals,"* 5th ed., New York: McGraw-Hill, 2012.
18. A. E. Fitzgerald, C. Kingsley, and D. Stephen, *"Electric Machinery,"* 6th ed., London: McGraw-Hill, 2012.
19. P.C. Krause, O. Wasynczuk, and S. D. Sudhoff, *"Analysis of Electrical Machinery and Drive Systems,"* 2nd ed., Hoboken, NJ: John Wiley & Sons, 2002.
20. A. Hughes, *"Electric Motors and Drives Fundamentals, Types and Applications,"* 3rd ed., Elsevier, 2006.
21. M. H. Rashid, *"Power Electronics, Circuits, Devices, and Applications"*, Upper Saddle River, NJ. Prentice, 2003.
22. P. Krause, O. Wasynczuk, S. Sudhoff, and S. Pekarek, *"Analysis of Electric Machinery and Drive Systems"*, 3rd ed., Hoboken, NJ: John Wiley & Sons, 2013.

Index

Note: Page numbers in italic and bold refer to figures and tables, respectively.